A Guidebook for Teaching
CONSUMER MATHEMATICS

PETER A. PASCARIS

Allyn and Bacon, Inc.
Boston • London • Sydney • Toronto

This book is part of A GUIDEBOOK FOR TEACHING SERIES

Copyright 1982 by Allyn and Bacon, Inc., 470 Atlantic Avenue, Boston, Massachusetts 02210.

All rights reserved. No part of the material protected by this copyright notice, except the Reproduction Pages contained within, may be reproduced or utilized in any form or by any means, electronic or mechanical, including photocopying, recording, or by any information storage and retrieval system, without written permission from the copyright owner.

The Reproduction Pages may be reproduced for use with this text, provided such reproductions bear copyright notice, but may not be reproduced in any other form for any other purpose without permission from the copyright owner.

Library of Congress Cataloging in Publication Data

Pascaris, Peter A.
   A guidebook for teaching consumer mathematics.

   Includes bibliographical references.
   1. Mathematics—Study and teaching. 2. Consumer education—Study and teaching. 3. Finance, Personal—Study and teaching. I. Title.
QA11.P37   513'.93   81-20650
ISBN 0-205-07388-3   AACR2

Printed in the United States of America
Printing number and year (last digits):
10 9 8 7 6 5 4 3 2 1   87 86 85 84 83 82

About the Author

Peter A. Pascaris has been a teacher of chemistry at Stevenson Senior High School in Livonia, Michigan. Prior to that position, Mr. Pascaris served as the Department Chairman of the Math/Science Department at Dickinson Junior High School for thirteen years. He received his B.S. and M.A. degrees from Wayne State University. He was a National Science Foundation Fellow at Michigan Technological University and is a member of the Mackenzie Honor Society at Wayne State and a member of the Omicron Delta Kappa Fraternity. Mr. Pascaris is also the co-author of two resource guides for teaching chemistry and mechanics at the junior high level.

Gene Stanford, *Consulting Editor for the Guidebook for Teaching Series*, received his Ph.D. and his M.A. from the University of Colorado. Dr. Stanford has served as Associate Professor of Education and Director of Teacher Education Programs at Utica College of Syracuse University and is a member of the National Council of Teachers of English and the International Council on Education for Teaching. Dr. Stanford is the author and co-author of several books, among them, *A Guidebook for Teaching Composition, A Guidebook for Teaching Creative Writing, A Guidebook for Teaching about the English Language*, and *Human Interaction in Education*, all published by Allyn and Bacon, Inc.

Contents

PREFACE ix

Chapter 1 MAKING SENSE OF DOLLARS AND SENSE 1
Introduction 1
Objectives 1
Content Overview 2
    Topic I: Expressing Figures in Dollars and Cents
          (The Decimal Point and Place Value) 2
    Topic II: Reading and Writing Money Figures 4
    Topic III: Comparing Money Figures 6
    Topic IV: Adding Money Figures 7
    Topic V: Subtracting Money Figures 8
    Topic VI: Rounding Money Figures and Estimating 10
Learning Experiences: The Diagnostic Survey and Independence
          Scale 10
Assessing Achievement of Objectives 13
Resources for Teaching About Dollars and Cents 13

Chapter 2 CHECKING ACCOUNTS 15
Introduction 15
Objectives 15
Content Overview 16
    Topic I: Why Use a Checking Account? 16
    Topic II: Terms 17
    Topic III: Forms 20
    Topic IV: Correct Usage and Spelling of Numbers 22
    Topic V: Endorsements 22
    Topic VI: Check Stubs and the Check Register 23
    Topic VII: Reconciling (Balancing) the Bank Statement 24

Learning Experiences: The Diagnostic Survey and Independence
　　　　　　　　　Scale　26
Assessing Achievement of Objectives　30
Resources for Teaching About Checking Accounts　32

Chapter 3　MULTIPLICATION AND MULTIPLE BUYING　35
Introduction　35
Objectives　36
Content Overview　36
　Topic I:　Review of Multiplication　36
　Topic II:　Estimating Products　40
　Topic III:　Multiple Purchases　41
　Topic IV:　Receiving Change　42
　Topic V:　Sales Tax　43
　Topic VI:　Using Multiplication to Compare Costs and Recognize
　　　　　　Value　47
　Topic VII:　Predicting Prices　48
Learning Experiences: The Diagnostic Survey and Independence
　　　　　　　　　Scale　49
Assessing Achievement of Objectives　52
Resources for Teaching About Multiplying and Multiple Purchases　54

Chapter 4　AVERAGES AND BUDGETS　57
Introduction　57
Objectives　58
Content Overview　58
　Topic I:　Division by Whole Numbers　58
　Topic II:　Averages　65
　Topic III:　Restaurant Math　67
　Topic IV:　Budgeting Food Costs　68
Learning Experiences: The Diagnostic Survey and Independence
　　　　　　　　　Scale　71
Resources for Teaching About Averages and Budgets　74

Chapter 5　TRANSPORTATION COMPUTATION　77
Introduction　77
Objectives　79
Content Overview　79
　Topic I:　Rounding and Estimating　79
　Topic II:　Odometer Readings　81
　Topic III:　Fuel Purchases　82
　Topic IV:　Fuel Efficiency　87
　Topic V:　Renting a Car　89
Learning Experiences: The Diagnostic Survey and Independence
　　　　　　　　　Scale　90
Assessing Achievement of Objectives　93
Resources for Teaching About Transportation Computation　94

CONTENTS   vii

Chapter 6   TRAVEL AND RECREATION   95
             Introduction   95
             Objectives   96
             Content Overview   96
                 Topic I: Fractions and Time   96
                 Topic II: Distance, Rate, and Time   100
                 Topic III: Map Reading   107
                 Topic IV: Bus, Train, and Plane Facts   107
                 Topic V: Vacation Trips   108
             Learning Experiences: The Diagnostic Survey and Independence
                                    Scale   108
             Assessing Achievement of Objectives   112
             Resources for Teaching About Travel and Recreation   112

Chapter 7   GETTING THE BEST VALUE   115
             Introduction   115
             Objectives   115
             Content Overview   116
                 Topic I: Measurement and Fractions   116
                 Topic II: More Thought for Food   122
                 Topic III: Do-it-Yourself Projects   126
                 Topic IV: Mail-order and Catalog Buying   128
             Learning Experiences: The Diagnostic Survey and Independence
                                    Scale   129
             Assessing Achievement of Objectives   132
             Resources for Teaching About Getting the Best Value   132

Chapter 8   PERCENTS PERCHANCE   137
             Introduction   137
             Objectives   138
             Content Overview   138
                 Topic I: Ratios, Proportions, and the Meaning of Percent   138
                 Topic II: Type One: A Certain % of Some Number is What
                             Number?   142
                 Topic III: Type Two: What % of a Certain Number is Another
                             Number?   145
                 Topic IV: Type Three: A Certain % of What Number is Another
                             Number?   147
                 Topic V: Mentally Estimating Percents   152
             Learning Experiences: The Diagnostic Survey and Independence
                                    Scale   152
             Assessing Achievement of Objectives   154
             Resources for Teaching About Percents Perchance   154

Chapter 9   PAYCHECKS, DEDUCTIONS, AND TAXES   157
             Introduction   157
             Objectives   157

Content Overview   158
   Topic I: Paychecks and Deductions   158
   Topic II: Methods of Determining Pay   164
   Topic III: Annual Income Tax Reports   169
Learning Experiences: The Diagnostic Survey and Independence Scale   175
Assessing Achievement of Objectives   180
Resources for Teaching About Paychecks, Deductions, and Taxes   180

Chapter 10   SPENDING AND SAVING   183
Introduction   183
Objectives   184
Content Overview   185
   Topic I: Credit and Loan   185
   Topic II: Savings and Investing   191
   Topic III: Insurance and Retirement   196
   Topic IV: Housing Costs   200
   Topic V: Buying and Maintaining a Car   205
Learning Experiences: The Diagnostic Survey and Independence Scale   207
Assessing Achievement of Objectives   212
Resources for Teaching About Spending and Saving   213

Appendix A   GENERAL INTEREST REFERENCE   217
Appendix B   ADDRESSES OF PRODUCERS OF RESOURCES   219
Appendix C   REPRODUCTION PAGES   223
Appendix D   FEEDBACK FORM   321

Preface

Although this book is written primarily as a professional resource book for teachers, it may also be used by individuals seeking to upgrade their own competency as consumers. A complete explanation of the topic, several examples, a multitude of Reproduction Pages ready to use for student assignments, and suggestions for evaluation are included in each chapter. My intentions are to provide direction, guidance, and examples so that teachers may plan and execute a program that is aimed chiefly at students in the following categories:

- A. Young people in the upper junior high school or senior high school who have an interest in or a need to learn basic consumer computation.
- B. High school seniors who have not demonstrated competency with basic arithmetic skills.
- C. Adults who:
    1. Have not completed high school.
    2. Need a refresher course in computation.
    3. May never have acquired the arithmetic skills necessary for effective money management.
    4. Are non-English-speaking or use English as a second language.
- D. Average general and specialized mathematics students who would benefit from supplemental experiences.

I have attempted to write the book in a manner that will appeal to both layman and professional so that any individual who is in one of the above categories will find the explanations, practices, and activities useful.

Emphasis is placed upon practical knowledge for everyday calculations. Personal economics is stressed, and considerations such as product evaluation and quality or business and corporate organization are secondary to the goal that students become proficient with the figures and calculations required to buy wisely and pay bills. This is not a home economics text; nor is it a business math primer. It is a survival kit. As such, I have begun with what I feel are basic consumer needs and have identified the various math skills associated with those needs.

Many of the people who seek this instruction have already attended classes in basic or general mathematics. Some may have had a great deal of difficulty with that type of study and may have experienced failure. Because of this, I have taken care to introduce skills in a developmental manner so as to build one computation skill upon the other. However, the particular consumer need is identified first. Thus, sections are entitled "Checking Accounts," "Food Expenses," "Taxes," and so forth rather than "Addition Skills," "Multiplication Skills," and the like.

In order to provide for the spiral development of skills, the sequence of consumer topics begins with those requiring the simplest arithmetic skills and builds toward the more complex. In this way, "Checking Accounts" (which requires addition and subtraction) precedes "Taxes" (which requires computation with percents). In other words, the *topics* are selected because a consumer need is identified, whereas the *sequence* is governed primarily by mathematical development.

The suggestions for culminating activities that are included both within each chapter and after a series of chapters reinforce the sequential development. For example, I suggest a vacation trip assignment after Chapter 7 because the skills in Chapters 1-7 are utilized in that activity. (Besides, it's a lot of fun to think about spending money in an enjoyable way. This particular unit was born from a log of expenses that I kept from a family camping trip to the Old West.)

On the other hand, each chapter can be used independently without the prerequisite of previous units. You may want to skip some chapters whenever the background, interest, and ability of the students warrants elimination of certain simpler assignments. In fact, teachers will be able to use the text as a guide to individualize instruction or to provide for independent study; I have often used the materials in that way. A chapter may be used by itself or out of the sequence presented, and the activities are not dependent upon one particular teaching strategy.

Calculators have become increasingly popular during recent years. Because some models are inexpensive and very compact, they lend themselves very well for use in a consumer math program. The degree to which they may be used depends upon the objective of instruction. If the goal is to achieve mastery of certain computational skills, calculators can be used as aids or tools in reaching that goal but not as ends in themselves. If, on the other hand, the objective is limited to attaining the facility to, say, balance a checkbook, then the operator must have only a basic understanding of computation: you must be able to recognize what operation is involved and what "buttons to push." Nevertheless, paper-and-pencil computation should not be sold short, nor should you ignore simple mental arithmetic. In addition, I think it's important to students' aspirations in that while a lower goal may be initially sighted, successful experience can encourage students to raise their goal toward higher achievement. I would suggest, therefore, that teachers and individuals strive toward mastery of computational skills as the ultimate or optimum goal. At the same time, you should realistically adjust your goals to take into account individual differences.

This brings us to the topic of evaluation. Suggestions and examples of evaluative tools are made at the end of each chapter. I have included sample tests as well as culminating activities and experiences that enable students to do some self-evaluation. However, it is appropriate for students to be examined or to conduct a self-examination prior to participating in the activities of the chapter. I call this "test before study" a *Diagnostic Survey*.

In order to diagnose skills, you must first begin with a clear set of performance objectives. These are provided at the beginning of each chapter. Next, instruments are suggested to take an accurate measure of the student's ability prior to doing the work of the chapter. Usually the results of such tests indicate whether a student can or cannot perform the given task. But this is not always a simple case of yes or no. In my experience, I have often heard a student say that he can do a certain task "with a little bit of help." That is to say, there are varying degrees of capability, and one way to indicate the variance is by the degree of independence exhibited by the performer. Therefore, I have designed an *Independence Scale* that can be used as a self-evaluation. Every chapter includes a Diagnostic Survey and an Independence Scale. Thus, a complete explanation of these instruments is given here so that they may be put to their most effective use.

The Diagnostic Survey. The Diagnostic Survey serves four main purposes: to introduce skills, to identify student competency prior to instruction, to facilitate planning, and to provide examples for final evaluation.

1. **Skills.** Remember, the survey is to be used as the initial activity of the chapter, prior to any lecture or instruction. You will want to use the survey to introduce specific skills as presented in the chapter. Upon close examination, you will note that each problem is associated with a specific performance objective and that usually there are several different types of problems associated with each objective.

2. **Student competency.** The survey will enable you to identify skill deficiencies that need to be improved and competencies that need little or no additional instruction. Each skill can be analyzed to its component parts so that the specific type of error the student is making can be identified.

3. **Planning.** Planning your teaching strategy is facilitated because both the teacher and the student know what to expect in the unit. The time and effort required to teach (and learn) each objective can be budgeted according to the needs assessed by the survey. Selection of topics is based upon demonstrated need. If the teacher wants to (and the teaching situation allows for it), the data collected from the Diagnostic Survey will facilitate planning a personally prescribed study for an individual.

4. **Final evaluation.** Finally, the survey serves as a fine example for a final test. (Pardon my play on words.) The final test, like the survey, ought to follow the performance objectives. In fact, if enough time has elapsed since it was first given, the survey itself may be used as the final test. The results of the final test, after all, represent what the student learned from the unit and can point to those areas that require further study. And, so, the process continues.

The Independence Scale. The Independence Scale is to be used directly with the Diagnostic Survey. Its purpose is simply to report the student's degree of competency to perform a specific skill. This is reported in terms of the amount of help a student feels he or she must have in order to be successful—in other words, his or her degree of independence. Each chapter has an Independence Scale to accompany the Diagnostic Survey, and you may wish to look ahead to see a specific example. (See Reproduction Page 2 for the Independence Scale used in Chapter 1.)

Note that the columns to the right indicate a decreasing need for assistance and instruction. Teachers can plan accordingly. If the same scale is used after instruction, students can compare their final performance level with their initial effort. More often than not, some growth takes place even though mastery may not have been attained.

In summary, use the Diagnostic Survey to estimate each student's degree of competence prior to instruction. Use the Independence Scale with the survey to identify those skills that need to be strengthened by further instruction and practice.

Both instruments are designed to:

1. Cause the student to think in terms of specific abilities for specific skills.

2. Motivate the student to strive *toward independence* by achieving mastery of the skills.

3. Cause the students to be realistic about their abilities and their needs for assistance in performing necessary skills.

4. Establish a set of skills as being important and necessary for survival in the modern world.

Of course, more traditional methods of evaluation are also suggested. This discussion only serves to point out that although the *Guidebook* is primarily aimed at providing a simplified, concise, and practical teaching program in and of itself, it also provides the flexibility that allows imaginative teachers to exercise their creative ability.

The Diagnostic Survey and the Independence Scale are the mainstays of an overall approach which involves identification of objectives, diagnosis of skills, estimation of independence, selection of topics and teaching strategy, study and practice, and final evaluation. All of these factors are to be considered in order to make the most effective use of the *Guidebook*. These features are incorporated into the six main sections of each chapter: "Introduction," "Objectives," "Content Overview," "Learning Experiences," "Assessing Achievement of Objectives," and "Resources."

Each chapter provides:

1. A brief introduction to the concepts the chapter deals with, along with a brief discussion of why these concepts might be important to study and suggestions as to what choices the teacher may need to make in organizing instruction.

2. A list of performance objectives that the methods and materials in the chapter help the student achieve, saving teachers the task of writing objectives of lesson plans.

3. An overview of the content of the chapter, which serves a dual purpose: first, to remind the teacher of the basic content of that particular topic; second, to provide the raw material for brief lectures by the teacher if he or she wishes to utilize this instructional method.

4. A wealth of learning experiences to involve the student actively in the study of the contents of the chapter; it is within this section that the Diagnostic Survey and the Independence Scale appear.

5. Sample tests and/or culminating activities to determine how well the student has achieved the objectives of the unit.

6. Annotated lists of materials including books, films, filmstrips, cassettes, slide-tape programs, and games.

1

Making Sense of Dollars and Cents

INTRODUCTION

This chapter introduces the concept of the decimal point and its significance in changing dollar figures to cents and vice versa; reviews place value for reading and writing dollar figures; explains comparison and the concept of greater or less than; shows methods (and pitfalls) of adding and subtracting dollar figures, including making change; and finally, explains the method for rounding off money figures.

The concepts presented here form the basis for understanding the meaning of all calculations the consumer will use. It is important, therefore, that these ideas be continually reinforced and developed, not only in this unit, but in succeeding ones as well. The practicality of the material speaks for itself, and even though some work may seem tedious, the student soon realizes the necessity of his or her labor.

Although this text is written primarily for teaching the student "everyday mathematics," it is important to be mindful of the higher-level understandings that can grow out of this study. Thus, the concept of mathematics as a system, well organized and logically developed, should permeate all instruction. The teacher who fails to utilize this thought, even in a subtle way, loses a valuable aid for instruction. The difference is a student who on the one hand sees mathematics as a tool and on the other sees mathematics as a troublesome burden. Furthermore, good teaching requires sharp perception not only of the immediate needs of the student but also of the student's potential to reach further than he or she may have ever dreamed. Consequently, success with an initial goal that is utilitarian in nature may be a prelude to more advanced undertaking. Ironically, this is also practical. What can be more practical than to seek the skills that provide the greatest opportunity for job placement?

Let us not lose sight, however, of the fact that most students who study consumer mathematics will have had some difficulty in previous encounters with arithmetic. The concepts are readily understood after the student repeats calculations time and time again through this and succeeding units. Therefore, it is most important that the teacher adopt the notion that habit often precedes understanding. Indeed, students can start and cultivate a habit in spite of their lack of understanding. We are talking here of economic survival. Students may well be saying: "Show me how, not why." It behooves us as teachers, then, to teach in such a way so as to utilize the mechanical, even the rote methods of learning, saying, "Here's how to do it. If you do it often enough and look for patterns that repeat in a similar way over and over, you will soon understand why you are doing it." In this chapter, for example, students ought to cultivate the habit of using a dollar sign and decimal point in all calculations (rather than a cent sign). Effort should be made to explain why, but it matters not that students understand all the logic of it or that they have a knowledge of alternative systems of numeration. This habit is a necessary one in order to correctly do the calculations that follow. Students do well to strive toward complete understanding, but if understanding is not attained, students will at least know how to perform the skill.

OBJECTIVES

If all the topics in this chapter are chosen by the teacher, the student should be able to:

1. Use the decimal point appropriately and correctly change "¢" figures to "$" figures and vice versa.

2. Read aloud or write in words money figures expressed to the tenth of a cent.

3. When money figures are expressed with a dollar sign, or a cent sign (or both) and written to a tenth of a cent, students should be able to compare money figures and rank them in order from greatest to smallest or smallest to greatest.

4. Add money figures.

5. Subtract money figures and make change.

6. Round off to the nearest cent.

7. Round off to the nearest tenth of a cent.

CONTENT OVERVIEW

Topic I: Expressing Figures in Dollars and Cents (The Decimal Point and Place Value)

Advertised prices are often misleading when written with a ¢ rather than a $. For example, to some persons "9¢ each" may appear to be less than "$.09 each." Consumers who fail to make the distinction between dollars and cents will be confused when attempting to calculate money figures. If they are not consistent when they write the figures or punch them on a calculator, they can make a serious mistake. That is why it is essential to recognize the importance of the unit of value (the sign), the decimal point, and the consequent place value.

In the United States the monetary units are dollars and cents. The cent is a fraction of a dollar; 100 cents equals a dollar. Therefore, the cent is 1/100 of a dollar or 0.01 of a dollar. Our monetary system follows the simple decimal or base 10 numeration system. This system uses ten symbols, 0, 1, 2, 3, 4, 5, 6, 7, 8, 9, to represent all numbers. To write numerals greater than nine, two or more number symbols are written next to each other. The position (or place) of these numerals determines the value of the number expression; thus the term *place value*. Compare this, for instance, with the Roman numeration system where a different symbol is used to represent various amounts. The value of each place in the decimal system is ten times the value of the place immediately to its right. Therefore, the value of each symbol in a decimal numeral depends upon: 1) what the symbol is and 2) in which position it appears. A "3," for example, means 3 tens in the numeral 38; while in the numeral 342, "3" means 3 hundreds.

When expressing parts of a whole unit, a fraction is used. A *decimal fraction* always has a denominator that is a power of ten(10, 100, 1000, etc.). It is important to recognize that we are concerned here with units between zero and one, *not* less than zero. This can be illustrated on a number line as a shaded area between, but not including, 0 to 1.

Furthermore, this shaded area can be divided evenly into 10, 100, 1000, or more parts. For purposes of most money calculations, 100 parts are enough, although 1000 parts or more are also used.

In decimal numerals a dot or *decimal point* is used to separate whole units and fractional parts of whole units. The "ones" place or unit place appears to the left of the decimal point, and the fractional parts of the unit appear to the right of the decimal point. Again the same ten symbols are used, and their value depends upon their position in the numeral. The first place to the right of the decimal point is 1/10 of a whole or 0.1, or "tenths"; the second place represents 1/100 of a whole or 0.01, or "hundredths"; the third is 1/1000 or 0.001, or "thousandths"; and so on.

These values hold true regardless of the units of measure, be they dollars or cents or any other unit. However, it is important to identify the unit and be consistent in using that unit throughout all calculations. Serious error can result if $34 is used when 34¢ was meant. Similarly, if $34 is added to 5¢ the sum is not $39 or 39¢, but $34.05.

The word "cent" and the cent sign mean that the units are "hundredths" of a dollar. Cents may be expressed with a cent symbol (¢) or with a dollar symbol ($) along with an appropriately placed decimal point. When a cent sign is used with a numeral, its equivalent value written as dollars is 1/100 of the cent figure. Since two places to the right of the point represent hundredths of a unit, cents can be expressed with a dollar symbol and numerals using two places to the right of the decimal point. Thus, 35¢ is $.35 and 5¢ is $.05. On the other hand, if the decimal point is used with a cent symbol, then the numerals to the right of the point represent *fractions of a cent*. Thus, 67.9¢ is 67 cents and 9 tenths of a cent. This will be discussed in more detail later in this chapter.

Each dollar represents 100 cents. Therefore, when the dollar sign is used with a numeral as $17, the same value expressed as cents is 100 times that figure or seventeen hundred cents (1700¢). Similarly, $17.53 becomes 1753¢.

In order to assume consistency and eliminate error as much as possible, one form of writing dollar figures should be insisted upon in all calculations. Because it is customary to write cent figures only at 99 cents or less (the expression 1753¢, for example, is not ordinarily used), then it is reasonable to use dollar expressions exclusively. This is necessary whether using paper and pencil *or* a calculator.

Expressions using one sign or unit can be written as expressions with the other sign or unit only after two adjustments are made: 1) the decimal point must be placed in a new position, and 2) the sign must be changed.

A. To change a figure from a cent expression to a dollar expression, e.g., 34¢ = $_____:

1. Place the decimal point two places to the left of where it appears in the cent expression. (If no decimal point exists, assume it to be immediately to the right of the last numeral. Zeros may be added as needed.)

2. Change the sign to a $ written to the *left* of the numeral. Thus, 34¢ becomes $.34; 66.9¢ becomes $.669; and 273¢ becomes $2.73.

B. To change a figure from a dollar expression to a cent expression:

1. Place the decimal point two places to the right of where it appears in the dollar expression. (If no decimal point exists, assume it to be immediately to the right of the last numeral. Zeros may be added as place holders as needed.)

2. Change the sign to a ¢ written to the *right* of the numeral.

In summary, because of the importance of consistency with units, it is recommended that teachers *insist that all calculations be done using the dollar sign ($) exclusively.*

Topic II: Reading and Writing Money Figures

The importance of reading and writing monetary figures is clear when one considers the frequency of the question, "How much does it cost?" Being able to express these amounts orally is most obvious. But the necessity of writing checks and receipts demonstrates the need to write dollars and cents in words, too. In order to facilitate the process of naming these figures, the following categories are suggested.

1. Naming dollars only (with no cents).

2. Naming cents only (with no dollars or fractions of a cent).

3. Naming dollars and cents with no fractions of a cent.

4. Naming cents and fractions of a cent.

1. Naming dollars only with no cents. These figures are essentially whole numbers, and one must merely be certain to remember that the value of each symbol depends upon what the symbol is and its position in the numeral. Thus, the place value must be known and memorized in the form shown in Figure 1-1.

MAKING SENSE OF DOLLARS AND CENTS 5

Figure 1-1
Whole Number Place Value

	trillions	hundred billions	ten billions	billions	hundred millions	ten millions	millions	hundred thousands	ten thousands	thousands	hundreds	tens	ones
	3,	5	4	2,	3	7	1,	2	6	8,	5	7	9

Place value for $3,542,371,268,579. The numeral is read: "Three trillion, five hundred forty-two billion, three hundred seventy-one million, two hundred sixty-eight thousand, five hundred seventy-nine dollars."

Note that the numeral is divided into groups of three starting from the right and moving left. The first digit at the right of each group names the group. Thus, from right to left, the first group is named ones, the second group, thousands; the third, millions; the fourth, billions; and the fifth, trillions. Within each group the middle digit represents *tens* of that group. Consequently, the middle digit of the first group names *ten* ones or tens; the middle digit of the second group names *ten* thousands; the middle digit of the third group names *tens* of that group. Thus, the left digit of the first group names *hundreds* of ones or hundreds; the left digit of the second group names *hundred* thousands; the left digit of the third group names *hundred* millions; and so on.

Other general rules to follow are:

The ones digits are named just as they would be if they appeared alone.

The numeral immediately following "nine" is "ten" followed by "eleven" and "twelve," and then by some form of the unit term used as a root with the suffix "-teen."

The remaining tens are named by some form of the unit term used as a root with the suffix "-ty."

The hundreds are all named by use of the unit term followed by the word "hundred."

The word "dollars" follows the naming of the numeral.

The word "and" should never be used when naming dollars with no cents.

2. Naming cents with no dollars or fractions of a cent. Whether using a cent sign (¢) or a dollar sign ($) with a decimal point, these numerals are read just as you would read a whole number followed by the word "cents." Only ones and tens are read, since the largest cent-only figure is 99¢ (ninety-nine cents).

3. Naming dollars and cents with no fractions of a cent. The dollars are read first and the cents last. The important idea here is that the decimal point that divides dollars from cents in the numeral is read "and" following the word "dollars." Thus, $15.72 is read "fifteen dollars *and* seventy-two cents."

4. Naming figures that include fractions of a cent. These are figures like $.679, or 67.9¢. Both $.679 and 67.9¢ are read the same way, with the unit named being "cents." In the case of

67.9¢ the digits to the left of the decimal point are read first; the word "and" is used at the decimal point followed by the next digit as "tenths of a cent" or simply the words "tenths cents." (If two digits followed the decimal point, they would be read "hundredths of a cent" or simply "hundredths cents.")

When examining the place values for $.679, it must be remembered that only the first two digits after the decimal point are whole cents. Therefore, the symbols that follow must be fractions of a cent. One must remember that the word "and" must follow the naming of the first two digits. Then, the fractions (tenths, hundredths, etc.) of a cent are named. Thus, $.679 and 67.9¢ are both read "sixty-seven and nine-tenths cents." (A variation is "sixty-seven cents and nine-tenths of a cent.")

Furthermore, the expressions $1.359 or 135.9¢ are to be read "one dollar and thirty-five and nine-tenths cents." When recalling that the place values from the decimal point to the right are tenths, hundredths, thousandths, and so on, confusion may arise. Using that pattern, $.679 can be read six hundred seventy-nine thousandths of a dollar. This type of a notation is not common, and the aforementioned method for naming figures is favored. Therefore, in order to avoid unnecessary problems, teachers should insist that all dollar-and-cent figures be written in the dollar sign and decimal point form and that the simplest form for naming dollars and cents be used consistently.

Topic III: Comparing Money Figures

Although no difficulty is experienced when comparing dollars with dollars, some instruction is necessary for comparing cents with cents and comparing mixed figures with dollars, cents, and other mixed figures.

Let us examine comparison of cents with cents. First of all, be sure that all figures are correctly written with a $. If simple cent figures are used (two digits to the right of the decimal point), the larger number is the larger value. If three digits to the right of the point appear in one numeral and the other has only two digits, add a zero to the right of the last digit so that each numeral contains the same number of decimal places, and then compare each figure. The larger number has the greatest value.

Examples:

    A. Which is more, $.65 or $.649?

        1. $.65 = $.650.

        2. $.650 is more than $.649. Therefore,

        3. $.65 is more than $.649.

    B. Which is more, $.0399 or 39¢?

        1. Change to $ figures: 39¢ = $.39.

        2. Rewrite so that both figures have the same number of digits: $.39 = $.3900.

        3. Compare: $.3900 is more than $.0399. Therefore,

        4. 39¢ is more than $.0399.

When comparing mixed figures (which contain both dollars and cents), compare the dollars first—if they are equal, then compare the cents as above.

More examples:

C. Which is more, $7.29 or 72.9¢?

1. Change to $ figures: 72.9¢ = $.729.

2. Compare dollars places: $7.29 contains seven dollars.
   $.729 contains no dollars.
   $7.29 is more than 72.9¢.

D. Which is more, 137.5¢ or $1.37?

1. Change to $ figures: 137.5¢ = $1.375.

2. Dollars places are the same; compare cents next.

3. Rewrite so that each contains the same number of digits: $1.37 = $1.370.

4. $1.375 is more than $1.370. Therefore, 137.5¢ is more than $1.37.

E. Which is more, 62.85¢ or $6.28?

1. Rewrite so that both figures show $: 62.85¢ = $.6285.

2. Compare: $.6285 with $6.28; $6.28 is more.

F. Arrange in order of size (largest first): $.57, $.573, $5.70, and 5.7¢.

1. Rewrite so that all figures show $: 5.7¢ = $.057.

2. $5.70 is the only figure with dollars; it is the largest.

3. Rewrite so that each contains the same number of digits: $.57 = $.570.

4. $.573 is larger than $.570; $.570 is larger than $.057.

5. Answer: $5.70, $.573, $.57, 5.7¢.

Topic IV: Adding Money Figures

When adding, all numerals must be written in columns so that the decimal points are directly under each other. In order that the number of places to the right of the decimal point be the same for each addend, zeros are used as place holders. A figure such as $8, without a decimal point, is to be written as $8.00. Addition then proceeds as with whole numbers. The decimal point in the sum is placed directly under the points in the addends. Again, be sure to change all figures to $ figures.

Examples:

A. 24¢ + $2.40 = ?   24¢ = $.24

Rewrite the problem and add:

$2.40
+ .24
─────
$2.64

B 73¢ + $7.30 + $7.03 + $7 + 7¢ + 3¢ + $3 = ?   *Rewrite the problem and add:*
    73¢ = $.73
    3¢ = $.03 $ .73
    7¢ = $.07   7.30
    $7 = $7.00   7.03
    $3 = $3.00   7.00
                                                          .07
                                                          .03
                                                       + 3.00
                                                       $25.16

Topic V: Subtracting Money Figures

Recalling the proper terminology:

    The number from which you subtract is the *minuend*.

    The number that you subtract is the *subtrahend*.

    The answer in subtraction is the *difference*.

    At the very start, be sure to write all figures correctly with $. The procedure to follow, then, is to write the subtrahend below the minuend being certain to line up the decimal points. As with addition, use zeros as place holders so that the number of places to the right of the decimal point in both numbers corresponds. Subtract as with whole numbers, placing the decimal point in the difference directly below the points in the subtrahend and minuend. Answers may be checked by adding the difference to the subtrahend; the sum ought to be the minuend. Some common errors to avoid are shown in the following examples.

Examples:

A. $42 − $.37.

    Correct solution: *Possible errors:*

        $42.00   1. Incorrect setup $42 $.42
        − .37   −.37   −.37
          1 9 $ 5 $.05

        4̷2.0̷0   2. No regrouping   42.00
          .37   − .37
        $41.63   42.37

                              3. Improper regrouping   1 99
                                                              4̷2.0̷0̷
                                                                .37
                                                             $41.62

MAKING SENSE OF DOLLARS AND CENTS

B. $4265 − $31.42.

Correct solution:

$4265.00
− 31.42

4 9

426$\cancel{5}$.$\cancel{0}$0
− 31.42
$4233.58

Possible errors:

1. Incorrect setup

 $4265
 − 31.42
 $11.23

2. No regrouping

 $4265.00
 − 31.42
 $4234.42

3. Improper regrouping

 4 99
 426\cancel{5}$.$\cancel{0}\cancel{0}$
 − 31.42
 $4233.57

C. $49 − 29¢.

Correct solution:

 8 9
 4\cancel{9}$.$\cancel{0}$0
 − .29
 $48.71

Possible errors:

1. Incorrect setup

 $49
 − 29
 $20

2. Improper regrouping

 8
 4\cancel{9}$.00
 − .29
 $48.29

D. $4 − 6¢.

Correct solution:

 3 9
 $4.$\cancel{0}$0
 − .06
 $3.94

Possible errors:

1. Incorrect setup

 $6 $.06
 − 4   or − .04
 $2 $.02

2. Incorrect regrouping

 3
 $$\cancel{4}$.00
 − .06
 $3.06

A careful analysis of the errors that students are making will enable a teacher to tailor the instruction for each student. Good teaching involves not only teaching what to do correctly but also how to avoid the most frequent errors. Note that several errors can be made if the cent figures are not correctly changed to the form using the dollar sign. The errors in the following section sometimes occur in checking account records.

E. $24.36 − $12.52.

Correct solution:

$24.36
− 12.52
$11.84

Possible errors:

1. Adding cent portion while subtracting dollars

 $24.36
 − 12.52
 $12.88

2. Adding dollars portion while subtracting cents

 $24.36
 − 12.52
 $36.84

Topic VI: Rounding Money Figures and Estimating

The importance of rounding numbers for estimation and quick calculation is evident in a trip to the shopping mall. Suppose the shopper is planning to buy Christmas gifts for 4 people and is limited to spending only $50. The wise consumer will round off the prices examined to the nearest dollar and estimate the total cost before making any final purchases. In business and industry as well, people are frequently asked to determine unit costs to the nearest cent, after more exact calculations have been made. For example, the owner of a local office supply store is considering the purchase of a small copy machine that is able to print copies at an average rate of $.0475 per sheet. The proprietor wants to make a small profit on the service, and she needs to determine exactly what to charge for 1 to 9 copies, 10 to 100 copies, and so on. It is likely that she will round off figures to estimate the approximate profit to be expected for each quantity printed.

In addition to writing all figures with $, remembering place values is very important for rounding figures. First of all, the number of digits needed to be retained must be determined. If rounding to the nearest dollar, all digits to the left of the decimal point (real or assumed) are retained. When rounding to the nearest ten cents, the first digit to the right of the decimal point is also retained. If rounding to the nearest cent, the first *two* digits to the right of the point (cents place) are retained. If the nearest tenth of a cent is desired, the third digit to the right is needed.

After determining the digits to be retained, rewrite only those digits needed and change the digits that appear to their right to zeros. Examine the first digit dropped. If it is a 5 or more, increase the preceding digit by one. If it is 4 or less, then the preceding figure remains unchanged. Thus:

$3.62 to the nearest dollar is $4. While $3.49 is rounded to $3.

$.659 to the nearest cent is $.66. While $.653 is rounded to $.65.

$.6687 to the nearest tenth of a cent is $.669. While $.6681 becomes $.668.

Furthermore:

$2.7384 to the nearest dollar is $3.

$2.7384 to the nearest cent is $2.74.

$2.7384 to the nearest tenth of a cent is $2.738.

LEARNING EXPERIENCES

THE DIAGNOSTIC SURVEY AND INDEPENDENCE SCALE

- Use the **Diagnostic Survey** on Reproduction Page 1 to estimate each student's degree of competence prior to instruction in the topics of this chapter.

- Upon completion of the survey, use the **Independence Scale** on Reproduction Page 2

MAKING SENSE OF DOLLARS AND CENTS

to identify those skills that need to be strengthened by further instruction and practice.

- Utilize the results of these two instruments to determine the appropriate assignments from the following list.

Topic I: Expressing Figures in Dollars and Cents (The Decimal Point and Place Value)

- Use Reproduction Page 3 to give students practice with changing signs of money figures.
- Identify grocery stores in your locality that show unit pricing on shelves and packages. Unit pricing shows the cost of the item purchased *and* the cost for each unit of measure (ounces, pounds, etc.). For example, if a can of baked beans costs 79¢ for 28 oz., the unit price is $.0282 per ounce. Use a form similar to the one shown here to record the survey.

GROCERY STORE SURVEY UNIT PRICING

Visit a grocery store that shows "Unit Pricing" on its shelves and packages. Find 15 different items of various measures. Complete this form with the appropriate information.

Item	Unit of Measure	Price Expressed In ¢	Cost Per Unit Expressed In ¢	Price Expressed In $	Cost Per Unit Expressed In $

- Ask students to research and report about the history of money. Include various ancient methods of exchange and recordkeeping as well as modern-day monetary systems used in other countries.

Topic II: Reading and Writing Money Figures

- Use a list of appropriate numbers so that students may learn to correctly spell words used for numbers (e.g. 1-20, 21, 31, 42, 53, 64, 75, 86, 97, 100, 150, 276, 1000, 1399).
- Use Reproduction Page 4, Part I, to give students practice reading and writing figures in $ (alone), ¢ (alone), and $ and ¢ together.
- Use Reproduction Page 4, Part II, to give students practice reading and writing money figures that include fractions of a cent.

Topic III: Comparing Money Figures

- Use Reproduction Page 5 for practice and drill with comparison of money figures.

Topic IV: Adding Money Figures

- Use Reproduction Page 6 to provide students with practice exercises adding money figures.
- For further exercises with addition of decimals and money figures, use *Stein's Refresher Mathematics* (Allyn and Bacon, Inc., 1980), pages 138-144.

Topic V: Subtracting Money Figures

- Use Reproduction Page 7 for practice problems in subtraction.
- Use Reproduction Page 8 for practice with making change.
- Use *Stein's Refresher Mathematics*, pages 146-152, for additional practice with subtraction of money figures.

Topic VI: Rounding Money Figures and Estimating

- Ask students to consider the many ways that rounding and estimating can help on a shopping trip. Plan such a trip and use Reproduction Page 9 as survey forms for recording prices and estimates. If a shopping trip is not possible, obtain catalogs from a catalog store or use newspaper ads and have students complete the forms in class.
- Use Reproduction Page 10 for practice in rounding and estimating. Also, drill students orally with problems like items 36-40 on that same page. Be sure students round and add mentally without the aid of a calculator or paper and pencil.

ASSESSING ACHIEVEMENT OF OBJECTIVES

Ongoing Evaluation

The extent to which students have mastered the concepts covered under the six topics in this chapter can be measured by any of the activities assigned to class members individually.

Final Evaluation

For an overall evaluation of the students' mastery of the concepts in this chapter, if all topics in the chapter have been taught, the **Diagnostic Survey** can be repeated.

RESOURCES FOR TEACHING ABOUT DOLLARS AND CENTS

Below is a selected and annotated list of resources useful for teaching the topics in this chapter, divided into audiovisual materials, games, and print materials. Addresses of publishers or distributors can be found in the alphabetic list in Appendix B.

Audiovisual Materials

For low ability and special education students.

Money Handling.

Understanding Addition and *Understanding Subtraction.* Kits include filmstrip, audio-cassette, workbooks, and teacher's guide. Interpretive Education, 1978.

How to Use a Pocket Calculator, filmstrip and audio-cassette. Interpretive Education, 1977.

Games

For low ability and special education students.

Cash in a Flash, Level 1 and *Level 2,* originated with Julie Jackels. The Math Group, Inc., 1977.

Making Cents. The Math Group, Inc., 1977.

The Disastrous Dollar. Creative Teaching Associates, 1978.

Print

A. FOR LOW ABILITY AND SPECIAL EDUCATION STUDENTS.

Money Makes Sense and *Using Dollars and Sense,* of the Pacemaker Practical Arithmetic Series by Charles H. Kahn and J. Bradley Hanna. Use entire *Makes Sense* book and pp. 3-74 of *Dollars and Sense* book for topics appropriate to Chapter 1. Fearon-Pitman Learning, Inc., 1973.

An Introduction to Everyday Skills and *Skills for Everyday Living, Book 1* and *Book 2,* by David H. Wiltsie. See especially *Everyday Skills,* pp. 9-10 and 17-21, as well as *Book 1,* pp. 13-23, and *Book 2,* pp. 8-9. Motivational Development, Inc., 1977, 1976, 1978, respectively.

Book II, Making Money Count, Book III, Buying Power, Book IV, Earning, Spending, and Saving, books of the Using Money Series, and *Useful Arithmetic, Volume II,* by John D. Wool. See especially all of *Book II,* pp. 1-34 of *Book III,* and pp. 1-11 of *Volume II.* Frank E. Richards Publishing Co., Inc., 1973.

It's Your Money, Book I and *Book II*, by Lloyd L. Feinstein and Charles H. Maley. See pp. 1-2 and 8-9 of each book. Steck-Vaughn Company, 1973.

Math for Today and Tomorrow by Kaye A. Mach and Allan Larson. Pp. 1-22. J. Weston Walch, Publisher, 1968.

Mathematics for Today, Level Orange, and *Level Green* by Wilmer L. Jones, Ph.D. See pp. 4-31, 152-165, 231-234, 247-250 in *Level Orange.* See pp. 4-29, 92-100, 231-234, 247-250 in *Level Green.* Sadlier-Oxford, 1979.

B. FOR GENERAL MATHEMATICS STUDENTS.

Stein's Refresher Mathematics, Seventh Edition, by Edwin I. Stein. See especially pp. 18-48 and 127-154. Allyn and Bacon, Inc., 1980.

Trouble-Shooting Mathematics Skills, Basic Competency Edition, by Allen L. Bernstein and David W. Wells. See especially pp. 3-27, 28-49, 99-115. Holt, Rinehart and Winston, 1979.

Activities Handbook for Teaching with the Hand-Held Calculator by Gary G. Bitter and Jerald L. Mikesell. See especially pp. 1-74, 177-190, 251-270. Allyn and Bacon, Inc., 1980.

Consumer Mathematics, Third Edition, by William E. Goe. Activities book available. See especially pp. 408-413 and 433-438. Harcourt Brace Jovanovich, 1979.

Mathematics in Life by L. Carey Bolster and H. Douglas Woodburn. See especially pp. 3-20 and 67-82. Scott, Foresman and Company, 1977.

Mathematics for Daily Living by Harry Lewis. See especially pp. 478-497 and 525-529. McCormick-Mathers Publishing Company, 1975.

Mathematics Plus! Consumer, Business & Technical Applications, by Bryce R. Shaw, Richard A. Denholm, and Gwendolyn H. Shelton. See especially pp. 2-11. Houghton Mifflin, 1979.

Consumer and Career Mathematics by L. Carey Bolster, H. Douglas Woodman, and Joella H. Gipson. See especially pp. 4-7. Scott, Foresman and Company. 1978.

Business and Consumer Arithmetic by Milton C. Olson and A. E. McVelly. See especially pp. 11-23 and 55-64. Prentice-Hall, Inc., 1974.

Business Mathematics for the Consumer by Mearl R. Guthrie, William Selden, and Delbert Karnes. See especially pp. 3-4 and 9-56. Fearon-Pitman Learning, Inc., 1975.

Developing Computational Skills by Marilyn N. Suydam and Robert E. Reys. See especially "Games: Practice Activities Basic Facts" by Robert B. Ashlock and Carolyn A. Washbon, pp. 39-50, and "Suggestions for Teaching the Basic Facts of Arithmetic" by Edward J. Davis, pp. 51-60. National Council of Teachers of Mathematics, 1978.

C. RESOURCE UNITS, PAMPHLETS, BROCHURES, ETC.

Money by Pamella Pruett. See especially pp. 1-33. Project Consumer, A Livonia Public Schools Project (with the Consumers' Education Office, Department of Health, Education, and Welfare, 1978).

Dollar Points by the Federal Reserve Bank of Boston, 1979.

2

Checking Accounts

INTRODUCTION

The chief aims of Chapter 2 are to provide information about various check using practices, including filling in forms, keeping records, and most importantly keeping a balanced checkbook. Specifically, this chapter explains why checking accounts are useful, identifies terms used in maintaining a checking account, demonstrates the importance of keeping records and filling in forms, reinforces the correct usage and spelling of numbers as used in writing checks, shows the correct methods of endorsement, explains the proper methods of maintaining a personal checking account record, introduces the bank statement, and finally demonstrates how to reconcile (balance) the account.

Maintaining a checking account is placed early in this text because of the type of math skills needed to be successful, primarily addition and subtraction. The notion of a checking account may seem far off for some students (depending on their age), but the relative simplicity of most of the tasks virtually assures success and will help motivate students to try more difficult problems when they arise.

Once the terms and bank routines are explained, the learning experiences are presented in a developmental fashion, the easiest appearing first. This approach allows for a sequential development of skills and provides the flexibility for selecting assignments according to individual needs.

OBJECTIVES

If all the topics in this chapter are chosen by the teacher, the student should be able to:

1. State three advantages for using a checking account.

2. Recognize various terms pertinent to checking, such as those listed in Topic II of this chapter.

3. Identify and complete forms such as: an application for a checking account and a signature card, a deposit slip, a check, a check stub, and a check register.

4. Show the correct usage and spelling of numbers for dollar figures as required on checks.

5. Understand and use the correct type of check endorsement.

6. Properly maintain a record of checks using a check stub or a check register.

7. Read and recognize the use of the bank statement.

8. Reconcile (balance) a bank statement.

CONTENT OVERVIEW

Topic I: Why Use a Checking Account?

Historically, some form of a check has been used in trade for many years. As early as 352 A.D., Romans were said to have accepted a written document to order the transfer of money from one holder to another. More general use did not occur until much later: around 1500 in Holland and the latter part of the 1600s in England. The "Fund at Boston in New England" was the forerunner of checking in America in 1681.

In effect, a check is a written order instructing someone to pay money. It is used in lieu of actual cash. Eventually real cash is transferred, but not "on the spot" when a purchase is made or a service is paid for. Checks, therefore, allow someone to make a purchase without the necessity of carrying large sums of money. The check can be cashed only by the person (or firm) it is written to and is of no value to someone else. Thus, one advantage of using checks is *security*. Other advantages include the ease of payment and the fact that the check provides a record of money spent and a proof of purchase.

Most businesses will accept a check in payment. This fact makes it easier to pay for items by mail, when pocket cash is short, or when an unexpected payment is necessary. However some form of identification is required, and there are a few businesses (e.g., restaurants and gas stations) that will not accept personal checks unless they know the writer personally.

The check register (or stub) and the cancelled check offer excellent records and proof of payment. The bank also gives a monthly report or bank statement that details all the checks written for that account. However, a fee or a minimum deposit is usually required. Even with safeguards, errors can be made and accounts can be overdrawn. These errors are often costly, and it is precisely because of these problems that this chapter has been written. Checking account errors are not limited to the incompetent; the most intelligent people are likely to err once in a while. Banks, too, can make a mistake, and it is important to know just how to spot mistakes and verify records.

Some knowledge about bank operation is also useful. How does a bank service a checking account? First of all, an individual who wishes to have an account must have an initial amount of money to make the first deposit, select the bank to deal with, and choose the type of account desired. Basically there are two types of account: regular and special.

A special account is for people who write only a few checks per month (say, 5 or 6). There is usually a service charge and a charge for each check that is written. The regular account has many variances: some with service and checkwriting charges and others with no charges if a minimum balance is maintained. It pays to investigate each choice carefully before making a final decision.

In addition, personal checking accounts can be either joint or individual. An individual account has only one holder, and only that person can write checks on that account. A joint account has more than one holder (usually two), and *either* one of the people can write checks on that account.

Once these choices are made, an application form and a signature card must be completed. Normally these are both on the same card. One side has information about the applicant; the other contains the agreement between the bank and the holder(s) signed by the holder(s). This signature is very important. It represents the legal manner in which the holder will sign all checks. It is used by the bank to verify the authenticity of a signature. A signature should be written and not printed.

With the completion of the card the applicant then deposits an amount of money in the bank and in effect tells the bank to "hold this money until such time as I give you written notice [a check] to give all or a portion of that sum to another party." As more funds are needed, the holder is responsible for making the appropriate deposits. The bank, therefore, serves as an agent to transfer money upon the written demand of the owner of that money. Because there are many different banks, this process can get complicated, so a system of banking known as the clearinghouse has been established to handle multiple transfers for many accounts at a time. There are local, regional, and national clearinghouses. In the United States this system was established by law in 1913 by the Federal Reserve Act; thus the name Federal Reserve System.

Each year the Federal Reserve System handles over 10 billion checks, equal to about $4 trillion. This does not include the transactions made by local clearinghouses. Some other interesting statistics are: There are over 100 million checking accounts in the United States today. About one half of 1 percent (1 out of 200) of the checks written are returned because of improprieties, such as insufficient funds, forgery, nonexistent accounts, stopped payment, or postdated (dated ahead of the current time). Each year the sum total of all checks increases at a rate of about 7%: about 30 billion checks were written in 1975. At this rate, by the year 2000 banks will be handling well over 165 billion checks per year!

Because of this increase banks are looking at alternative methods of money transfer. Electronic transfer that eliminates paper notes and utilizes credit cards may be a partial solution. Paying a bill or making a deposit may become as easy as picking up the telephone. However, a cautionary note should be sounded. As there is greater reliance upon electronic transfer, overall paper checks and reports will decrease, but the individual responsibility for accurate recordkeeping and accounting will probably increase, since it will be incumbent upon the individual to record all transactions carefully in order to verify the periodic electronically generated reports from the bank.

Topic II: Terms

Following is a glossary of terms used with checking. Anyone who intends to maintain a personal checking account should be familiar with all of them.

Cash — Currency, coin or paper money, used as legal exchange in a country.

Check — A written demand or order instructing a bank to pay money.

   Blank Check — A check form that has not been completed. Blank check also refers to one that is not preprinted with depositor's name and account number.

   Cancelled Check — A check that has been paid. It is marked paid and returned to the checkwriter. Its value is subtracted from the checking account.

   Cashier's Check — A check purchased by an individual or party (the drawer) and issued by a bank. The bank itself guarantees that payment will be made and is solely responsible for it.

   Certified Check — Similar to the cashier's check, but both the drawer and the bank are liable.

   Draft — A check. A bank draft is a cashier's check. A credit union draft is a check issued by a credit union drawn from an account of a credit union depositor.

   Forged Check — A fraudulent check, an imitation passed off as genuine.

   Money Order — A check issued for a fee, usually on a one-time basis. Banks, post offices, loan associations, credit unions, and stores sell money orders. The seller itself guarantees that payment will be made and is solely responsible for it.

   Personal Check — A check drawn from an individual's own account at a bank. The individual (writer) guarantees that payment will be made and is solely responsible for it.

   Traveler's Check — A special check, usually sold in sets, issued by a financial institution and sold at a small cost to individuals for use when traveling. The issuing institution guarantees that payment will be made. Its security lies in the fact that each check must be signed by the purchaser in the presence of a bank officer when they are issued. They are then signed again, witnessed by the payer, at the time the check is used.

Checkbook — A book containing detachable blank checks and deposit slips issued by the bank in which the depositor has an account.

   Checkbook Balance — The exact amount of money on deposit in a checking account at any given time.

Checking Account — An account into which money is deposited for the purpose of writing checks. Also called "a demand deposit."

   Individual Account — A checking account held (owned) by one person.

   Joint Account — A checking account held (owned) by more than one person. Any of the holders may write checks without the consent of the other(s).

   Regular Account — A checking account with normal banking features.

   Special Account — A checking account with special features, usually for those who write very few (less than 6) checks per month.

Check Register — A checking account record, a record of checks that have been written, usually separated from the blank checks in a checkbook.

Check Stub — A record of checks that have been written; like a register but is attached to the blank check and remains in the same section as the checks.

Credit — To add to one's account; e.g., to add $100 to a checking account is to *credit* that account for $100; each time a deposit is made, a credit is entered for the amount of deposit.

Currency — Coin or paper money used as legal exchange in a country.

Debit — The opposite of credit, to subtract from one's account; e.g., to debit an account $100 is to subtract $100 from that account; each time a check is written a debit is entered for the amount of the check. Debits may also be charges against an account.

Deduction — A subtraction; a debit is a deduction.

Demand Deposit — A checking account.

Deposit — The amount of money placed into (added to) a checking account.

> *Deposit Form, Deposit Slip, or Deposit Ticket* — A form on which a precise, itemized list of all that is deposited is shown in detail: coins, dollars, and checks.
>
> *Deposit Receipt* — A record and proof of deposit showing when and how much was deposited into a particular account.

Drawer — The person writing the check; the drawer orders the bank to pay the amount shown on the check from his account.

Endorse — (also spelled "indorse") To write one's name (signature for individuals) on the back of a check in order to cash it or, with additional written instructions, pass it on to another party. The first endorsement must be made by whomever is named on the check; after that, whomever is given the check may endorse it.

Entry — The writing down (on a check stub or register) of a record of a check written on an account, a deposit made, a debit, or a credit.

Fee — A charge or expense for using a checking service.

Legal Tender — Money or currency that, by law, must be accepted to pay debts. Checks are *not* legal tender.

Money — That which serves as an acceptable or common medium of exchange.

Outstanding — Those checks or deposits not shown on a bank statement. Outstanding checks are those that have not yet been cancelled. Outstanding deposits are those that have not yet been credited.

Overdraft or Overdraw — The situation that exists when a check is drawn for more than the balance on deposit in an account. That check can be returned to the drawer or to the payee, and usually no funds are transferred. A fee is charged against the drawer's account and entered as a debit.

Payee — The person or organization to whom a check is written and made payable.

Signature — A person's name written by himself.

Statement — A financial account.

>*Reconciling or Balancing a Bank Statement* — Procedure followed to verify that the checkbook balance and the bank statement agree and the correct balance is recorded on the stub or register.

>*Bank Statement* — A monthly accounting, issued by the bank, showing all transactions (checks, deposits, and other debits and credits) made through a checking account; all cancelled checks are returned with the statement.

Stop Payment — Instruction by a drawer of a check that he does *not* want the payee of a check to receive the money previously ordered; the drawer instructs the bank to refuse payment. A fee is usually charged, and a debit is entered on the drawer's account.

Verify — To prove that something is true or correct. To verify a balance is to prove it is the correct balance.

Void — To annul or nullify a check. When written upon a check, the check is of no value.

Withdraw or Draw — To take out; a check is a means to withdraw money on deposit in a checking account.

Topic III: Forms

APPLICATION AND SIGNATURE CARD

A local bank will supply a copy of a checking account application and signature card. Note that the information requested is of a general nature and includes the applicant's age and taxpayer identification number (social security number). Most banks do not stipulate an age for the applicant, but if the applicant is under eighteen a parent or guardian signature may be required, and the applicant must be able to write his or her own name. Some banks may require that a youthful applicant have a driver's license. Also included on the application form is a choice of the type of account desired and important statements as to the obligations and responsibilities of both the bank and the depositor.

The signature itself is very important. Although banks will generally accept a signature unless challenged, it is advisable that the applicant sign the signature card in the same manner that he or she is likely to repeat over and over again as checks are written. No one can be held liable for any document if his or her signature does not appear on that document. A preprinted check with all information but no signature is of no value to the payee. This also means that in a joint account, if one party overdraws the account upon signing a check (or checks) that is greater than the balance, the other party in the joint account (who did not sign the check) cannot be held responsible.

Often a person will attempt to sign in a unique way in order to discourage forgery. If a personal check is cashed with a forged signature, some inconvenience and possible embarrassment is likely. However, if the forgery is the result of the depositor's negligence and not the bank's, then the depositor is held liable. This is the case when a depositor carelessly leaves a checkbook lying about, so that it is available to unauthorized persons, or fails to report incorrect or not-received monthly bank statements.

CHECKING ACCOUNTS

DEPOSIT TICKET

A facsimile of a deposit ticket is shown on Reproduction Page 13. The deposit ticket is simply an order to place money into an account. It is important, therefore, that the correct amounts be shown and the appropriate account be credited. To assure credit to the proper account, usually a printed form is used that has the name and address of the depositor, the bank name, the bank number, and the account number. If a blank form is used, only the bank name and number appear. The depositor must carefully write his or her name and address (exactly as they appear on the signature card) as well as the correct account number. The correct account number is especially important.

Figure 2-1
Listing Cash and Checks

CASH →	27	00 38	
LIST CHECKS SINGLY 04-567	15	01	
01-234	37	29	88-13
02-345	2	13	533
06-789	10	00	USE OTHER SIDE FOR ADDITIONAL LISTING
TOTAL FROM OTHER SIDE	203	15	ENTER TOTAL HERE
TOTAL ITEMS TOTAL	215	28	BE SURE EACH ITEM IS PROPERLY ENDORSED

The responsibility to be sure that the amount credited for deposit is correct rests with the depositor. Focusing on the boxed section at the upper right of the ticket (see Figure 2-1), several items are identified.

Cash — Refers to the coins or currency deposited.

Checks — This section contains several lines on the front *and* the back. Each check is to be listed separately. Recording the *bank number* (*not* the check number) on the left and the amount of the check in the appropriate dollars column and cents column on the right.

Total From Other Side — Add the total of all checks listed on the *back* of the ticket and record that total here.

Total — Add Cash, Checks, and Total from Other Side and record the total here.

Less Cash Received — This is the amount of cash the depositor wishes to retain and *not* deposit. It is to be subtracted from the previous total.

Net Deposit — Subtract Cash Received from Total and enter the difference here.

THE DEPOSIT RECEIPT

This is a printed record of the deposit made via the ticket. It is a proof of deposit showing when, where, how much, and into what account a deposit has been entered.

CHECK

A check is a written demand or order instructing a bank to pay money. It tells who and how much is to be paid. Preprinted checks are preferred and usually will be more widely accepted. A preprinted check will have all the features of a standard check form, plus the name and address of the account owner, the account number, and the check number. In either form, a check, once written, should not be altered. If a correction is to be made, the old check should be marked "void" (or better yet, torn up and discarded). Then a new check can be written. Also, the check register or stub should have "void" recorded for the check number corresponding to the check destroyed.

Topic IV: Correct Usage and Spelling of Numbers

When a check is written, the amount to be drawn is written both in numbers and in words. Bank rules stipulate that the words are given consideration over figures unless the words are not clear. Furthermore, handwritten terms are given priority over typewritten or printed terms, and typewritten terms control printed terms. For these reasons it is important that correct usage of the written words for dollar figures be learned.

In general, blank spaces are to be avoided anywhere on a check. This means that wherever figures are written, spaces before, between, and after each digit must be so small that another figure cannot be inserted without being obtrusive. Since words govern figures, the same rule holds for spaces between words for dollar amounts. Thus, the use of the word "and" to separate the whole dollar words and the fraction of a dollar as well as the use of the scribbled line from the fraction figure to the printed "dollar" are means to control otherwise blank spaces. The first word on the line is written as near to the left edge of the check as possible in order to prevent any insertions. The word "and" is to be used *only* to separate dollars and cents, that is, at the position of the decimal point. The cents are always written as a fraction of a dollar (denominator 100) immediately following the word "and." All of these procedures help to prevent alterations by unscrupulous persons. Words should be written clearly and in the same hand as the signature. Typewritten amounts are, of course, acceptable, but remember that handwritten terms have control over typewritten terms. Checkwriters are, therefore, obliged to practice a legible, unique style and learn the correct spelling of numbers.

Topic V: Endorsements

When someone is issued a check (the payee), it is necessary for that payee to sign the check in order to cash it. This is called an *endorsement*. In addition, the endorsement may carry additional written instructions to transfer the check to another party. That party must also endorse the same check in order to cash it. The person named on the check must make the first endorsement, however. An endorsement, therefore, will transfer the check from one party or another. The endorsement normally appears on the back of the check and may take one of several forms, the most likely of which are: blank, special, and restrictive.

BLANK ENDORSEMENT

A blank endorsement simply carries the signature of the payee designated on the check. It has the effect of converting the check to cash. The check is then negotiable and can be used by anyone. If it is lost or stolen, the finder can cash it as if the check were written to him in the first place.

SPECIAL ENDORSEMENT
In order to restrict who can cash the check, the payee may write an endorsement that designates a specific party to be paid. The words "pay to the order of . . ." are written prior to the name of the person to be paid and then signed by the payee named on the check. With this *special endorsement*, the party named must also endorse the check before receiving payment, and only the named party may do so.

RESTRICTED ENDORSEMENT
If a check is to be deposited in a bank, an endorsement that restricts its use to that purpose only may be made by writing the words "for deposit only," naming the bank and account number into which the check is to be deposited, and then signed by the payee named on the check. The check cannot be used for any other purpose and is safe to be sent in the mail.

OTHER ENDORSEMENT RULES
If the payee's name is misspelled on the front of the check, the endorsement should first be signed as written on the check and then signed again, correctly spelled. If there are two or more parties named as payees, endorsement rules are as follows. If the payees are named jointly with the word "and" connecting each name, then each party named must endorse the check. On the other hand, if the parties named are named alternatively with the word "or" connecting each name, then any one party may sign the endorsement and receive payment.

Topic VI: Check Stubs and the Check Register

Each time a check is written or a deposit is made, a record of that transaction must be kept. This can be recorded on either a check register or a check stub. When a checking account is opened, the depositor has a choice as to which method of recordkeeping is to be used.

CHECK REGISTER
A check register is not attached directly to the checks but is part of the checkbook. Registers appear in several styles, one of which is illustrated on Reproduction Page 22. Other forms are simply variations of this style.

CHECK STUBS
A record that is attached to the end of each check is called a check stub. An example of a check stub appears on Reproduction Page 23.
    The following rules apply to check stubs.

1. A new stub is completed each time a new check is written.
2. When a stub is completed, the last entry ("Bal For'd") is to be carried over to the first line of the next stub (also marked "Bal For'd").
3. If a deposit is made, it should be entered immediately on the new (unused) stub. More than one deposit may be entered.

4. The first line ("Bal For'd") is then added to the next lines ("Deposits") and the sum entered on the line marked "Total." If there is no deposit, the "Total" is the same as line one.

5. The amount of the current check is then entered.

6. The amount of other deductions is entered next. These may be bank charges for checking account services.

7. Finally, the amount of the check is subtracted from the total, and then the other deductions are subtracted. The final difference is entered as "Bal For'd," and the process continues on the next stub.

Topic VII: Reconciling (Balancing) the Bank Statement

Each month the bank at which the account is kept will send a report showing all transactions (checks, deposits, and other debits and credits) made through a checking account. This is called a *bank statement*. (See Reproduction Pages 28-34.) At this time the bank will also send all of the cancelled checks drawn from that account. The balance shown on the bank statement and the balance shown on the check register (or stub) must agree. However, there usually are many checks shown on the register but not on the statement, because some checks are written after the statement is printed. Therefore, the balances must be verified by a process that takes the outstanding checks into consideration. The process of verifying the balances is called *reconciling*. To reconcile a statement, the following steps should be completed.

1. Sort all of the cancelled checks and arrange them consecutively by number. Verify that there is a cancelled check for each amount identified on the statement.

2. Go through the register (or stubs) and verify that there is a record of each cancelled check reported. Place a "√" mark at the appropriate spot for each verification. All checks not reported are considered *outstanding checks*.

3. Go through the register (or stubs) and verify each of the deposits shown on the statement. Place a "√" mark in the appropriate spot for each verification. Deposits not reported are considered outstanding.

4. On the register (or stub) enter any debit or credit shown on the statement but not on the register (or stub).

At this point procedures differ slightly, depending upon the form used by a particular bank. Using the sample form shown on Reproduction Page 28, the steps are as follows.

5. On the "Checks Outstanding" form, list and total all checks that have been drawn but are not shown on the statement. Complete steps 6-10 on the "Balance Form."

6. On line 1 of the "Balance Form," enter the balance shown on the front of the statement.

7. On line 2 enter all outstanding deposits.

CHECKING ACCOUNTS 25

8. Add lines 1 and 2 and enter the total on the line that follows.

9. Enter the total of all outstanding checks (last line of the "Checks Outstanding" form).

10. Subtract the checks outstanding from the previous entry. Enter the difference on the final line. This amount should agree with the checkbook (or stub) balance.

If the final line balances (agrees) with the checkbook, there is nothing more to be done (except, perhaps, to initial or "OK" the last balance reconciled on the register).

If, however, the lines are not the same, then the following actions must be taken. First, be sure that all credits and debits shown on the statement are correct and have been entered in the register (or stub). Second, go back to steps 1 and 2 and scrutinize each entry and each cancelled check carefully, double-checking to verify that the amount written on the check is the same as the amount shown in the register and the statement. Third, carefully examine the deposits shown on the statement and the register. Fourth, when a check is first cashed at a bank, that bank will print the amount for which the check is cashed at the lower-right-hand corner of the face of the check. Verify that the amount in that corner agrees with the amount written on the check.

Finally, check the adding and subtracting. Initially, verify the arithmetic on the "Checks Outstanding" form, then the "Balance Form," and then (and most painstakingly) verify the arithmetic for each entry in your record, beginning with the last balance that was reconciled with the previous statement. Remember that subtraction can be checked by adding in reverse (see Chapter 1).

POSSIBLE COMPUTATION ERRORS

Figure 2-2 shows several examples of some of the types of arithmetic errors people make when keeping a checking account record.

Figure 2-2
Possible Computation Errors

Ex. 1

NUMBER	DATE	DESCRIPTION OF TRANSACTION	PAYMENT/DEBIT (−)	✓ T	FEE (IF ANY) (−)	DEPOSIT/CREDIT (+)	BALANCE $ 214 26
X	7/13	DEPOSIT (pay)	$		$	$ 372 35	372 35 / (586) 61

Error: Addition Fact: 3 + 2 = 5
not 3 × 2 = 6

Ex. 2

							BALANCE $ 214 26
X	7/13	DEPOSIT (pay)	$		$	$ 372 35	372 35 / 586 (5)1

Error: Addition, regrouping. Failure to "carry the one" to dimes column. (Similar errors occur when someone "carries the one" when there is no regrouping necessary; this is the case in example 3, below.)

Ex. 3

							BALANCE $ 214 26
X	7/13	DEPOSIT (pay)	$		$	$ 372 35	372 35 / 596 61

Ex. 4

193	9/22	Otto Finance	$176 00	$	$		BALANCE $627 91
							176 00
							551 91

Error: Subtraction; incorrect regrouping. Note that this could also be a reversal error. (2-7) is correct. (7-2) is a reversal error.

Ex. 5

193	9/22	Otto Finance	$176 00	$	$		BALANCE $627 91
							176 00
							803 91

Error: Adding when subtracting is required. Note that cents equal 91 either way.

Ex. 6

X	9/25	DEPOSIT (birthday gift)	$		$	$10 00	BALANCE $426 83
							10 00
							416 83

Error: Adding when subtracting is required.

Ex. 7 Using the above (example 6), if this is the last line on this page, the balance must be brought forward to the top of the next page. Another error may occur there. For example, one might mistakenly write:

BALANCE $461 83

Incorrectly copying numbers, transposing positions.

Ex. 8

216	3/4	Hab's Dashing Store (sport coat)	$	$	$		BALANCE $354 86
							42 51
							396 35

Error: Subtracting cents columns but adding dollars columns. (Similar error could be to add cents and subtract dollars.)

Ex. 9

1801	6/23	P&U Market (groceries)	$	$	$		BALANCE $342 71
							18 01
							324 70

Error: Check number incorrectly entered as amount of check.

Often the method of entry will contribute to the error. The mistakes below are not uncommon when a single line entry is used.

Ex. 10

1315	3/7	Hurts Car Rental	$		$	$27 10	BALANCE $346 21
							319 11
1316	3/8	All Inn Motel				39 73	39 73

Error: Amount of check entered as balance instead of subtracting the check amount and entering the difference. Similarly, a deposit may be incorrectly entered as the balance when it should be added instead.

LEARNING EXPERIENCES

THE DIAGNOSTIC SURVEY AND INDEPENDENCE SCALE

- Use the **Diagnostic Survey** on Reproduction Page 11 to estimate each student's degree of competence prior to instruction in the topics of this chapter.

- Upon completion of the Survey, use the **Independence Scale** on Reproduction Page 12 to identify those skills that need to be strengthened by further instruction and practice.

- Utilize the results of these two instruments to determine the appropriate assignments from the following list.

Topic I: Why Use a Checking Account?

- Assign research reports (oral or written) on the following topics.
    1. The history of checks and checking. Show how the need for security, convenience, and records of transaction led to checks and are, today, the three main advantages of checks.
    2. How does a bank work?
    3. What is the Federal Reserve System?
    4. What will future banking be like?

Topic II: Terms

- Have students do the following exercise to learn terms and definitions (sometimes called "memory" or "concentration"). Select all or some of the terms defined in the "Content Overview" and write each term on a small (3" X 5" or smaller) card, one term per card. On another set of cards write the definition of each term, one definition per card. Write on one side of the card only. Mix the sets together and shuffle the cards. Place each card face down on the table and have a student turn up one card and then another. If a term is matched with the correct definition, the student keeps the pair and takes another turn. If there is no match, then both cards are turned over and another student takes a turn. The object is to remember where each revealed card is and match it with its pair when the opportunity arises.

- Use the list of terms given in the "Content Overview" for student practice or to test students on recall of definitions. Select only a few terms at a time. (Additional practice with definitions appears on certain other Reproduction Pages.)

- For student practice or to test students on recognition of definitions, list a few of the more important definitions and ask students to match each one with the appropriate term. (See **Diagnostic Survey**, Part II.)

Topic III: Forms

- Ask students to practice signing their name until they have a signature they wish to keep as their standard signature. Then, on a sheet of loose-leaf paper, ask all students to write their signature 25 times. Check to be sure that each signature is very nearly the same. Have students exchange papers and try to "forge" another student's signature. Point out the need to write legibly but uniquely.

- Obtain application forms from a local bank and have students complete a sample checking account application form and signature card.

- Refer to Reproduction Page 13 and ask students to identify each part of a deposit ticket.

- Use Reproduction Pages 13 and 14 and ask students to complete a deposit ticket for each of the deposits described.

- Use Reproduction Page 15 and ask students to identify each part of a check.

Topic IV: Correct Usage and Spelling of Numbers

- Use Reproduction Page 16 and ask students to complete the page by correctly writing the words for the numbers given.

- Using Reproduction Pages 17 and 18, students should complete each check correctly for the examples shown. (See also assignments that follow for "Endorsements" and on checking account records. For this section, write checks only.)

- Using Reproduction Pages 13, 17, and 19, ask students to write checks and make deposits as directed. (See also assignment that follows in checking account record.)

- Ask students to examine checks on Reproduction Page 20 and identify as many errors or careless entries as possible. If the check can be easily altered, have students demonstrate the changes. The purpose here is not to teach forgery but to help students realize that if they fail to write a check correctly, someone else may alter their check and cash it for a greater amount. (This assignment may be done with the whole class, using an overhead transparency.)

Topic V: Endorsements

- Use Reproduction Page 21. In Part One, ask students to identify the type of endorsement shown and explain the advantages or disadvantages of each. In Part Two, have students properly endorse the checks given the conditions described.

- Using the checks written in the "Checking Accounts" assignment in Topic IV (Reproduction Page 19), ask students to endorse correctly each check according to the following situations. Students are to assume that they are the person named as payee in the corresponding problem as it appears on Reproduction Page 19.

    1. You wish to receive cash for the check written to you.

    2. You want to deposit all of the check in your checking account.

3. You want to use your check to pay Miss Dee Boat.

4. You want to deposit only part of your check in your checking account.

5. You want to use your check to pay Al O. Wishus, and he wants to cash it. (Show both endorsements.)

6. You want to use your check to pay Red E. N. Able, and he wants to place the entire check amount in his checking account.

7. You want to use part of the check to pay Ann Ditover the $15.00 you owe her. What could you do and how should you endorse the check?

8. You want to use all of your check to pay Tanks A. Lot. She wants to use it to pay Tona Bricks, who in turn wants to cash it. Show *all* endorsements

Topic VI: Check Stubs and the Check Register

- Use Reproduction Page 13 for deposit tickets, Reproduction Page 17 for checks, Reproduction Page 22 for the check register, and Reproduction Page 23 for check stubs as needed for the following assignments. Having completed checkwriting assignments in Topic IV, students are now ready to keep a checking account record for the same assignments. If these assignments (from Topic IV) were not assigned previously, they may be given now so that they include checks, deposits, and records. Although bank statements are not yet a part of these assignments, the teacher serves the same purpose when checking papers and reporting to students. Therefore, when pointing out errors, the instructor ought to help the student make corrections by following procedures similar to reconciling a statement.

- Using the checks already written for Reproduction Page 18, show entries in a check register. Begin with a balance of $170.

- Show entries on check stubs for those checks and deposits already written for Reproduction Page 19. Begin with a $1000 balance.

- The following Reproduction Pages contain problems for checkwriting in the categories shown. They appear in order of increasing difficulty. Select those assignments according to the goals desired to be attained.
    - Reproduction Page 24, checks with deposits made in specified order.
    - Reproduction Page 25, checks with deposits as needed. (The answer to "Test Your Wits" is 40 sows and bucks.)
    - Reproduction Page 26, make deposits as needed but in specified order.
    - Reproduction Page 27, same as the preceding item.

Topic VII: Reconciling (Balancing) the Bank Statement

- Use Reproduction Page 28 and ask students to identify the sections of a bank statement. Questions about the use of a statement may be posed to the students as part of a class discussion.

- Reproduction Page 29 is a modified (partly completed) bank statement for Reproduction Page 18. After students have completed the checking account register for Reproduction Page 18, ask them to reconcile the account by verifying each entry on the statement and the register and entering any debit or credit as necessary. Assume that a cancelled check is returned for each check reported on the statement.

- Reproduction Page 30 is a modified bank statement for Reproduction Page 19. After students have completed the check stubs for Reproduction Page 19, ask them to reconcile their account.

- Reproduction Page 31 is a modified bank statement for Reproduction Page 24. Ask students to reconcile the account.

- The following Reproduction Pages are modified bank statements. Ask students to reconcile the statement with the check record from previous assignments as indicated. Each assignment is of increasing difficulty, as noted in the comments below.

    - Reproduction Page 32, from Reproduction Page 25. Forms partly completed; outstanding checks, no outstanding deposits.

    - Reproduction Page 33, from Reproduction Page 26. Nothing completed; start from scratch.

    - Reproduction Page 34, from Reproduction Page 27. Nothing completed; start from scratch.

ASSESSING ACHIEVEMENT OF OBJECTIVES

Ongoing Evaluation

The extent to which students have mastered the concepts covered under the seven topics in this chapter can be measured by any of the activities assigned to class members individually.

Culminating Activity

The following activity may be done in small groups or as a whole class activity and may be used as an evaluative instrument or as a practice activity prior to the final test. It is designed as a simulation game whereby students do some role playing while carrying out responsibilities in keeping a checking account record. The rules for the game follow. Figure 2-3 lists 40 jobs and their fees.

1. Assign each student one job from the list of 40 jobs available.

2. Each job has a fee for a service performed or a product produced and sold. (If students are allowed to choose their own job, do not reveal the fee until after the job has been selected.)

3. The object is for each student to simulate a real-life situation by attempting to receive as much income and make as many purchases as possible.

4. Students are to pay by check and keep a checking account record (register or stub).

5. Each student may make only one purchase or receive only one fee from the same student but is not required to purchase or receive from everyone.

6. Everyone is limited by the income received. This means that some (those with a higher income) can purchase more than others.

7. There is no single item or service that everyone must buy, but all are expected to continue to buy as long as they are able.

8. Each student is to begin with $52.37. In order to continue to make purchases, each student must collect as many fees as possible.

9. The total accumulated wealth for each student may be determined by adding the value of all purchases plus the balance in the checking account.

10. The activity may be carried out a step further by appointing a banker (or several bankers) who will cancel checks and issue bank statements after all purchases and deposits are made.

11. If a less difficult activity is appropriate, round off the fees and the beginning checking account balance to the nearest dollar.

12. While the activity is going on, circulate among the students and check to see that they are keeping their records up to date and are not overdrawing their accounts.

Figure 2-3
List of Forty Jobs and Fees

Job	Fee	Job	Fee
Secretary	$ 6.15	Welder	$ 8.77
Stenographer	4.75	Machinist	11.18
Store Clerk	4.10	Medical Aide	8.85
Nursing Aide	3.90	Mechanic	8.41
Health Technician	12.15	Farm Mechanic	7.65
Laboratory Clerk	5.75	Carpenter	9.00
Computer Assistant	15.35	Construction Worker	14.25
Office Clerk	4.15	Data Processor	7.95
Waitress	6.32	Governess	8.75
Chef's Assistant	6.38	Electrician	16.85
Hairdresser	7.50	Farmworker	3.75
Draftsperson	9.45	Plumber	17.35
Dental Assistant	12.25	Gardener	10.25
Printer	11.75	Artist	18.50
Cooling Expert	13.85	Electronics Aide	14.32
House Painter	9.45	Clothing Manager	9.65
Advertiser	4.80	Appliance Repairperson	12.90
Auto Painter	13.10	TV Repairperson	11.75
Metalworker	13.18	Woodworker	8.95
Aviator	16.65	Business Machine Repairperson	17.20

Final Evaluation

For an overall evaluation of the students' mastery of the concepts in this chapter, if all topics in the chapter have been taught, a test constructed directly from the "Objectives" listed at the beginning of the chapter can be used. As an alternative, one might consider using the **Diagnostic Survey** as a final test.

RESOURCES FOR TEACHING ABOUT CHECKING ACCOUNTS

Below is a selected and annotated list of resources useful for teaching the topics in this chapter, divided into audiovisual materials, games, and print materials. Addresses of publishers or distributors can be found in the alphabetic list in Appendix B.

Audiovisual Materials

A. FOR LOW ABILITY AND SPECIAL EDUCATION STUDENTS

Banking Series, 5 filmstrips, 5 audio-cassettes, and 20 student workbooks. Interpretive Education, 1977.

Bills—Why We Have to Pay Them, filmstrip and audio-cassette. Interpretive Education, 1977.

B. FOR GENERAL MATHEMATICS STUDENTS

"Proper Checkbook Management," color and sound slides and workbooks. Bankers Systems, 1974.

"How's Your Checkbook Math?" *Consumer Math Cassettes*, produced by F. Lee McFadden. Filmstrip and cassette. The Math House, 1977.

Paying by Checks, cassette, teacher's guide, student worksheets and tests. Media Materials, Inc., 1975.

Reconciling Your Checking Account, cassette, teacher's guide, student worksheets, and tests. Media Materials, Inc., 1975.

Money, Checks, and Banks, set of 6 filmstrips. Universal Education and Visual Arts, 1970.

Checking Accounts, film, 12 minutes. Consumer Reports Films, Consumers Union of the United States, 1975.

Checking Accounts, film, 12 minutes. Consumer Reports Films, Consumers Union of the United States, 1977.

Games

Bank Account by Arthur Wiebe. Creative Teaching Associates, 1976.

Cutting Corners developed by John Schatti. The Math Group, Inc., 1977.

"Managing Your Money," Credit Union National Association Mutual Insurance Society, 1969, 1970.

Print

A. FOR LOW ABILITY AND SPECIAL EDUCATION STUDENTS

Working Makes Sense by Charles H. Kahn and J. Bradley Hanna. See especially pp. 59-78. Fearon-Pitman Learning, Inc., 1973.

Useful Arithmetic, Volume II, by John D. Wool. See especially pp. 73-78. Frank E. Richards Publishing Company, Inc., 1972.

Banking, Budgeting, and Employment by Art Lennox. See expecially pp. 1-21. Frank E. Richards Publishing Company, Inc., 1979.

The Bank Book by John D. Wool. See especially pp. 28-64 and 74-89. Frank E. Richards Publishing Company, Inc., 1973.

Skills for Everyday Living, Book 2, by David H. Wiltsie. See especially pp. 44-82. Motivational Development, Inc., 1978.

Using Money Series, Book IV, Earning, Spending, and Saving, by John D. Wool. See especially pp. 40-45. Frank E. Richards Publishing Company, Inc., 1973.

Michigan Survival by Betty L. Hall and David Landers; also available for all states. See especially pp. 67-99. Holt, Rinehart and Winston, 1979.

Mathematics for Today, Level Red, by Saul Katz, Ed. D.; Marvin Sherman; Patricia Klagholz; and Jack Richman. Sadlier-Oxford, 1976.

B. FOR GENERAL MATHEMATICS STUDENTS

Stein's Refresher Mathematics, Seventh Edition, by Edwin I. Stein. See especially pp. 584-585. Allyn and Bacon, Inc., 1980.

Trouble-Shooting Mathematics Skills, Basic Competency Edition, by Allen L. Bernstein and David W. Wells. See especially pp. 356-360. Holt, Rinehart and Winston, 1979.

Consumer Mathematics with Calculator Applications by Alan Belstock and Gerald Smith. See especially pp. 159-170. McGraw-Hill Book Company, 1980.

Consumer Math, A Guide to Stretching Your Dollar, by Flora M. Locke. See especially pp. 36-52. John Wiley and Sons, Inc., 1975.

Consumer Mathematics, Third Edition, by William E. Goe. Activities book available. See especially pp. 303-316 and 321. Harcourt Brace Jovanovich, 1979.

Mathematics for Today's Consumer by Jack Price, Olene Brown, Michael Charles, and Miriam Lien Clifford. Compiled from selections from *Mathematics for Everyday Life* and *Mathematics for the Real World*. See especially pp. 33-50. Charles E. Merrill Publishing Company, 1979.

Mathematics for the Real World by Jack Price, Olene Brown, Michael Charles, and Miriam Lien Clifford. See especially pp. 33-50. Charles E. Merrill Publishing Company, 1978.

Mathematics in Life by L. Carey Bolster and H. Douglas Woodburn. See especially pp. 75-79. Scott, Foresman and Company, 1977.

Mathematics for Daily Living by Harry Lewis. See especially pp. 248-261. McCormick-Mathers Publishing Company, 1975.

Mathematics Plus!, Consumer, Business & Technical Applications, by Bryce R. Shaw, Richard A. Denholm, and Gwendolyn H. Shelton. See especially pp. 122-127. Houghton Mifflin, 1979.

Consumer and Career Mathematics by L. Carey Bolster, H. Douglas Woodman, and Joella H. Gipson. See especially pp. 88-97. Scott, Foresman and Company, 1978.

Business and Consumer Arithmetic by Milton C. Olson and A. E. McVelly. See especially pp. 149-161. Prentice-Hall, Inc., 1974.

Business Mathematics for the Consumer by Mearl R. Guthrie, William Selden, and Delbert Karnes. See especially pp. 165-178. Fearon-Pitman Learning, Inc., 1975.

Brady on Bank Checks by Henry J. Bailey. Warren, Gorham, and Lamont, 1978.

C. RESOURCE UNITS, PAMPHLETS, BROCHURES, ETC.

Personal Banking Services by James Cooper. Project Consumer, A Livonia Public Schools Project (with the Consumers' Education Office, Department of Health, Education, and Welfare, 1979).

Commercial Banking Services and an Overview of Credit by William R. McQuesten. Project Consumer, A Livonia Public Schools Project (with the Consumers' Education Office, Department of Health, Education, and Welfare, 1978).

Buying Furniture by Patricia Arent. Project Consumer, A Livonia Public Schools Project (with the Consumers' Education Office, Department of Health, Education, and Welfare, 1978).

"I bet you thought . . ."; "The Story of Banks." Federal Reserve Bank of New York, 1980.

"Checkbook Management," 1973. "Now Account Management," 1977. "A Mini-Course on Proper Checkbook Management," 1974. Bankers Systems, Inc.

3

Multiplication and Multiple Buying

INTRODUCTION

This chapter reviews multiplication and introduces the concept of multiple purchases (whereby more than one item of the same cost is purchased), demonstrates and provides practice with making change, shows how to figure sales tax, and offers some suggestions for comparing costs and recognizing savings.

In later chapters other types of purchases are included, and the concepts are expanded to include unit pricing, percent discount, and so forth. The ideas introduced here reinforce the mathematics skills already developed and build upon them by expanding from addition and subtraction to simple multiplication and division. The notion of percent is only partially developed here, and calculations are shown in a utilitarian way. This is in keeping with the practical theme of the *Guidebook*. The consumer's most common use of percent, the sales tax, is treated in a specific way. The generalizations regarding the concept of percent are explained in greater detail later in the text, where it is more necessary to understand such things as commissions, discounts, interest rates, and the like.

Certain other themes are recurrent throughout the text, particularly comparing values and decision making. This chapter begins to explore each, utilizing some fundamental skills that are continually sharpened and broadened. In the case of comparing prices, for example, this chapter explains some of the hidden costs involved with mail-order and catalog sales. Later chapters demonstrate more sophisticated means for recognizing value, such as unit pricing.

Likewise, choosing and selecting items from a catalog offer some practice with decision making. This chapter attempts to make the consumer more conscious of the decision-making process so that more important decisions (auto and home purchases, job and career decisions, etc.) may be made more wisely.

> **OBJECTIVES**
>
> If all the topics in this chapter are chosen by the teacher, the student should be able to:
>
> 1. Multiply whole numbers and dollars-and-cents numbers by one- and two-digit numerals.
> 2. Estimate products in simple multiplication problems.
> 3. Calculate the total cost when buying more than one of the same item.
> 4. Identify and use the symbol "@."
> 5. Give (or verify receipt of) correct change, using the minimum number of coins and bills.
> 6. Calculate simple sales tax.
> 7. Compare costs and recognize value in simple purchases.
> 8. Predict prices and verify their accuracy.

CONTENT OVERVIEW

The students have already had experience adding money figures (Chapter 1). In this chapter the teacher should review addition simultaneously as simple multiplication is introduced. Initially, multiplication facts can be reinforced, as necessary, followed by multiplication of dollars and cents by whole numbers (one- and two-digit multipliers). The practical use is obvious when one considers that consumers often encounter an example, such as: "If one blouse costs $7.98, how much do three blouses cost?"

The operation called for is multiplication, and a brief presentation will assist those students who demonstrate a need for more instruction. This chapter offers examples of problems wherein specific types of multiplication skills are identified. The explanation that follows each example can serve not only as a basis for direct teacher instruction (lecture, etc.) but also to help the teacher pinpoint individual needs. These needs are met when the teacher is able to say, "Johnny needs help with three-digit multiplicands that contain a central zero." That is a more analytical appraisal than the indictment, "Johnny can't multiply."

Topic I: Review of Multiplication

The essential terms used in multiplication are:

The *multiplicand* ... the number being multiplied.

The *multiplier* ... the number by which the multiplicand is multiplied.

The *product* ... the answer after multiplication.

Whenever the multiplier is a two-digit number or greater, each of the digits must be used as a multiplier and each answer is called a *partial product*. Thus, a three-digit multiplier will yield three partial products. Each number used to form a product is also called a *factor*. Of course, the *times sign* (X) is the symbol used to indicate multiplication.

Another useful term in discussing multiplication is *array*, an orderly grouping of numbers. Numbers of an array may take the form of groups of 2, 3, 4, and so on to show various multiplication patterns. When the grouping is arranged so that numbers are listed beginning with zero from left to right and top to bottom (from the upper left), it is called a *multiplication table* (also called a *times table* or a *multiplication grid*). The numbers of the multiplication table are arranged so that all the basic facts of multiplication can be easily determined. The basic facts of multiplication are the products of all factors, one through ten. These facts should be memorized, and the table offers an excellent tool to help students learn them. The rows and columns correspond to the factors in the problem. When examining seven rows of three columns, the last number is 21; thus: 7 X 3 = 21. Note that multiplication tables may be made to any size.

All multiplication utilizes the basic facts. Numbers larger than ten are, in fact, expressed with the same symbols (0, 1, 2, 3, 4, 5, 6, 7, 8, 9). It is their placement in the numeral that determines their value. The symbol "27" is read "twenty-seven" and it means two tens and seven. (See Chapter 1 for complete explanation of place value.) Understanding place value and the basic multiplication facts is essential to forming the partial products that lead to the final solution. Below are examples and brief explanations of some specific types of problems encountered in multiplication.

Example 1 One-digit multiplier, multiplicand ending in zeros.

a) 40　　b) 400　　c) 4000　　etc.
　X 2　　　X 2　　　 X 2
　 80　　　 800　　　8000

Here the basic facts are 2 X 0 = 0 and 2 X 4 = 8. But it is the product of zero that often causes errors. For instance, students may neglect the zero product or fail to write all the appropriate zeros. The product should have at least the same number of zeros as the multiplicand. Another common error is to say 2 X 0 = 2. This is often the result of carelessly adding instead of multiplying.

Example 2 Multiplying by multiples of ten.

a) 78　　b) 78　　c) 78　　d) 78　　etc.
　X 10　　X 100　　X 1000　　X 10000
　 780　　 7800　　 78000　　 780000

The rule of thumb is to write the multiplicand as it originally appears, then write additional zeros equivalent to the number of zeros in the multiplier.

In the examples that follow, the process of regrouping or "carrying" is shown in its short form because it is the most widely used *and* the method most likely to cause errors. "Carrying" is a means to facilitate multiplication when the product has two digits. For a more detailed explanation of regrouping, see Figure 3-1.

Figure 3-1
Regrouping Explained

1) 64   64 = 60 + 4   4   60
   × 2   × 2   × 2
                             8   120   8 + 120 = 128

2) 46   46 = 40 + 6   6   40
   × 2   × 2   × 2
                             12   80   12 + 80 = 92

3) 76   76 = 70 + 6   6   70
   × 2   × 2   × 2
                             12   140   12 + 140 = 152

(Similar steps are followed for larger numbers.)

Example 3 Two-digit multiplicand.

a) 24   Both products are one digit with   b) 64   Last product has two digits.
   × 2   no regrouping.   × 2
    48   128

c) ^{1}46 First product has two digits. Re- d) ^{1}76 Both products are two digits. Re-
   × 2   grouping is required. (Note the   × 2   grouping is required.
    92   "1" raised to left of the "4.")   152

Example 4 Three-digit multiplicand.

a) 423   Each product is only one digit.   b) 713   Only final product has two
   × 2   × 2   digits. Regrouping not
    846   1426   necessary.

c) 3^{1}17 Only first product has two digits. d) 8^{1}17 First and last products are two
   × 2   Regrouping once is required.   × 2   digits. Regroup only once.
    634   1634

e) ^{1}1^{1}78 Two consecutive two-digit   f) ^{1}6^{1}78 Three consecutive two-digit
   × 2   products. Regroup twice.   × 2   products. Regroup twice.
    356   1356

g) 403   Zero middle term. No regrouping.   h) 4^{1}09   Zero middle term, first product
   × 2   Common error is to write: 403   × 2   has two digits. Regroup required.
    806   × 2   818   Tendency is to write:
                                      826
                                                         4^{1}09
                                                         × 2   (2 × 1 = 2)   or
                                                         828

                                                         4^{1}09
                                                         × 2   (2 + 1 = 3)
                                                         838

MULTIPLICATION AND MULTIPLE BUYING 39

Example 5 Multiplicand greater than three digits. Many errors already mentioned in Examples 3 and 4 also occur here. In addition, some special cases are noted below.

a) 4003   Two center zeros. No regrouping.   b) 4007   Two center zeros with regrouping.
    X 2   Error may be: 4003   X 2   Error may be:
    8006   X 2   8014   40¹07   4007
                           8226   X 2   or   X 2
                                                                 4024   8004

c) 5007   Two center zeros, last product
    X 2   requires another zero. Error may
    10014   be: 5007
                 X 2
                 1014

Example 6 Multiplier is greater than one digit. Recall from Chapter 1 that a two-digit number represents tens and units; a three-digit number represents hundreds, tens, and units; a four-digit number represents thousands, hundreds, tens, and units; etc. Thus:

a) 7859 means 7 thousands + 8 hundreds + 5 tens + 9 ones.

b)   34 (no regrouping). 34 means 3 tens (or 30) + 4 ones (or 4) = 30 + 4
    X 12 means 1 ten (or 10) + 2 ones (or 2) = 10 + 2.

Long Method   *Short Method*

              34   34   Each line represents a
            X 12   X 12   partial product. The 2d-line
    2 X   4 = 8   2 X 34 =   68   partial product must end
    2 X 30 = 60   10 X 34 =   340   in zero.
   10 X   4 = 40   408
   10 X 30 = 300
             408

                    Be sure columns are lined up.

c) Two-digit multiplier with regrouping.   d) Three-digit multiplier.

                    368   368
                  X 34   X 349
    4 X 368 =   1472   9 X 368 =   3312
   30 X 368 = 11040   40 X 368 =   14720 ← 2d line ends
               12512   300 X 368 =   110400   in zero.
                                                                128432 ↖ 3d line ends in
                                                                ↑↑↑↑↑↑ at least 2 zeros.

                    Be sure columns are aligned.

e) Central zero in multiplier.   f) Central zero in both factors.

        368   308
      X 309   X 207
       3312   2156
       0000   0000
     110400   61600
     113712   63756

g) Multiple central zeros.   h) Multiple ending zeros.

       7008   7000
     X 5009   X 413
      63072   21000
      00000   70000   one "extra" zero.
     000000   2800000   two "extra" zeros.
    35040000   2891000
    35103072

Example 7 Multiply dollars and cents by whole numbers. Multiplication of this type proceeds in the same manner as with whole numbers. The product will have the same number of digits to the right of the decimal point as the multiplicand. Be sure to change all cent figures to dollar figures before multiplying.

a) 91¢ $ .91
   X 3   X 3
           $2.73

b) 47¢ $ .47
   X 9   X 9
           $4.23

c) 83¢ $ .83
   X 17   X 17
            581
            830
          $14.11

d) $2.34
   X 2
   $4.68

e) $ 6.34
    X 2
   $12.68

f) $ 6.37
    X 2
   $12.74

g) $ 6.87
    X 2
   $13.74

h) $ 2.34
    X 32
    4 68
   70 20
   $74.88

i) $ 6.87
   X 32
   13 74
  206 10
  $219.84

j) $ 6.07
   X 49
   54 63
  242 80
  $297.43

k) $ 4.06
   X 509
   36 54
   00 00
  2030 00
  $2066.54

Topic II: Estimating Products

A consumer's ability to think quickly and mentally estimate calculations is indispensable. Dollars can be saved when the buyer is able to compare prices of various products by quickly estimating their respective costs. In order to accurately estimate products, two skills are necessary: the ability to round off figures and the knowledge of basic multiplication facts. A brief explanation and examples of estimating products follow.

Generally, dollars-and-cents products are estimated to the nearest dollar, and therefore initial figures are rounded to the nearest dollar. One-digit numbers are usually not rounded off. However, numbers 6, 7, 8, and 9 can be rounded to 10, depending upon the precision desired. Two-digit numbers are usually rounded to the nearest ten. The exceptions are 95, 96, 97, 98, and 99, which are rounded to 100.

The methods for rounding numbers are discussed in greater detail in Chapter 1. At this point, suffice it to say that after determining the digits to be retained, rewrite only those digits needed and change the other digits to zero. If the first digit changed was a 5 or more, increase the preceding digit by one. If the first digit changed was a 4 or less, then the preceding figure remains unchanged. After rounding off, multiply the rounded-off figures. To check the accuracy of your estimate, multiply the numbers completely.

Example 8 Estimating products by rounding factors.

a) 214   round to   200
   X 6   X 6
                       1200

Actual product is 1284

b) 214   round to   200
   X 8   round to   X 10
                       2000

Actual product is 1712

c) 214   or   200
   X 28   X 30
                  6000

Actual product is 5992

d) 91¢ is $.91 or $1.00
  X 3   X 3   X 3
                          $3.00

Actual product is $2.73

e) 47¢ is $.47 or $ .50
  X 9   X 9   X 10
                       $5.00

Actual product is $4.23

f) 83¢ is $ .80 or $ 1.00
  X 17   X 20   X 17
               $16.00 $17.00

Actual product is $14.11

g) $6.34   or $ 6.00
   × 7   × 7
                   $42.00

   Actual product is $44.38

h) $18.27   or $ 18.00
   × 18   × 20
                  $360.00

   Actual product is $328.86

i) $23.57   or $ 24.00
   × 31   × 30
                  $720.00

   Actual product is $730.67

j) $80.51   or $ 81.00
   × 12   × 10
                  $810.00

   Actual product is $966.12

k) $354.49   or $ 354.00   or $ 350.00
   × 28   × 30   × 30
                  $10620.00 $10500.00

   Actual product is $9925.72

Topic III: Multiple Purchases

Consumers often buy more than one of the same item. When considering the cost of multiple purchases, it is more expedient to multiply rather than to add. Another expedient is the use of the business form "@" to abbreviate the phrase "at the cost of," with the price that follows designating the cost per unit.

Example 9 Multiple purchases.

a) The purchase of 4 cans of corn when each can costs 39¢, may be written 4 @ $.39. (Be sure to use $ figures.) The total is 4 × $.39 = $1.56.
(Note that the @ has the same effect as the ×.)

b) The purchase of 3 pairs of socks at 93¢ each becomes 3 @ $.93, which is 3 × $.93 or $2.79 total.

c) 17 gallons of paint that costs $11.99 each is 17 @ $11.99, which is 17 × $11.99 or $203.83.

Further clarification is needed when the quantity purchased is confused by the packaging. For instance, three dozen eggs is 36 eggs, but eggs are priced by the dozen, not per egg. Therefore, the expression that reads eggs, 3 @ $.97, means that each dozen (not each egg) costs $.97. The expression "@ $.97" still means "ninety-seven cents per unit," but the unit of measure here is the dozen. Other units can also be signified—units may be case, carton, pack, package, and the like—whatever is appropriate for the item. Careful buyers will check price labels to be sure they are getting the best buy.

Consider this consumer's plight. The sign at the store read: "3 for $2.89." The basket of socks to which the sign referred held several packages, none of which was marked with the price. Each package contained 3 pairs of socks. The buyer assumed that the sign referred to the packages (3 packages for $2.89) and picked up 3 packages. Having made several other purchases, it was not until the consumer arrived home that she realized she had spent $8.67 for 9 pairs of socks. The sign was misinterpreted as meaning:

3 *packages* for $2.89

when in reality the sign meant:

3 *pairs* for $2.89 or

3 packages $8.67

Example 10

Suppose that in planning for a family trip a consumer bought 3 cartons of individual serving boxes of cereal. Each box is priced at 12¢. Is it correct to indicate the purchase as: 3 boxes of 10 cartons @ 12¢?
No: The correct way to express the purchase is:

$$30 \text{ boxes @ } 12¢$$

for a total of 30 × $.12 or $3.60.

Topic IV: Receiving Change

Store clerks are just like any of the rest of us; they can make mistakes. So it is not only appropriate but also very wise to count your change immediately upon receiving it. This requires a good deal of mental arithmetic and quick calculation. A good way to do this is to figure out the exact change (type of coins, etc.) before you receive it. Because there are various denominations of coins (pennies, nickels, dimes, quarters, etc.), checking to be sure that the change is correct requires more than merely subtracting. A knowledge of coin and bill values and their possible combinations is necessary. Therefore, instruction ought to begin with a review of coin values and should be followed by lessons demonstrating how many of each coin are need to make a certain total. Finally, students should be given exercises that require verifying change with the least amount of bills and coins.

Example 11 Counting change.

a) Use at least one of each of the following coins to make 68¢: pennies (p), nickels (n), and quarters (q).

Solution:

Begin with the largest coin.

$$2\,q \text{ is } 2 \times 25¢ \text{ or } 50¢$$
$$3\,n \text{ is } 3 \times 5¢ \text{ or } 15¢, \text{ running total is } 65¢$$
$$3\,p \text{ is } 3 \times 1¢ \text{ or } 3¢, \text{ running total is } 68¢.$$

b) Make a total of 85¢ using at least one of each coin, nickels, dimes (d), and quarters.

Solution:

$$2\,q \text{ is } 2 \times 25¢ \text{ or } 50¢$$
$$3\,d \text{ is } 3 \times 10¢ \text{ or } 30¢, \text{ running total is } 80¢$$
$$1\,n \text{ is } 1 \times 5¢ \text{ or } 5¢, \text{ running total is } 85¢.$$

c) Use four standard U.S. coins to make 70¢.

Solution:

$$2\,q \text{ is } 2 \times 25¢ \text{ or } 50¢$$
$$2\,d \text{ is } 2 \times 10¢ \text{ or } 10¢, \text{ running total is } 70¢.$$

d) Use nine standard U.S. coins to make $1.15.

Solution:

$\quad\quad\quad\quad$ 3 *q* is 3 × 25¢ or 75¢

$\quad\quad\quad\quad$ 2 *d* is 2 × 10¢ or 20¢, running total is 95¢

$\quad\quad\quad\quad$ $\underline{4\ n}$ is 4 × 5¢ or 20¢, running total is $1.15.

$\quad\quad\quad\quad$ 9 coins

e) Use the least number of coins and bills named to obtain the designated total. Select only those coins and bills that are needed from a selection of:

Bills of $20, $10, and $2. $\quad\quad\quad$ Coins of *$1*, *q*, *d*, and *p*.

1) Make 35¢. $\quad\quad\quad\quad\quad\quad\quad\quad$ 2) Make $28.42.

Solution: 2 coins, one *q* and one *d*. $\quad$ *Solution:* 5 bills,

$\quad\quad\quad\quad\quad\quad\quad\quad\quad\quad\quad\quad\quad\quad\quad$ 1 @ $20 and 4 @ $2 plus
$\quad\quad\quad\quad\quad\quad\quad\quad\quad\quad\quad\quad\quad\quad\quad$ 6 coins,
$\quad\quad\quad\quad\quad\quad\quad\quad\quad\quad\quad\quad\quad\quad\quad$ 4 *d* and 2 *p*.

f) Tell what your change would be for each of these purchases for the amount given to the clerk. Name the change specifically (how many standard U.S. bills and coins) so that the smallest number of bills and coins are used.

Amount of Purchase	Amount Given to Clerk	Change Received
$ .48	$ 1.	2 *p* and 1 *h-d* = $.52
$33.56	$50.	Bills of 1 @ $10, 1 @ $5, 1 @ $1; coins of 1 *q*, 1 *d*, 1 *n*, and 4 *p* = $16.44

Topic V: Sales Tax

Most states (and some cities) now have a sales tax on consumer purchases. Some states restrict this tax so that food and drugs are not taxed. Others do not tax labor or service charges. Some have a special rate for certain purchases, such as motels, hotels, restaurant meals, and entertainment. In nearly every case, however, the tax is determined as a percent of the purchase price. In this chapter we shall introduce percent calculations in terms of the sales tax and limit our use of percents to utilitarian methods of figuring sales tax. A more in-depth presentation of the meaning of percent and its other uses in made in another chapter. At this point, it is sufficient that the student be able to convert a given percentage figure to a decimal number and multiply.

The sales tax rate varies from state to state, and although all rates may not be specifically represented, we will concentrate on the following sales taxes:

$\quad\quad\quad\quad$ 1%, 1 1/2%, 2%, 2 1/2%, 3%, 3 1/2%, 4%, 4 1/2%, 5%.

A percent means "per 100" or "for each 100," 3% means "3 for each 100." Since there are 100 cents in a dollar, a 3% tax means "3 cents for each dollar." Therefore, if there is a purchase of $4, a 3% tax would amount to $.12. Mental logic or intuition works well for full dollar figures or for estimating but breaks down when considering the exact sales tax on, say, $6.78. In order to calculate this tax accurately, you must multiply. And in order to multiply, the percent figure must be changed into a decimal fraction.

Percents can be very easily converted to decimal fractions because the expression "per 100" can be interpreted as a fraction whose denominator (bottom number) is 100; thus, 4% become 4/100. A decimal fraction is simply a special fraction whose denominator is a multiple of 10 (10, 100, 1000, etc.). (See Chapter 1.) If the denominator is 100, the decimal numeral is written two places to the right of the decimal point. Consider these conversions.

$$1\% = 1/100 = .01$$
$$2\% = 2/100 = .02$$
$$3\% = 3/100 = .03$$
$$4\% = 4/100 = .04$$
$$5\% = 5/100 = .05$$

This process can be seen as simply moving a decimal point. Since each percent figure can have, in fact, a decimal point at its right (2% is 2.%, etc.), then the procedure is to move the decimal point two places to the left, using a zero as a place holder. Why two places? Because two decimal places represents hundredths: Thus,

2% = 2.%   Move the point two places to the right; use a zero as a place holder; and drop the percent sign.

.02.% = .02

Percent figures with fractions are best understood by initially changing the fraction to a decimal part of the percent and then converting the entire percent to a decimal numeral by the method just described. Consider the examples shown in Figure 3-2.

Figure 3-2
Converting Percents to Decimals

½% = .5%	.00.5% = .005
1½% = 1.5%	.01.5% = .015
3½% = 3.5%	.03.5% = .035
7¼% = 7.25%	.07.25% = .0725
10¼% = 10.25%	.10.25% = .1025

To calculate the amount of tax owed on a purchase when the tax rate is known,

1. Change the % to a decimal and
2. Multiply.

You will recall that in Chapter 1 it was recommended that all money figures be changed to dollars and cents (decimals). This practice should continue throughout the text. The numeral "$.93" should be used instead of "93¢." This is necessary both when using paper and pencil and when using a calculator.

MULTIPLICATION AND MULTIPLE BUYING

Multiplication of two-decimal numerals proceeds in the same manner as the multiplication already described. The partial products are obtained by multiplying as though the multiplier and multiplicand are whole numbers. The sums of the partial products are written in the same manner. However, the final answer must have as many places to the right of the decimal point as the sum total of digits to the right in the multiplicand *and* the multiplier. Thus, in the problem $4.83 X .035 there are two places to the right of the point in the multiplicand and three places to the right in the multiplier. The total number of digits to the right of the point is 5; therefore, the final product must show 5 places to the right of the point.

```
                          $  4.83
                          X .035
          483 X  5  =      2 415
          483 X 30  =     14 490
                          $.16905
```

Zeros as place holders. When counting the number of digits in the final product, begin from the right and count to the left. If not enough digits are present, use as many zeros as necessary and place them to the left of the product obtained. For example:

```
           $  4.83
           X .015
             2 415
             4 830
           $.07245
```

Rounding off. In this section we will be rounding off to the nearest cent. (See Chapter 1 for more explanation.) The cent figure is at the second position to the right of the decimal. To round off to the nearest cent, examine the digit to the right of the cent figure.

   a. If it is 5 or more, increase the cent figure by one and drop the digits to the right.

   b. If it is less than 5, leave the cent figure unchanged and drop the digits to the right.

To illustrate, the answers to the two previous examples are rounded as follows:

$.16905 is rounded to $.17.

$.07245 is rounded to $.07.

Calculating sales tax. Note in the examples that follow, the product is only the amount of *tax*. In order to calculate the total cost of the purchase, add the tax to the purchase price.

Examples involving sales tax.

Directions: Figure the tax and total cost.

Example 12.

   1% sales tax on a purchase of $7.86.

   1% = 1.%   .01.% = .01

```
   $ 7.86
   X .01
   $.0786  is rounded to  $ .08  Tax
                         + 7.86  Purchase
                          $7.94  Total
```

Example 13.

    4-1/2% sales tax on a purchase of $42.73.

    4-1/2% = 4.5%   .04.5% = .045

```
                    $ 42.73
                    X  .045
     4273 X  5 →    21 365
     4273 X 40 →   170 920
              $1.92285 is rounded to  $ 1.92  Tax
                                     +42.73  Purchase
                                     $44.65  Total
```

Use common sense. Recall that a 4% tax means only $.04 is added for each dollar spent and that *tax can never be greater than the purchase price.* This knowledge will assist in checking the answer for the correct placement of the decimal point. The multiplication may be accurate, but if the decimal point is misplaced, the answer is far off. An advantage of calculation with dollars-and-cents figures is that one can verify the answer by asking, "Does this make sense? Is the answer reasonable?" The following examples illustrate some possible errors in placement of the decimal point.

Example 14   4% sales tax on a purchase of $9.37.

$ 9.37   *(Error is in not changing 4% to the proper decimal number.)*
  X 4
$37.48 Tax

Is it reasonable to pay a $37 tax on a $9 item? Of course not! Go back and redo the problem.

Example 15   2-1/2% tax on a purchase of $2.50.

2-1/2% = 2.5%   .02.5% = .025

$2.50   *(Error is counting places only in the multiplier instead of counting places in*
X .025   *multiplier and multiplicand.)*
1 250
 5 00
$6.250

Does it make sense to pay $6 tax on a $2.50 item? No! Redo the problem.

Example 16   3% tax on a purchase of $234.87.

3% = 3.%   .03.% = .03

$214.87   *(Error is in counting digits to the left of the decimal instead of to the right.)*
 X .03
$ 64.461

This one could get by students unless they are good at mental calculations. Three cents for each dollar is also $3 for each $100. Since the purchase is a little more than $200, the tax should be a little more than $6. Redo this one, too.

MULTIPLICATION AND MULTIPLE BUYING

Topic VI: Using Multiplication to Compare Costs and Recognize Value

Producers and storekeepers often package items in larger quantities to encourage people to "buy more and save." But is it always the case that the larger quantity is cheaper? In order to determine whether or not a larger package is less expensive, the buyer can multiply the price on the smaller package to determine what it would cost for the same amount as the larger package. That is, if the same item sells for 35¢ in a package of one and $1.00 in a package of three, multiply the price on the package of one by 3 and compare with the larger package.

For each of the following examples, tell which package is the best buy.

Example 17 A package of 3 bars of soap for $1.00 or a package of one for 35¢.

Solution: 35¢ = $.35 $ .35
                              X 3
                            ──────
                            $1.05   *Therefore, the package of 3 is the best buy.*

Example 18 A 28-oz. bottle of ketchup for $.93 or a 14-oz. bottle for $.46.

Solution: 28 oz. is 2 times 14 oz., so multiply the smaller by 2.

$.46
X 2
──────
$.92   *Therefore, the 14-oz. bottle is cheaper.*

Example 19 A 510-g. box of cereal for $.97 or an 85-g. box of cereal for $.16.

Solution: 510 is how many times larger than 85? (Although division is an appropriate means to solve this problem, encourage students to use multiplication, since division skills have not yet been reviewed.) You can determine this by trial and error. The number 85 is near 100; 500 is 5 times 100, so try 5: 85 X 5 = 425. Too small, try 6: 85 X 6 = 510. Therefore, multiply the price of the smaller by 6.

$.16
X 6
──────
$.96   *The smaller is cheaper.*

Example 20 A package of 3 16-oz. cans for $.89 or a package of 3 4-oz. cans for $.25.

Solution: How many total ounces in each?
              3 16-oz. cans is 3 X 16 or 48 oz.
              3 4-oz. cans is 3 X 4 or 12 oz.
              48 is how many times 12? 4!

Therefore, multiply the price of the smaller by 4.

$ .25
X 4
──────
$1.00   *The larger package is cheaper.*

Example 21 A package of 3 what-d'you-callits for $1.00 or a package of 8 for $2.59.

Solution: 3 cannot be multiplied by a whole number to obtain 8. Therefore, this problem requires that both numbers (and their corresponding prices) be multiplied to obtain the same product. This product is called the *least common multiple* (LCM). The LCM for 3 and 8 can be determined by listing the multiples of each side by side and then identifying the smallest product that appears in both lists. The multiples of 3 are 3, 6, 9, 12, 18, 21, ㉔. . . . The multiples of 8 are 8, 16, ㉔, 32, 40 . . . 24 is the LCM of 3 and 8. Change 3 to 24 by multiplying by 8 (multiply the price by 8 also). Change 8 to 24 by multiplying by 3 (multiply the price by 3 also).

3 for $1.00 becomes 24 for $8.00
8 for $2.59 becomes 24 for $7.77

$2.59
× 3
―――
$7.77   Clearly the 8 for $2.59 is the better buy *if* you need eight what-d' you-callits.

This method of comparing prices is advantageous because multiplication is used rather than division (as is the case with unit pricing). Multiplication is more easily grasped, and students will likely find success with this method. However, certain comparisons become cumbersome if not impossible without using division. The multiplication and price comparison techniques demonstrated in this chapter will contribute to a better understanding of division and unit pricing when they are introduced later. Keep in mind the developmental approach that is essential to successful teaching. Good teaching requires that skills be taught sequentially, beginning with the simplest and moving to the more difficult. Success is assured not only because students are better equipped to handle the more complicated skills but also because they are building upon successful experiences.

Topic VII: Predicting Prices

Consumers should cultivate an ability to estimate the price of an item in order to plan expenses. For instance, a shopper who can predict the approximate cost of the week's groceries is able to plan a budget effectively. The ability to predict prices accurately is a skill that can be developed only through trial and error and is directly dependent upon how informed the consumer is. Therefore, some practice with predicting and then verifying prices is suggested. A simple way to verify prices without going to the store is through newspaper ads. Catalogs also serve as convenient shopping guides; however, a word of caution should be noted. The cost of shipping and handling is usually added to the catalog price, making the actual cost of the item slightly higher. This procedure is described in detail in another chapter; shipping costs can be ignored for now, and catalogs can serve as excellent pricing guides. At this point shipping costs are unimportant, since the main focus is on approximating prices. Students can use a form similar to the one shown in the "Learning Experiences" section for Topic VII to practice predicting prices and comparing the estimates to actual prices. The column for "Difference" is obtained by subtracting the higher price from the lower price. If the actual price is higher than the predicted price, the difference is marked with a "(+)." Conversely, if the actual price is lower, the difference is marked with a "(−)."

Note that these comparisons are dependent upon the type and quality of the items compared. For example, if someone predicts that a price for a shirt will be $4.98, it is likely that a shirt can be found at that price. But it may not be the type and quality originally sought. Therefore, when identifying an item, the type and quality should be well defined.

On the other hand, a shopper may be working from a strict budget, for Christmas gifts, for instance. Then the buyer might state a predicted price and shop around until an item that is close to that price is found. In this case, the consumer may be guided more by the price than by the quality.

LEARNING EXPERIENCES

THE DIAGNOSTIC SURVEY AND INDEPENDENCE SCALE

- Use the **Diagnostic Survey** on Reproduction Page 35 to estimate each student's degree of competence prior to instruction in the topics of this chapter.

- Upon completion of the Survey, use the **Independence Scale** on Reproduction Page 36 to identify those skills that need to be strengthened by further instruction and practice.

- Utilize the results of these two instruments to determine the appropriate assignments from the following list.

Topic I: Review of Multiplication

To provide practice and drill in the specific types of multiplication problems most needed by your students, select from any or all of the following.

- Use problems 1 and 2 of the **Diagnostic Survey** and write exercises for drill and practice with multiples of ten. Include a variety of problems: some that require no regrouping, some that regroup only the last product, some that regroup only the first product, and others that require several regroupings. Include problems that contain central zeros.

- For additional drill and practice with multiplication, use *Stein's Refresher Mathematics* (Allyn and Bacon, Inc., 1980), pages 155-161 and 173 and 174.

Topic II: Estimating Products

- Use Reproduction Page 37, Part One, for review practice of rounding off numbers.
- Use Reproduction Page 37, Part Two, for drill and practice with estimating products.

50 GUIDEBOOK FOR TEACHING CONSUMER MATHEMATICS

Topic III: Multiple Purchases

- Present the following assignment as an oral exercise for class discussion.

    A. Rewrite each of these phrases using the "@" symbol. Calculate the total price. Use the $ and decimal point rather than the ¢.

    1. Buy 3 cans of beans when each can costs $.79.
    2. Buy 4 boxes of cereal when each box costs 88¢.
    3. Buy 5 pairs of socks when each pair costs $1.19.
    4. Buy 11 gallons of paint when each gallon costs $14.47.
    5. Buy 3 dozen pencils when each pencil costs 8¢.

    B. Interpret these price labels and calculate the total cost of the purchase.

    6. Buy 12 items @ $1.20
    7. Buy 17 items @ 38¢.
    8. The sign reads "4 for $5.97." Each package contains 4 items. Buy 4 items.
    9. The sign reads "2 for $8.31." Each package contains 2 items. Buy 6 items.
    10. The sign above a display of carriage bolts says 36¢ each. Below the sign are loose bolts and bolts boxed in boxes of 6. Buy 12 bolts.

- Use the table on Reproduction Page 38 to provide practice with multiple purchases, following these directions.
  In this exercise various purchases of some articles of clothing are listed.

    1. Figure the cost of each purchase.
    2. Figure the sales tax for each purchase.
    3. Figure the cost plus tax.
    4. Figure the amount of change you should receive if you paid the amount indicated.
    5. On a separate sheet of paper, indicate the specific denominations of the change to receive. Use the least number of bills and coins possible.
    6. Add the columns for "Total Cost," "Sales Tax," and "Cost Plus Tax." Verify the totals by the method shown below the columns.
    7. Use Reproduction Pages 13, 17, 22, and 23 and start a checking account record with $75. Write a separate check for the cost plus tax amount for each purchase. Add a $75 deposit whenever necessary. Make all checks payable to J. C. Seawards Department Store.

Topic IV: Receiving Change

- Use Reproduction Page 39, Parts I and II, to provide students with practice in counting out change.

- Use Reproduction Page 39, Part III, to provide students with further practice in counting change. Note the bonus problem and end of assignment. Although it is presented as a puzzle, its practical value is clear.

- Use Reproduction Page 40 for cumulative practice in receiving change.

MULTIPLICATION AND MULTIPLE BUYING

Topic V: Sales Tax

- Use Reproduction Page 41, Part I, for drill and practice in changing a % figure to a decimal fraction and vice versa.

- Use Reproduction Page 41, Part II, for review practice in rounding off dollars and cents.

- Use Reproduction Page 41, Part III, to sharpen intuitive recognition of correct sales tax calculations.

- Use Reproduction Page 41, Part IV, for drill and practice with sales tax calculations.

Topic VI: Using Multiplication to Compare Costs and Recognize Value

- Use Reproduction Page 42 for practice in using multiplication to compare prices.

- Ask students to visit a supermarket or a shopping area to survey the stores and make a list of items sold in packages of different quantities. Ask students to check labels and packaging for inducements that encourage the consumer to buy more (e.g., size markings like "giant economy size," etc.). After checking the same item in several different-sized packages, ask students to determine the best buy. Use a form similar to the one below to complete the survey.

Item	Size or Quantity Inducement	Actual Size or Quantity	Price	Best Buy (Check)

Topic VII: Predicting Prices

- Use a form similar to the one on page 52 and give students the following directions.
    Identify a list of items that you can find advertised in a newspaper or a catalog. Be specific about the type and quality of the item desired. After listing the item on the form, predict the price you expect to pay for it. Then look through the newspapers and/or catalogs until you find the comparable item and write down its actual cost. Find the difference by subtracting the lower cost from the higher. If the actual price is higher than the predicted price, mark the difference with a "(+)." If the actual price is the lower cost, mark the difference with a "(−)."

Item	Predicted Price	Actual Price	Difference (+) Or (−)	Is This Type And Quality Originally Desired?

- Use the same form but ask students to visit a shopping area to verify prices.

ASSESSING ACHIEVEMENT OF OBJECTIVES

Ongoing Evaluation

The extent to which students have mastered the concepts covered under the seven topics in this chapter can be measured by any of the activities assigned to class members individually.

Culminating Activities

Use the following form and directions to provide practice with multiple purchases, sales tax, making change, estimating products, and writing checks. Teachers may wish to assign only parts of the assignment. For example, checkwriting could be omitted in order to shorten the assignment.

Shopping for Best Buys

Complete this form using instructions from your teacher.

Item	Predicted Cost	Actual Price	Sales Tax	Total Cost	Difference (+) Or (−)

A. SHOPPING BY NEWSPAPER ADS AND CATALOGS

Use the form provided by your teacher and make a list of items that can be found advertised in newspapers or catalogs.

1. Predict the cost of each item, including tax, for a particular style and quality that you prefer.

2. Look through newspaper ads, catalogs, or visit a shopping area to find a comparable item. Record the actual cost.

3. Figure a 4-1/2% sales tax on each item.

4. Add the tax to the purchase price to obtain the "Total Cost."

5. Find the "Difference" between the total cost and the predicted cost.

6. Add to get the totals for each column and verify correctness by comparing "Total Cost," "Predicted Cost," and "Difference."

7. Start a checking account with a balance of $250. Write a separate check for each purchase and keep a record or check register. Add deposits of $100 whenever necessary. Make checks payable to N. E. Store and sign your own name. Use the form provided by your teacher and the appropriate Reproduction Pages from Chapter 2.

B. SHOPPING FOR CHRISTMAS GIFTS

Use the same form as used in Part A to make lists of Christmas gifts you might purchase. Use two forms: one for "fun" gifts (toys, games, nonsense items), and another for "sensible" gifts. Complete the forms according to the following rules.

1. Buy at least two presents for each of ten people: one for "fun" and one that is "sensible."

2. Spend as near to exactly $200 as possible.

3. Record a predicted cost for each item.

4. Record the actual cost for each item.

5. Figure the sales tax on each purchase. Use the current sales tax rate of your own state and city.

6. Add the tax to the actual price to obtain "Total Cost."

7. Find the difference between "Total Cost" and "Predicted Cost."

8. Add to get the totals for each column and verify correctness by comparing "Total Cost," "Predicted Cost," and "Difference."

9. Start a checking account with a balance of $45. Write a separate check for each purchase and keep a record or check register. Add deposits of $50 whenever needed. Make checks payable to N. E. Store and sign your own name.

> **Final Evaluation**
>
> For an overall evaluation of the students' mastery of the concepts in this chapter, if all topics in the chapter have been taught, a test constructed directly from the "Objectives" listed at the beginning of the chapter can be used. As an alternative, one might consider using the **Diagnostic Survey** as a final test.

RESOURCES FOR TEACHING ABOUT MULTIPLYING AND MULTIPLE PURCHASES

Below is a selected and annotated list of resources useful for teaching the topics in this chapter, divided into audiovisual materials, games, and print materials. Addresses of publishers or distributors can be found in the alphabetic list in Appendix B.

Audiovisual Materials

A. FOR LOW ABILITY AND SPECIAL EDUCATION STUDENTS

Using Arithmetic When Shopping for Groceries, filmstrip and audio-cassette. Interpretive Education, 1977.

Sales Tax, filmstrip and audio-cassette. Interpretive Education, 1977.

B. FOR GENERAL MATHEMATICS STUDENTS

"How's Your Shopping Math?" filmstrip and cassette. *Consumer Math Cassettes* produced by F. Lee McFadden. The Math House, 1977.

Learning to Use Money, 10-1/2 minutes. Coronet Instructional Films, 1973.

Money: How Its Value Changes, 13 minutes. Coronet Instructional Films, 1977.

Games

Paying the Cashier. McGraw Hill Book Company, 1975.

Print

A. FOR LOW ABILITY AND SPECIAL EDUCATION STUDENTS

Using Dollars and Sense by Charles H. Kahn and J. Bradley Hanna. See especially pp. 27-86. Fearon-Pitman Learning, Inc., 1973.

Working Makes Sense by Charles H. Kahn and J. Bradley Hanna. See especially pp. 14-25 and 53-56. Fearon-Pitman Learning, Inc., 1973.

Useful Arithmetic, Volume I, by John D. Wool and Raymond J. Bohn. See especially pp.1-14, 18-24, 39-41. Frank E. Richards Publishing Company, Inc., 1972.

Useful Arithmetic, Volume II, by John D. Wool. See especially pp. 1-4 and 14-28. Frank E. Richards Publishing Company, Inc., 1972.

An Introduction to Everyday Skills by David H. Wiltsie. See especially pp. 8-21 and 136-145. Motivational Development, Inc., 1977.

Skills for Everyday Living, Book 1, by David H. Wiltsie. See especially pp. 13-23 and 84-96. Motivational Development, Inc., 1976.

Skills for Everyday Living, Book 2, by David H. Wiltsie. See especially pp. 31-39 and 42-44. Motivational Development Inc., 1978.

Using Money Series, Book III Buying Power, by John D. Wool. See especially pp. 5-62. Frank E. Richards Publishing Company, Inc., 1973.

It's Your Money, Book 1, by Lloyd L. Feinstein and Charles H. Maley. See especially pp. 2-4, 9-11, 44-56. Steck-Vaughn Company, 1973.

It's Your Money, Book 2, by Lloyd L. Feinstein and Charles H. Maley. See especially pp. 2-4 and 9-11. Steck-Vaughn Company, 1973.

Math for Today and Tomorrow by Kaye A. Mach and Allan Lawson. See especially pp. 5-7 and 20-23. J. Weston Walch, Publisher, 1968.

Scoring High in Survival Math by Tom Denmark. See especially pp. 10-12, 14-16, 24-30. Random House, Inc., 1979.

Mathematics for Today, Level Orange, by Wilmer L. Jones, Ph.D. See especially pp. 32-49 and 235-238. Sadlier-Oxford, 1979.

Mathematics for Today, Level Green, by Wilmer L. Jones, Ph.D. See especially pp. 30-61, 101-115, 251-254, Sadlier-Oxford, 1979.

B. FOR GENERAL MATHEMATICS STUDENTS

Stein's Refresher Mathematics, Seventh Edition, by Edwin I. Stein. See especially pp. 49-57, 155-162, 557-568. Allyn and Bacon, Inc., 1980.

Trouble-Shooting Mathematics Skills, Basic Competency Edition, by Allen L. Bernstein and David W. Wells. See especially pp. 50-73, 116-129, 274-375. Holt, Rinehart and Winston, 1979.

Activities Handbook for Teaching with the Hand-Held Calculator by Gary G. Bitter and Jerald L. Mikesell. See especially pp. 80-88. Allyn and Bacon, Inc., 1980.

Consumer Mathematics, Third Edition, by William E. Goe; an activities book is available. See especially pp. 433-450. Harcourt Brace Jovanovich, 1979.

Mathematics for Today's Consumer by Jack Price, Olene Brown, Michael Charles, and Miriam Lien Clifford. Compiled from selections from *Mathematics for Everyday Life* and *Mathematics for the Real World.* See especially pp. 160-175. Charles E. Merrill Publishing Company, 1979.

Mathematics in Life by L. Carey Bolster and H. Douglas Woodburn. See especially pp. 21-40 and 83-89. Scott, Foresman and Company, 1977.

Mathematics for Daily Living by Harry Lewis. See especially pp. 99-114, 498-515, 525-529. McCormick-Mathers Publishing Company, 1975.

Mathematics Plus!, Consumer, Business & Technical Applications, by Bryce R. Shaw, Richard A. Denholm, and Gwendolyn H. Shelton. See especially pp. 12-21. Houghton Mifflin, 1979.

"Childrens Spending," *Money Management Library,* 12 booklets. Household Finance Corporation, Money Management Institute.

Consumer and Career Mathematics by L. Carey Bolster, H. Douglas Woodman, and Joella H. Gipson. See especially pp. 8-13 and 324-335. Scott, Foresman and Company, 1978.

Business and Consumer Arithmetic by Milton C. Olson and A. E. McVelly. See especially pp. 19-30, 63-68, 254. Prentice-Hall, Inc., 1974.

Business Mathematics for the Consumer by Mearl R. Guthrie, William Selden, and Delbert Karnes. See especially pp. 57-82 and 225-236. Fearon-Pitman Learning, Inc., 1975.

"Teaching Multiplication and Division Algorithms" by Donald Hazekamp. "Estimation and Mental Arithmetic: Important Components of Computation," by Paul R. Trafton; see especially pp. 196-213. "Computation and More" by Diane J. Thomas; see especially pp. 214-225. "Teaching Computation Skills with a Calculator" by Edward C. Beardslee; see especially pp. 226-242. *Developing Computational Skills* by Marilyn N. Suydam and Robert E. Reys. National Council of Teachers of Mathematics, 1978.

C. RESOURCE UNITS, PAMPHLETS, BROCHURES, ETC.

Money by Pamella Pruett. See especially pp. 20-25 and 32-44. Project Consumer, A Livonia Public Schools Project (with the Consumers' Education Office, Department of Health, Education, and Welfare, 1978).

4

Averages and Budgets

INTRODUCTION

Several new topics are introduced in this chapter, including simple division; mean, median, and mode; budget planning; and restaurant menus and service tips. In addition, the concepts of recordkeeping, sales tax, and comparative costs are reinforced and elaborated upon. Division of whole numbers and dollars-and-cents figures by one-, two-, and three-digit whole numbers is presented here primarily to enable students to find averages. In keeping with the developmental approach of this *Guidebook*, topics are introduced according to the degree of mathematics skill needed for the computations involved. Thus, it is not until Chapter 5 that topics requiring division by decimal fractions are introduced.

Finding averages (or the mean) is an important tool in budget planning. Consider the fact that most people are paid weekly or biweekly and that they must, therefore, plan the use of their money on a per day or per week basis. The cost per day for food, for example, is an average. Unfortunately, many people spend their money day by day as they see fit, and their "average cost per day" is whatever they have already spent, not necessarily what they planned to spend. It is the objective here to demonstrate how planning ahead is beneficial for the average consumer. Planning ahead often involves looking back at the past. Appropriately, the ideas of median and mode are also presented here so that these methods of statistical analysis may be applied to consumers' budgets.

Today's consumers, at almost all income levels, find room in their budgets for restaurant dining. For this reason and because restaurant menus are convenient and often appealing, they are used as a means to practice budget and meal planning. In addition, percent computation can be reinforced with sales taxes and service tips. Comparative costs of meals are easily discovered on a menu, and the student is able to transfer this knowledge to meal planning at home.

Ordering from a menu means that choices and decisions are made that often reflect a particular value. Teachers may attempt to point this out to students, and encourage them to be introspective. For example, on what basis are choices made—cost? nutritional value? taste? It is hoped that students who master the objectives of this chapter will be better able to make wise choices.

> ### OBJECTIVES
>
> If all the topics in this chapter are chosen by the teacher, the student should be able to:
>
> 1. Divide dollars-and-cents figures by whole numbers (up to three digits).
> 2. Find the range and approximation for division problems.
> 3. Recognize that the *mean* is the same as the *average*.
> 4. Distinguish among and determine mean, median, and mode.
> 5. Use restaurant menus for selection and meal planning.
> 6. Keep records in a neat and orderly manner.
> 7. Plan personal meals for a week based upon budget restrictions.
> 8. Determine sales tax for restaurant purchases.
> 9. Determine service tip for restaurant service.
> 10. Compare food expenses on a per serving, per meal, and per day basis.

CONTENT OVERVIEW

Topic I: Division by Whole Numbers

This section will deal specifically with division of whole numbers and dollars-and-cents figures by whole numbers. The process of rounding off numbers has been covered in previous chapters, and students are expected to be able to further develop those skills as they apply to division. Several terms are used in describing division. The number that you divide is the *dividend*; the number by which you divide is the *divisor*. The answer in division is the *quotient*; when the quotient is not exact, the number left is called the *remainder*. Division may take two forms: twelve divided by four may be written $12 \div 4$ or $4\overline{)12}$. In either case, 12 is the *dividend*, 4 is the *divisor*, the *quotient* is 3, and there is no *remainder*.

DIVISION OF WHOLE NUMBER DIVIDENDS

Students must known and be proficient with the ideas of numeration (place value), subtraction, and multiplication before taking on division. These topics were presented earlier and may need to be reinforced for certain individuals. There are two basic methods (called "algorithms") for dividing. These algorithms, the subtractive and the distributive, are shown in Figure 4-1. For the purpose of this text, the distributive algorithm is used.

In both examples of Figure 4-1, 384 is the *dividend*, 12 is the *divisor*, 32 is the *quotient*, and the *remainder* is zero. When using the distributive algorithm, the numeration (place value) of the quotient digits corresponds to that of the dividend.

Figure 4-1
Division Algorithms

Subtractive	Distributive
12)̄384	
−240 20 (20 × 12)	32
144	12)̄384
	−360
120 10 (10 × 12)	24
24	24
24 2 (2 × 12)	0
0	

quotient is: 20 + 10 + 2 or 32

One-digit divisors. Multiplication and division are opposite operations, and knowledge of the basic multiplication facts is a necessity, since the division facts follow directly from multiplication. In its simplest form, division proceeds with a multiplication question; for example:

27 ÷ 3 or 3)̄27 asks "what number times 3 equals 27?"

Another way to interpret the same example is:

27 ÷ 3 or 3)̄27 asks "how many groups of 3 equal one group of 27?"

This, of course, can also be related to multiplication, since "9 × 3" is the same as "9 sets of 3." Indeed, if instruction proceeds in a developmental manner, this notion is presented when teaching multiplication skills.

Memorization of basic multiplication facts, through 100, enables students to divide by one-digit divisors easily. This is true even for dividends greater than 100. The notion of dividing a three-digit number by a one-digit number can be understood in terms of place value or numeration. A two-digit number really represents a certain number of tens and ones combined. A three-digit number also represents tens and ones; for example:

270 is 27 tens and 0;
276 is 27 tens and 6.

Thus, when dividing 270 ÷ 3, the problem may be rephrased 27 tens ÷ 3.

Since 27 ÷ 3 = 9,
then 27 tens ÷ 3 = 9 tens or 90.

Similarly, 276 ÷ 3 becomes (27 tens and 6) ÷ 3.

Since 27 tens ÷ 3 = 9 tens and 6 ÷ 3 = 2
(27 tens and 6) ÷ 3 = 9 tens and 2 or 92.

When applying the algorithmic solution, some of this work needs to be done mentally. Therefore, the solution of 276 ÷ 3 using the algorithm becomes:

```
    92
3) 276      (Note the corresponding place
  -270      values in quotient and dividend.)
    6
   -6
    0
```

However, a sequential development of thinking skills, leading up to the algorithm, will assist the learner in applying the algorithm. Such a sequence may result in the following explanation.

```
   2          6 tens ÷ 3 = 2 tens or 20. Place the 2 above the 6
3) 69         because you are dividing into 6 tens, and "2" is in
              reality "20" and belongs in the tens place.

   2          Multiply "20" X 3 = 60 and subtract to get the
3) 69         remainder.
 -60
   9

  23          9 ÷ 3 = 3
3) 69
 -60          3 X 3 = 9
   9
  -9          9 - 9 = 0
   0
```

When dividing larger numbers, the basic skills are expanded upon and place value continues to be important. For example, 7254 has 72 hundreds; 13,472 has 13 thousands; and so on. The question is: Which numbers are used to find the first digit in the quotient? In the case of a number like 7254, there are 72 *hundreds*, to be sure; but there are also 7 *thousands*. Is the 72 or 7 used to find the first digit of the quotient? Of course, it depends upon the size of the divisor. If the divisor is seven or less, then the 7 *thousands* is used. If the divisor is greater than seven, the 72 *hundreds* is used. Since the initial step determines the first digit of the quotient, a zero is not needed as a place holder. (An exception occurs with certain decimal fraction dividends discussed later.) This process continues for the next digit, and the next, and so on, until division is complete. However, succeeding steps require special attention because zeros may be needed as place holders. See Figure 4-2 for a summary of division with larger dividends.

Figure 4-2
Dividing Larger Numbers

```
         2418
a) 3)7254            7 thousands ÷ 3 = 2 thousand. Write the 2 in the thousands place.
      6000
      1254           12 hundred ÷ 3 = 4 hundred. Write the 4 in the hundreds place.
      1200
        54           5 tens ÷ 3 = 1 ten. Write the 1 in the tens place.
        30           24 ÷ 3 = 8. Write the 8 in the ones place.
        24
        24
         0


          806
b) 9)7254            72 hundreds ÷ 9 = 8. Write the 8 in the hundreds place.
     7200
       54            Tens place is the next vacant place. However, no number X 9 = 5.
       00            Therefore a zero is written in the tens place.
       54
       54            54 ÷ 9 = 6. Write the 6 in the ones place.
        0
```

(Note the declining sequence from thousands, to hundreds, to tens, to ones.)

Two-digit divisors. As divisors increase in size, students need other skills to be successful. *Approximating* the quotient and *finding the range* of the quotient are two methods that help students estimate answers prior to division. Both skills require knowledge of place value, and they not only serve to assist actual computation but also enable consumers to estimate costs and other data when full computation is impractical.

Finding the range establishes a "ball-park" figure for the quotient. Using the divisor and the dividend, the quotient is determined to be either more than or less than a certain number. Since we use the base ten numeration system, the range is established in terms of multiples of ten, one hundred, one thousand, and so on, depending upon the size of the quotient and the degree of precision desired. The *range* of the quotient for 27)162 is "less than 10." This is because 10 X 27 is 270 and 162 is less than 270; therefore, the quotient is "less than 10." This range is very broad, but it does give a general idea as to the final answer.

In order further to pinpoint the answer, an *approximation* is needed. To approximate a quotient you begin by approximating the divisor and the dividend in terms of multiples of ten, hundred, and so on. These approximate figures are then divided to obtain an approximate quotient. Thus 27)162 is approximately 30)150 or 5. Note that in order to divide, the student must utilize place-value concepts.

$$\phantom{30\overline{)150}\text{ is the same as }}\overset{5\text{ ones}}{3\text{ tens}\overline{)15\text{ tens}}}$$
$$30\overline{)150}\text{ is the same as }3\text{ tens}\overline{)15\text{ tens}}.$$

Once the range and approximation have been established, the actual quotient is determined.

$$\begin{array}{r}5\\27\overline{)162}\\135\\\hline 27\end{array}$$ *Choose 5, since it is an approximation. The 5 is placed over the 2 because 27 cannot be divided into the first two digits alone.*

If the remainder is greater than or equal to the divisor, the division is incomplete; the next larger number should be tried as a quotient.

$$\begin{array}{r}6\\27\overline{)162}\\162\end{array}$$

Larger dividends are also handled in a similar manner:

$$69\overline{)276{,}207}$$

Find the range:　　　　*Approximate:*

69 × 1,000 = 69,000　　　69 is near 70

69 × 10,000 = 690,000　276,207 is near 280,000

The quotient lies between

1,000 and 10,000　　　　$$\begin{array}{r}4{,}000\\70\overline{)280{,}000}\end{array}$$

$$\begin{array}{r}4{,}003\\69\overline{)276{,}207}\\276{,}000\\\hline 207\\000\\\hline 207\\000\\\hline 207\\207\\\hline 0\end{array}$$

Place the 4 above the 6 because of the range and approximation already stated. (Also, 69 does not divide into 27.)

69 will not divide into 2; place a zero above 2.

69 will not divide into 20; place another zero in the quotient.

69 is near 7 tens. 207 is near 21 tens.

21 tens ÷ 7 tens = 3 ones.

AVERAGES AND BUDGETS

Handling remainders. Often remainders other than zero occur that cannot be further divided. There are several ways to handle this result. One is to write the quotient, then write a small letter *r* to the right, and then write the remainder. Another is to write the remainder as a fraction of the divisor and write the fraction to the right of the quotient. In this text, we shall use the remainder either to round off the quotient or to carry out further division by inserting zeros after the decimal point of the dividend. Rounding off quotients can be accomplished simply by determining if the remainder is at least one half of the divisor. If it is, the last digit of the quotient is increased by one. If it is not, the quotient remains unchanged. Although rounding off in this manner is expedient, it is often necessary to continue division before considering a rounded-off quotient. This is particularly true with decimal figures such as dollars and cents.

Consider the following practical problem. A young adult is attempting to make it on his own living in his own apartment and paying all expenses from a weekly take-home pay of $186. What is the average amount of money available each day? Obviously, the average can be determined by dividing $186 by 7. Thus:

```
         $26
      _____
    7)$186
       140
       ___
        46
        42
        __
         4
```

The answer is somewhat more than $26 because there is a remainder of 4. If division is continued two places beyond the decimal point, the cents can be determined. Recall that any whole number has an imaginary decimal point at the right of the last digit, and zeros can be written thereafter. So the division can continue.

```
       $ 26.57
      _____
    7)$186.00
       140 00
       _____
        46 00
        42 00
        _____
         4 00
         3 50
         ____
           50
           49
           __
            1
```

Note that the decimal point in the quotient corresponds to that in the dividend. This follows because of corresponding place values.

Division proceeds as with whole numbers.

Although this remainder may be handled as described earlier, another method is to continue division (write another zero in the dividend) one more place beyond the place value desired. If this "extra" digit in the quotient is five or greater, the previous digit is increased by one. If not, the previous digit remains unchanged. Regardless of the result, the "extra" digit is not written in the final answer. In the case of the previous example:

```
    $ 26.571   = $26.57
  7)$186.000   The remainder was one, so
    .  - - -   another zero is written and
    .          division continues. The
    .  - - -   final remainder (3) is
       500     ignored.
       490
        10
         7
         3
```

Division of dividends with decimal fractions. The previous discussion leads directly to problems involving decimal fractions. This section will explore decimal fraction *dividends* only. The presentation involving divisors with decimal fractions is made in Chapter 5. As has already been seen, whole number dividends can become decimal dividends by merely writing zeros after the decimal point. Although division continues in the same manner as with whole numbers, much is left to mental or even intuitive thinking. However, the thinking patterns established with simple division will assist the learner in applying the distributive algorithm to more complex problems. Ultimately, success with division rests with mastery of basic multiplication and subtraction skills, and the development of a thinking pattern that involves understanding place value, approximating and finding the range of a quotient, following a sequence of steps when applying the algorithm, and dealing with the remainder as required by the problem.

Although decimal dividends present special problems, they are not unlike those faced with whole numbers. Take, for example, approximating and finding the range of the quotient in this problem.

$$278 \overline{)17.514}$$

Find the range:

278 × 1 = 278

278 × .1 = 27.8 ⎫
278 × .01 = 2.78 ⎬ 17.514

The range is between .01 and .1

(This leaves .02, .03, .04 . . . , .08, .09)

Approximate:

278 is near 300

17.514 is near 18.000

(be sure the zeros are included)

```
        .06
  300)18.000
       18 00
```

The approximation and the range must support one another. Although the digit for the approximation is clearly a "6," it is the range that positively places its value at ".06." Thus:

```
       .063
 278)17.514    The zero can be placed immediately
     16 680   after the decimal point before division.
        834   A "6" can then be tested in the
        834   quotient.
```

Examples And Solutions. The following is a representative selection of sample problems along with their solutions and explanatory comments.

```
         349.3                                              $ 371.43
1) 5)1746.5         The decimal point should be    2) 8)$2971.44
    1500 0          placed before dividing.            2400 00
     246 5                                              571 44
     200 0                                              560 00
      46 5                                               11 44
      45 0                                                8 00
       1 5          Note that the decimal point           3 44
       1 5          does not appear in the sub-           3 20
         0          traction.                              24
                                                           24
                                                            0

         .1282                        $ .26                $ .116
3) 7).8974                  4) 38)$9.88            5) 732)$84.912
    7000                         7 60                     73 200
    1974                         2 28                     11 712
    1400                         2 28                      7 320
     574                            0                      4 392
     560                                                   4 392
      14                                                       0
      14
       0

         .013                                              .092
6) 3).039           Note the placement of          7) 8).736
    000             zeros.                            000
     39                                               736
     30                                               720
      9                                                16
      9                                                16
      0                                                 0

         .009                       .08                    .0006
8) 8).072                   9) 27)2.16             10) 214).1284
```

Topic II: Averages

It is common for consumers to refer to the "average" amount for several different things: money spent, weekly pay, quantity purchased, and so forth. The term is also used to describe what is normal or usual or common, as expressed in an "average person." What is really expressed is a quantity or description that best generalizes a given set of data. The average is sometimes called a *central tendency of data.* This quantity can be calculated in three different ways; each method has its own value depending upon the purposes intended. What is most often referred to as average is the *arithmetic mean* or simply *mean.* Other methods of expressing central tendency are *median* and *mode.* Although this text shall most often refer to the mean (and use *mean* as synonymous with *average*), we shall explore all three methods.

MEAN

If the sum of the numbers in a set is divided by the number of items in that set, the *arithmetic mean* (or *mean*) is determined. Since this text shall follow common practice and use average and mean interchangeably, whenever an average of a given group of numbers is to be determined, the numbers must be added and the sum must be divided by the number of items in the group. Thus, the average or mean of $.62, $1.18, $3.27, $.93, $.81, and $1.05 is found in the following manner.

$$\frac{\$.62 + \$1.18 + \$3.27 + \$.93 + \$.81 + \$1.05}{6}$$

$$= \frac{\$7.86}{6} \quad \text{or} \quad 6\overline{)\$7.86}$$

$$= \$1.31$$

MEDIAN

When a set of numbers is arranged in order of size, the middle number of the set is the *median*. Thus, the median of 42, 21, 46, 35, and 53 is determined by writing the numbers in order—21, 35, 42, 46, 53—and then counting toward the middle from either end; 42 is the median of this set.

If the number of items is even, there are two middle terms, and the median is found by dividing the sum of those numbers by two. The median of 28, 17, 13, 36, 31, and 22 is discovered in the following manner.

1. Arrange in order: 13, 17, 22, 28, 31, 36

2. Add the middle terms: 22 + 28 = 50

3. Divide by 2: 50 ÷ 2 = 25

The median is 25.

MODE

The number that occurs most frequently in a set of numbers is the *mode*. The surest way to determine the mode when a large number of items are considered is to arrange the numbers in order of size (marking off each number as it is written to be sure none is missed), and then count the occurrences of each number. For example, find the mode of:

13, 22, 33, 51, 22, 29, 33, 19, 15, 51, 23, 29, 19, 29, 3

1. Arrange in order (Note: numerals in the original list may be checked (√) as they are rewritten in order.)

    3, 13, 15, 19, 19, 22, 22, 23, 29, 29, 29, 33, 33, 51, 51

2. Count to find the most frequent number.

    29 is the mode.

When the mean, median, and mode are compared, there is some similarity, but they do not measure exactly the same thing. Each concept reports something different, and a comparison often suggests a caution for interpretation of data, especially when there is a wide range between the greatest and the least numbers. Using the above example, 29 is the mode, 23 is the median, and the average is 26, even though 26 never appears on the list.

When all three central tendencies and the extremes (a low of 3 and a high of 51) are considered, the data take on more meaning. If these were dollar figures showing amounts spent for food on a 15-day vacation, an interpretation that "somewhere near $26 was spent every day" would be false. The average of $26 must be recognized to include many items more and less than $26. This knowledge helps in budgeting expenses, because the buyer can plan his spending according to a "target" average. Whenever the amount exceeds the average, an effort needs to be made to reduce spending below the average.

Averages offer a basis for comparison, and consumers often use them in food planning by determining the average cost on the basis of time (day, week, etc.), meal, or number of people served. The term "per" is used to denote the basis on which the average is calculated. It is used to stand for the phrase "on the basis of." Thus, we refer to the average cost *per* day, *per* meal, *per* serving, and so on. The meaning of the word *per*, therefore, is "divided by." When calculating the cost of an item on the basis of each day (per day), the cost is divided by the number of days. When units of measure form the basis of cost (cents per ounce, per gram, etc.), the cost is divided by the number of ounces, grams, and so on. Some supermarkets display the cost per unit on the shelf. This "unit pricing" enables consumers to compare various brands of the same food item. However, all stores do not display unit prices, and it is a wise buyer who is able to divide units into price quickly and accurately.

The concept of the term "per" can be expanded in a way that is helpful to gain a better understanding of percents. Recall that the term "percent" is derived from *per* and *centum*. "Centum" means 100. Percent means "on the basis of 100" or "divided by 100." It is a kind of average whereby the numbers are compared with 100. If, for example, a consumer spends 19% of his total income on food, it means that $19 out of every $100 he earns ($19 per $100) covers his food costs. Further discussion about percents and unit pricing appears in later chapters as more complex topics are introduced.

Topic III: Restaurant Math

This topic is included here because it is highly motivational and offers students the opportunity to make decisions while planning meals. Many activities can be designed into the experience, and the student is able to see the real practical value of learning the division and averaging skills introduced in this chapter. Other skills such as wise selection and decision making (based on nutritional and economic value), living within a budget, calculating sales tax and service tip, keeping neat and orderly records, and comparison of prices (restaurant versus restaurant as well as restaurant versus home) are all included in this topic. These skills, important to the consumer and essential for certain jobs, can be presented in stages leading up to the student's selecting, planning, and calculating several restaurant meals on his own. At first, assignments might begin with a sample school cafeteria menu with no tax, no service tip, and a small selection. Fast-food dining follows and is similar, except that there is a sales tax. Family dining and more elegant restaurant entertainment may be presented next.

The service tip at a restaurant is a percent of the total bill not including tax. The amount of tip may vary according to the quality of service rendered by the waitress or waiter, but 15% is the acceptable amount. This is usually calculated mentally and rounded off. Let us examine the process of figuring service tip.

There are two percent figures that must be understood: 10% (for smaller tips) and 15%. The relationship between percents and decimal fractions was introduced in Chapter 3 with regard to sales tax. Recall that a whole number percent can be represented by a decimal fraction written with two digits to the right of the decimal point. Thus:

10% becomes .10 and
15% becomes .15.

Once the percent is changed to a decimal, the service tip is calculated by multiplying that figure times the total food cost (excluding tax). For example, if the food cost totals $4.20:

A 10% tip is:		A 15% tip is:
$4.20	*Note: Decimal point*	$4.20
X .10	*is 4 places to the left.*	X .15
$ .4200		2100
or $.42		420
		$ .6300
		or $.63

Only two places to the right of the decimal are used for cents, and the previously discussed rules for rounding off are still applied.

Since a service tip is usually calculated mentally, it is wise to learn some shortcuts. First, multiplying by .10 gives the same result as moving the decimal point of the multiplicand one place to the left: $4.20 becomes $.4₂20 or $.42.

Second, 15% is 10% plus one half of 10% (10% + 1/2 of 10% = 10% + 5% or 15%). Therefore, since it is so easy to determine 10% of a number, use this method.

15% of $4.20

1. Find 10% of $4.20 ──────────────→ $.42
2. Find 1/2 of 10% of $4.20 ──→ 1/2 of $.42 ──→ .21
3. Add to obtain 15% of $4.20 ──────────→ $.63

Of course, some students should continue to use the methods of full calculation.

Topic IV: Budgeting Food Costs

This topic is an extension of Topics II and III and offers a preliminary step to unit pricing. Here the notion of cost per serving may be presented in an elementary way with sample menus for home planning. Budgeting food costs by determining serving size and cost is facilitated by package labels on many products. Cans, boxes, and other packages often carry information regarding the number of servings per package. Initially, students are asked to respond to a preplanned list of items with servings identified for them. In later assignments (See Chapter 7, in the topic entitled "More Thought for Food") students will be asked to consider factors like nutritional value, calories, balanced meals, weights and measures, unit pricing, and the like and plan their own food budget.

AVERAGES AND BUDGETS

At this time students may be introduced to budgeting food costs that require simple division and little or no knowledge of units of weights or measures. For example, given the following information, ask students to determine the cost per serving. Note that the amounts in customary units (and metric units in parentheses) are not used in the calculations but are presented to acquaint the students with weights and measures.

Item	Cost Per Item	Servings Per Item
1 qt. or 32 oz. Juice (.95 l.)	$ .88	8
20 oz. Cereal (567 gm.)	$1.89	20

The cost per serving is determined by dividing cost per item by servings per item. A good way to show this is:

$$\frac{\text{cost per item}}{\text{servings per item}}$$

The "per items" cancel (or divide) out, leaving cost divided by serving. Recall that "per" means "divided by," so that we are left with cost per serving.

For the juice this becomes:

$$\frac{\$.88}{8} \quad \text{or} \quad 8\overline{)\$.88} = \$.11 \text{ per serving}$$

For the cereal it is:

$$\frac{\$1.89}{20} \quad \text{or} \quad 20\overline{)\$1.890} = \$.09 \text{ per serving}$$

Once the cost per serving is determined, several things become apparent. First, the cost for serving several persons can be obtained by multiplying. Second, comparison of home-prepared food with restaurant food can be made. (e.g., What does a small glass of juice cost at a restaurant?) Third, comparisons can be made of the cost for different food items prepared at home. (e.g., Which is cheaper on a per serving basis, ham or beef?)

Another computation is required when planning for a large group. Suppose, for example, a party is being planned for eight people and the following items are offered. How is the amount to be purchased and the cost for each purchase determined?

Item	Servings Per Item	Cost Per Item
8-lb. or 128-oz. Ham (3.63 kg.)	64	8 lb. for $23.84
5-lb. or 80-oz. Potato Salad (2.27 kg.)	16	5 lb. for $4.45
1-gal. or 128-oz. Ice Cream (3.79 l.)	28	1 gal. for $4.73

The cost per serving can be obtained by dividing the cost per item by the servings per item:

$$\left(\frac{\text{cost per item}}{\text{servings per item}}\right)$$

This figure can be multiplied by the number of people (or the number of servings) expected. Thus:

$\underline{\$.37}$
$64) \overline{\$13.84}$ is the cost per serving of ham

$\underline{\$.278}$
$16) \overline{\$4.450}$ or $.28 is the cost per serving of potato salad

$\underline{\$.176}$
$28) \overline{\$4.930}$ or $.18 is the cost per serving of ice cream

To serve eight people (assuming one serving each), it would cost:

$ .37	$ .28	$ .18
× 8	× 8	× 8
$2.96 for ham	$2.24 for potato salad	$1.44 for ice cream

This would be the cost if the items could be purchased in the quantity needed. However, most often the items are not available in the precise quantities desired. On the other hand, if the servings per item are accurate, the actual cost of the party will be correct. The additional quantities are "leftovers" and not considered party expenses. To calculate the amount needed to be purchased, first divide the number to be served by the servings per item and then multiply by the package size of the item. Another way to state this is to set up a ratio between the number to be served and the servings per item. When eight are to be served, this becomes:

$\frac{8}{64}$ for ham; $\frac{8}{16}$ for potato salad; and $\frac{8}{28}$ for ice cream

When two numbers are compared by division, we have a ratio. Since a division problem is also a fraction, it is clear that the above figures represent fractions of the quantities shown for each item. Thus, by multiplying the fraction by the package size, the quantity needed is discovered.

$\frac{8}{64} \times 8$ lb. of ham $= \frac{1}{8} \times 8$ or 1 lb. of ham

$\frac{8}{16} \times 5$ lb. of potato salad $= \frac{1}{2} \times 5$ or $2\frac{1}{2}$ lb. of salad

$\frac{8}{28} \times 1$ gal. of ice cream $= \frac{2}{7} \times 1$ or $\frac{2}{7}$ gal. of ice cream

(Note that each fraction is reduced.)

It is clear that in order for these quantities to make sense, the student will need to understand fractions as well as have a good idea of weights and measures. At this point in the developmental sequence of the *Guidebook*, it is recommended that these concepts be touched upon now but not fully developed until later chapters.

LEARNING EXPERIENCES

THE DIAGNOSTIC SURVEY AND INDEPENDENCE SCALE

- Use the **Diagnostic Survey** on Reproduction Page 42 to estimate each student's degree of competence prior to instruction in the topics of this chapter.

- Upon completion of the Survey, use the **Independence Scale** on Reproduction Page 43 to identify those skills that need to be strengthened by further instruction and practice.

- Utilize the results of these two instruments to determine the appropriate assignments from the following list.

Topic I: Division by Whole Numbers

- Part I of the **Diagnostic Survey** identifies the kind of division problems most important at this time. Separate worksheets may be written for A. *Simple division with no remainder*, B. *Rounding off quotients*, C. *Finding the range and approximation*, and D. *Division requiring zeros as place holders*. Place particular emphasis on the latter two concepts.

- For further practice appropriate to this section, see *Stein's Refresher Mathematics* (Allyn and Bacon, Inc., 1980). See especially pages 58-68 and 162-166.

Topic II: Averages

- Use Reproduction Page 44 to give students practice with defining and determining mean, median, and mode.

- Use Reproduction Page 45, Part I, for practice with calculating average expenses per day when the total expense is given.

- To help clarify the difference between average and median, assign Reproduction Page 45, Part II.

- Refer to *Stein's Refresher Mathematics* for further examples appropriate to this section. See especially pages 68-70, 175-176, and 532-534.

Topic III: Restaurant Math

- Use the following assignment for students to develop a facility with sales tax and service tip calculations.

TAXES AND TIPS

Determine the tax, tip, and total cost for each food expense given below.

a. $4.00   b. $3.95   c. $2.50   d. $6.55   e. $9.21

Problems 1-5, determine 5% tax, 10% tip, and total cost for each food expense.

Problems 6-10, determine 4% tax, 15% tip, and total cost for each food expense.

Problems 11-15, determine 5.75% tax, 20% tip, and total cost for each food expense.

Problems 16-20, determine 3.125% tax, 15% tip and total cost for each food expense.

- Obtain a school menu and price list. Give the following assignment using the form on Reproduction Page 47. Write in the % sales tax and % service tip that is appropriate for your region and give students the following directions.

    1. Use the school menu and price list provided by your teacher to plan ten different lunches for ten school days. You must plan a balanced meal with an entree and/or a sandwich, fruit, vegetable, plus a dessert and a beverage for each day. You may add as much extra as you wish. Change all prices from ¢ to $. For example:

    $$5¢ = \$.05;\ 45¢ = \$.45$$

    2. Determine the tax and service tip you might expect to pay at a restaurant.

    3. Add the total cost for all ten days and find the average cost per day.

    4. Use the form on Reproduction Page 47 to record this assignment.

- Obtain menus from various restaurants in your community. (Old menus are periodically replaced. Perhaps you can obtain those directly from the restaurant or from the menu printer.) Use the form on Reproduction Page 47; write in the appropriate % figures for tax and tip; and give the students the following directions.

Eating Out

Use the menus in the room and plan a different breakfast, lunch, and dinner for yourself for 5 days. You must order a beverage, a main dish (and a dessert for lunch and dinner).

1. Order for *5 days*.

2. *Add* the totals for all 5 days and

3. Find the average cost per day.

For *Extra Credit* find the average cost for breakfast, the average for lunch, and the average for dinner.

- For practice with planning expenses on a strict budget, have students do the following assignment using the form on Reproduction Page 47, with the appropriate % figures for tax and tip.

Eating out on a budget

Repeat the "Eating Out" assignment, but this time you must spend only $68.00, including tax and tip. You must spend as close to *exactly* $68.00 as possible—no more, no less. Again, you are to buy breakfast, lunch, and dinner for 5 days, a total of 15 meals.

You must order a main dish (breakfast could be cereal; lunch and dinner could be a sandwich) and a beverage. Dessert is not necessary but is recommended.

Topic IV: Budgeting Food Costs

- Draw up an assignment similar to Part IV, problems 25 and 26, of the **Diagnostic Survey**. Students will gain experience determining cost per serving and cost per meal when eating at home.

- Have students compare the costs per serving calculated above with the prices for similar servings at a restaurant.

- Ask students to visit a supermarket and survey the package labels for information regarding the number of servings per package. Use a form similar to the one below and determine the cost per serving.

Supermarket Survey

Item	Package Size	Cost Per Package	Servings Per Package	Cost Per Serving

- Select the best buys from the above survey and plan a full day's menu (breakfast, lunch, and dinner) for one person. Make each a balanced meal and compare the cost with similar meals at a restaurant.

- Use Reproduction Page 48 for students to practice planning for a large party. Have students suggest ways to save money on a dinner party. Party foods often cost more than everyday foods, for instance, and perhaps leftovers can be reduced by careful planning.

ASSESSING ACHIEVEMENT OF OBJECTIVES

Ongoing Evaluation

The extent to which students have mastered the concepts covered under the four topics in this chapter can be measured by any of the activities assigned to class members individually.

Final Evaluation

For an overall evaluation of the students' mastery of the concepts in this chapter, if all topics in the chapter have been taught, a test constructed directly from the "Objectives" listed at the beginning of the chapter can be used. As an alternative, one might consider using the **Diagnostic Survey** as a final test.

RESOURCES FOR TEACHING ABOUT AVERAGES AND BUDGETS

Below is a selected and annotated list of resources useful for teaching the topics in this chapter, divided into audiovisual materials, games, and print materials. Addresses of publishers or distributors can be found in the alphabetic list in Appendix B.

Audiovisual Materials

A. FOR LOW ABILITY AND SPECIAL EDUCATION STUDENTS

Sales Tax, filmstrip and audio-cassette. Interpretive Education, 1977.

Selecting and Eating at Restaurants, filmstrip and audio-cassette. Interpretive Education, 1977.

Understanding Averages, filmstrip, audio-cassette, and 20 student workbooks. Interpretive Education, 1977.

Using Arithmetic When Shopping for Groceries, filmstrip and audio-cassette. Interpretive Education, 1977.

B. FOR GENERAL MATHEMATICS STUDENTS

"How's Your Budget Math?" filmstrip and cassette. *Consumer Math Cassettes,* produced by F. Lee McFadden. The Math House, 1980.

Budgeting, film, 11 minutes. Aetna Life and Casualty, 1977.

Managing Your Money, 4 filmstrips and 4 cassettes. Teaching Resources Films, 1974.

Print

A. FOR LOW ABILITY AND SPECIAL EDUCATION STUDENTS

Using Dollars and Sense by Charles H. Kahn and J. Bradley Hanna. See especially pp. 27-112. Fearon-Pitman Learning, Inc., 1973.

Working Makes Sense by Charles H. Kahn and J. Bradley Hanna. See especially pp. 25-40 and 53-56. Fearon-Pitman Learning, Inc., 1973.

Useful Arithmetic, Volume I, by John D. Wool and Raymond J. Bohn. See especially pp. 60-63. Frank E. Richards Publishing Company, Inc., 1972.

Useful Arithmetic, Volume II, by John D. Wool. See especially pp. 11-14. Frank E. Richards Publishing Company, Inc., 1972.

An Introduction to Everyday Skills by David H. Wiltsie. See especially pp. 2-11 and 131-135. Motivational Development, Inc., 1977.

Skills for Everyday Living, Book 1, by David H. Wiltsie. See especially pp. 93-96. Motivational Development, Inc., 1976.

Using Money Series, Book IV Earning, Spending, and Saving, by John D. Wool. See especially pp. 34-38. Frank E. Richards Publishing Company, Inc., 1973.

It's Your Money, Book 1, by Lloyd L. Feinstein and Charles H. Maley. See especially pp. 2-4, 9-11, 44-56. Steck-Vaughn Company, 1973.

It's Your Money, Book 2, by Lloyd L. Feinstein and Charles H. Maley. See especially pp. 2-4 and 9-11. Steck-Vaughn Company, 1973.

Math for Today and Tomorrow by Kaye A. Mach and Allan Larson. See especially pp. 5-14 and 20-25. J. Weston Walch, Publisher, 1968.

Scoring High in Survival Math by Tom Denmark. See especially pp. 6-8 and 10-12. Random House, Inc., 1979.

Mathematics for Today, Level Red, by Saul Katz, Ed.D.; Marvin Sherman; Patricia Klagholz; and Jack Richman. See especially pp. 27-50. Sadlier-Oxford, 1976.

Mathematics for Today, Level Orange, by Wilmer L. Jones Ph.D. See especially pp. 50-73, 152-177, 235-238. Sadlier-Oxford, 1979.

Mathematics for Today, Level Green, by Wilmer L. Jones Ph.D. See especially pp. 30-61, 101-115, 251-254. Sadlier-Oxford, 1979.

B. FOR GENERAL MATHEMATICS STUDENTS

Stein's Refresher Mathematics, Seventh Edition, by Edwin I. Stein. See especially pp. 58-73, 163-177, 557-568. Allyn and Bacon, Inc., 1980.

Trouble-Shooting Mathematics Skills, Basic Competency Edition, by Allen L. Bernstein and David W. Wells. See especially pp. 74-98, 130-147, 374-375. Holt, Rinehart and Winston, 1979.

Activities Handbook for Teaching with the Hand-Held Calculator by Gary G. Bitter and Jerald L. Mikesell. See especially pp. 89-270. Allyn and Bacon, Inc., 1980.

Consumer Mathematics, Third Edition, by William E. Goe. Activities book available. See especially pp. 433-450. Harcourt Brace Jovanovich, 1979.

Mathematics in Life by L. Carey Bolster and H. Douglas Woodburn. See especially pp. 41-66 and 85-104. Scott, Foresman and Company, 1977.

Mathematics for Daily Living by Harry Lewis. See especially pp. 498-529. McCormick-Mathers Publishing Company, 1975.

Mathematics Plus!, Consumer, Business & Technical Applications, by Bryce R. Shaw, Richard A. Denholm, and Gwendolyn H. Shelton. See especially pp. 22-32. Houghton Mifflin, 1979.

Consumer and Career Mathematics by L. Carey Bolster, H. Douglas Woodman, and Joella H. Gipson. See especially pp. 8-13 and 324-341. Scott, Foresman and Company, 1978.

Business and Consumer Arithmetic by Milton C. Olson and A. E. McVelly. See especially pp. 28-40, 67-79, 254. Prentice-Hall, Inc., 1974.

Business Mathematics for the Consumer by Mearl R. Guthrie, William Selden, and Delbert Karnes. See especially pp. 83-104 and 225-236. Fearon-Pitman Learning, Inc., 1975.

"Teaching Multiplication and Division Algorithms" by Donald Hazekamp, 1978.

—"Estimation and Mental Arithmetic: Important Components of Computation" by Paul R. Trafton, pp. 196-213, 1978.

—"Computation and More" by Diane J. Thomas, pp. 214-225, 1978.

—"Teaching Computation Skills with a Calculator" by Edward C. Beardslee, pp. 226-242, 1978.

Developing Computational Skills by Marilyn N. Suydam and Robert E. Reys. National Council of Teachers of Mathematics, 1978.

5

Transportation Computation

INTRODUCTION

This chapter focuses on some of the expenses and calculations involved with driving an automobile. Topics include odometer readings, gasoline purchases, understanding fuel economy, and renting a car. Other topics, such as map reading and recreational travel, are treated elsewhere in this book.

The activities suggested here are designed to help the student become a wiser consumer while at the same time develop certain mathematics skills such as multiplying, dividing, and rounding off decimal numerals.

Because the United States will soon be using the metric system, it is important that all consumers become acquainted with metric units. One must be cautious, however, not to go into more calculation than necessary. For example, once the metric system is adopted there will be no need to "convert" from the customary system to the metric, because only the metric units will be used. Thus, when fuel purchases are considered in the customary system, gallons are used and the unit price is per gallon. In the same manner, with the metric system fuel is sold by the liter, and the unit price is per liter. For this reason, it is recommended that very little time be spent on conversion factors. Instead, a conversion table is offered for teacher use only with instructions on how to rewrite the assignments from customary to metric units. The suggested activities, therefore, offer a choice between assignments using each system, allowing the teacher to select the appropriate units necessary for specific objectives.

However, during the transition period in which our nation moves toward full adoption of metrics, most consumers will be predisposed to the customary system. Therefore, knowledge of the corresponding units of each system is necessary, and the following list should be committed to memory.

If the unit in the Customary system is:	The corresponding Metric unit is:
miles	kilometers
gallons	liters
miles per gallon (mpg)	kilometers per liter (kpl)

Note that only those units pertaining to this chapter are shown. Also keep in mind that this list does not mean that miles are the same as kilometers or that gallons are the same as liters. They are not identical.

As an illustration of the relationship between the two system, let's take a hypothetical case. Suppose Mr. Hanks, driving an American car, and Herr Hans, driving a German auto, are traveling in the same direction, starting from the same spot, each with a full tank of gas. After traveling nonstop along the same route for many hours, the two men stop to purchase the same kind of fuel at the same station. They discover that each man pays the exact same amount for the fuel: $14.58. Therefore, each man must have purchased the same amount of fuel. However, Herr Hans calculates his purchase in terms of the metric system, whereas Mr. Hanks still uses the customary system. Their figures look like this.

	Distance Traveled	Amount Purchased	Cost Per Unit	Total Cost
Mr. Hanks	235.5 miles	10 gallons	$1.458 or 145.8¢ per gallon	$14.58
Herr Hans	379 kilometers	37.85 liters	$.385 or 38.5¢ per liter	$14.58

Therefore, paying 38.5¢ per liter is the same as paying 145.8¢ per gallon. The consumer is not paying more (or less) for the fuel in either system. The fuel is simply measured differently. Fuel economy is identical also. The German gets 10 kilometers per liter, while the American gets 23.55 miles per gallon, but these are equivalent amounts measured in different units.

In order to facilitate instruction with metric units, two steps have been taken. First, wherever appropriate, examples appear in two versions—one customary, one metric. Second, the conversion table mentioned earlier appears within the "Learning Experiences" section for the exclusive use by the teacher. The examples utilize the conversion factors so that the metric example is equivalent to the customary except for the units. It must be noted, however, that the conversion factors offer only approximate equivalency, and the examples will not be completely identical.

Whatever system is used, it is important that students show their work in all calculations. By carefully examining a student's work, a teacher is better able to assess the degree of understanding that has been achieved. Furthermore, the specific type of error the student made can be readily identified by close examination of the student's work. Therefore, it is recommended that the teacher regularly admonish students to show all their work neatly.

TRANSPORTATION COMPUTATION

OBJECTIVES

If all the topics in this chapter are chosen by the teacher, the student should be able to:

1. Read an automobile odometer.

2. Determine the number of miles (kilometers) traveled when given the starting and ending odometer readings.

3. Round off to the nearest mile (kilometer).

4. Round off to the nearest tenth of a cent.

5. Round off to the nearest tenth of a gallon (liter).

6. Mentally estimate sums, products, and quotients after rounding to the nearest mile (kilometer), the nearest cent, and the nearest gallon (liter).

7. Determine the total cost of a fuel purchase if given the number of gallons (liters) purchased and the cost per gallon (liter).

8. Determine the cost per gallon (liter) if given the total cost and the number of gallons (liters) purchased.

9. Determine the number of gallons (liters) purchased if given the total cost and the cost per gallon (liter).

10. Determine the miles per gallon (or kilometers per liter) if given the miles (kilometers) traveled and gallons (liters) purchased.

11. Calculate the amount of fuel used, given the fuel efficiency (miles per gallon or liters per kilometer).

12. Identify all the likely charges when renting a car.

CONTENT OVERVIEW

Topic I: Rounding and Estimating

This topic is covered in detail in Chapter 1 for dollars-and-cents figures. As recommended there, the concepts should be reinforced regularly. Of course, the basic rules apply for all rounding operations, whether using dollars, cents, miles, or whatever. The important thing to remember is place value. Therefore, when rounding to the nearest mile (kilometer), retain all digits to the left of the decimal point; if the first digit to the right of the point is 5 or more, add 1 to the miles. If the first digit to the right of the decimal point is less than 5, do not change the miles. Thus:

37,521.4 miles is rounded to 37,521 miles, and
37,693.5 miles is rounded to 37,694 miles.

Similarly, 60,409.4 kilometers is 60,409 km. and 60,686.5 kilometers is 60,687 km. Automobile odometers usually show mileage to the tenth of a mile (kilometer), but common usage is to refer only to the miles (kilometers) traveled.

On the other hand, fuel prices and gallons (liters) purchased are posted on the pump in *tenths* of a cent and tenth of a gallon (liter), respectively. So, when rounding these figures, one must be sure to retain the proper digits and use the appropriate digit for determining whether or not to change the figure.

When rounding to tenths of a gallon, for example, all digits to the left of the decimal point and the first digit to the right of the point are retained. The "rule of 5" is then applied using the second digit to the right of the point.

Although rounding to the nearest tenth of a cent is similar, the use of the "$" rather than the "¢" may complicate the problem for some students. However, because greater confusion results when using the ¢ in calculations, teachers should insist that students always use dollar figures. Therefore, when rounding a figure after calculation is completed, the student will already have the answer in terms of the dollar sign, and the following rules will apply.

1. Retain *the first three digits* to the right of the decimal point.

2. Apply the "rule of 5."

The above discussion pertains to precise calculations. However, when making mental estimation, less precise figures are used. Distance is still rounded to miles (kilometers), but the quantity of fuel is rounded to gallons (liters) and the cost is rounded to the nearest cent. Thus, when mentally estimating the total cost of:

10.4 gallons of fuel at $1.399 per gallon

the figures are rounded to:

10 gallons @ $1.40

for an estimated total of $14.00. (Compare this with the actual total of $14.55.)

In the same manner, when estimating miles (kilometers) per gallon (liter), round to the nearest gallon (liter) first, and then estimate mentally. Thus, 760.3 kilometers traveled on 37.6 liters of fuel is 760 kilometers on 38 liters of fuel for an estimated fuel economy of 20 liters per kilometer. (Compare this with the more precise calculation of 20.2 liters per kilometer.)

Of course, one might round off to the nearest cent, ten cents, or dollar, depending upon the degree of accuracy desired. Likewise, distance can be rounded to the nearest mile (kilometer), ten miles (ten kilometers), or hundred miles (hundred kilometers).

Usually when the purpose is to achieve a more precise answer, only the final figure (the answer) is rounded after all calculations are completed. When estimating, the figures are rounded *before* making a mental calculation. Estimating is a fundamental tool for planning and budgeting. With practice, a wise consumer will be able to anticipate expenses before they occur.

Topic II: Odometer Readings

Although they may not realize what it is called, most students will have seen the odometer dial on an automobile. It is located along the bottom half of the speedometer. In most autos this dial reads in tenths of a unit, and when the car is moving the last digit on the dial moves most rapidly. As this last digit moves from 9 to 0, the units digit changes. The same sequence occurs for each digit from right to left.

The odometer, of course, operates continuously, and only the total distance the car has traveled since its manufacture is shown (although some prestige autos have special optional features). In order to determine the distance traveled in one trip, one must record the starting reading and subtract this figure from the reading at the end of the trip. For some, this is often confusing. However, if one remembers that he or she must subtract the smaller number from the larger number, the matter is simplified. To find how far you have traveled, you must:

a. Record the reading at *start*.

b. Record the reading when you reach your destination and *stop*.

c. Subtract the start from the stop.

Note that in the following example the units are not shown, because the procedure is the same for kilometers and for miles. Also, commas do not appear on the odometer, but they are written here for clarity.

a. Start ... 37,951.7
b. Stop ... 38,240.4

c. Subtract
Stop   38,240.4
Start 37,951.7
       ‾‾‾‾‾‾‾‾
         288.7

In addition to informing the driver about the distance he has traveled, the odometer serves other purposes. For example, the fuel efficiency of the auto can be determined. By recording the distance traveled and the fuel purchased, the consumer can determine the miles per gallon or the kilometers per liter. Also, when the fuel efficiency rate is known, the driver can budget his or her expenses and plan for a long trip or estimate the probable cost for fuel over a period of time. (See section on "Fuel Efficiency.") Map reading and following directions on the highway (which are presented in a later chapter) are also dependent upon accurate odometer calculations.

Perhaps the most exploited misuse of the odometer is in the sale and purchase of used cars. Used-car salespersons often make exaggerated claims about the "low mileage" shown on an auto they wish to sell. One must be cautious, however, because the odometer reading can be tampered with by unscrupulous owners and salespersons. The odometer cable is connected to the wheel, and if it is disconnected it will not record the distance traveled. If this is done over a long period of time, the inaccuracy is significant. The odometer can be reversed by disconnecting the cable and running the dial backward by some mechanical means. That is why it is advised that a used car be checked carefully for any unlawful tampering. It is best to buy from a reputable owner or dealer.

The effect of tampering with the odometer is illustrated by an old joke. It seems that an owner of an older car was discussing his problems with a friend. He related that his car had 87,000 miles on it and he wanted to sell it, but he was afraid he would not get much because of the high mileage. "But," his friend advised, "why don't you turn the mileage back a little? Then, you'll get more money for it." A few days later the two friends met again. One asked the other if he had sold his car yet. The owner of the old clunker replied, "Why should I sell it? It's only got 10,000 miles on it."

Topic III: Fuel Purchases

The cost of gasoline has skyrocketed in the past several years, and increases are likely to continue. This is one automobile expense most consumers reluctantly accept. But choices do exist. There are, of course, various grades of gasoline (including gasoline and alcohol mixtures), and recently the popularity of Diesel fuel for Diesel engines has increased. Prices do vary from station to station, and self-service stations usually charge less than full-service stations. So it pays to shop around.

While checking prices one is likely to find the display of prices somewhat deceiving, because the charge per unit (gallon or liter) is shown to the tenth of a cent. The cents are often prominently displayed in large-sized numerals, while the tenth (usually 9/10) is somewhat smaller. This extra 9/10¢ is nearly a full cent more and realistically ought to be considered as such. In some states laws have been passed that require the fraction to be shown clearly and be at least half the size of the cent numerals. Some states also require that the display price include all taxes.

Sometimes the price is shown with a cent sign; for example, 138-9/10¢ or 138.9¢. However, for the purpose of calculation, all figures should be written with a dollar sign. Thus, the correct expression to use for calculation is $1.389.

There are three types of problems dealing with fuel purchase.

Type I: Computing the total cost when the gallons (liters) purchased and the cost per gallon (liter) are known.

Type II: Computing the cost per gallon (liter) when the total cost and the number of gallons (liters) are known.

Type III: Computing the number of gallons (liters) when the total cost and the cost per gallon (liter) are known.

In each of these problems, the cost per gallon (liter) is likely to be shown with tenth of a cent (as explained above). Furthermore, the number of gallons (liters) purchased will often also be shown with tenth of a unit; for example, 12.3 gallons or 46.7 liters. The following examples and explanations are offered to illustrate each of the three types of gasoline purchase problems.

TYPE I: FUEL PROBLEMS

12.3 gal. @ 139.8¢

Compute the total cost.

First of all, note the use of the symbol "@." As mentioned in an earlier chapter, this is a business notation that means "at a cost for each unit of . . . ." In the case of this example, it is read "at the cost of 139.8¢ for each gallon." More simply, it may be read: "at 139.8¢ per gallon."

Next, be sure to change the cent figure to one with a dollar sign: 139.8¢ is $1.398. Insist that this be done for *all* calculations.

Third, all Type I problems require multiplication. Set up the problem as multiplication without lining up the decimal points and multiply.

Example 1a:

$ 1.398
X 12.3
─────
4194
2796
1398
─────
$17.1954

Be sure to place the decimal point correctly. Round to the nearest cent. Recall the discussion entitled "Common Cents" in an earlier chapter. Does the answer look appropriate for a purchase of gasoline?

rounded to $17.20

Example 1b: (metric measure)

46.6 liters @ 36.9¢

Compute the total cost.

36.9¢ is $.369

$ .369
X 46.7
─────
2214
2214
1476
─────
$17.2323 rounded to $17.23

Again, teachers should insist that students show all their work. This will enable a diagnosis of the specific type of error the student is making. Some of the skill deficiencies to look for are:

    a. Multiplication facts.

    b. Place-value recognition.

    c. Neatness in setting up columns of figures.

    d. Addition facts.

    e. Placement of the decimal (remember "common cents").

    f. Rounding off.

TYPE II: FUEL PROBLEMS

13.6 gal. @ ___?___ = $19.56

Find the cost per gallon

Both Type II and Type III problems require division. With Type II, no figures need to be changed before dividing. However, the first decision encountered is: "Which number is the divisor?" One way to remember this is to think in terms of the units. The answer will be cost (dollars) per gallon. Remember that the word "per" means "for each" or "divided by." Therefore, the answer is dollars divided by gallons (or, in the case of metric measures, dollars divided by liters). So, the setup for this Type II example is:

$$\frac{\$19.56}{13.6} \quad \text{or} \quad \$19.56 \div 13.6 \quad \text{or} \quad 13.6 \overline{)\$19.56}$$

In order to divide, the divisor must be changed to a whole number. Mechanically, that can be done simply by moving the decimal point one place to the right in the divisor *and* in the dividend. This has the effect of multiplying both by 10.

$$\frac{\$19.56}{13.6} \times \frac{\times 10}{\times 10} = \frac{\$195.6}{136}$$

Note that the division problem is multiplied by 10/10 or 1; and multiplication by 1 does not change the actual value of the quotient. In the most common form we indicate the new position of the decimal points by a *caret* (∧), and cross out the old position.

$$13_{\times}6_{\wedge}\overline{)\$19_{\times}56_{\wedge}}$$

Note also the position of the point in the quotient. Division then proceeds according to the rules for division of whole numbers. (See Chapter 4.)

Example 2a:

Carry out to four places beyond the decimal point.

```
           $1.4382
13ₓ6∧)$19ₓ5∧6000
           13 6
            5 9 6
            5 4 4
              5 20
              4 08
              1 120
              1 088
                 320
                 272
                  48
```

Be sure to place the decimal point correctly and divide to the fourth place as explained for division of whole numbers. Round to the nearest tenth of a cent, since gasoline prices are displayed that way. Recall the "Common Cents" discussions: Does the answer look appropriate for the cost of a gallon of gasoline?

Round off to $1.438 per gallon

Example 2b: (metric measure)

51.6 liters @ ___?___ = $19.56

Find the cost per liter.

```
           $.3790
51ₓ6∧)$19ₓ5∧6000  = $.379 per liter
         15 48
          4 0 80
          3 6 12
            4 680
            4 644
              360
```

TRANSPORTATION COMPUTATION

TYPE III: FUEL PROBLEMS

$$\underline{\quad ?\quad} \text{ gal. @ } 145.8¢ = \$16.32$$

Find the number of gallons purchased.

As previously stated, Type III problems are also division. Here, as with Type I, change the ¢ figure to a $ figure: 145.8¢ = $1.458. Next, determine which number is the divisor. One might be able to reason this out by recalling a simpler example, such as: How many candies can I purchase if each costs 5¢ and I have 30¢ or __?__ candies @ 5¢ = 30¢. The answer is obviously 6, and it is obtained by dividing 5¢ into 30¢ or $5\overline{)30}^{\,6}$

Example 3a:

$$\underline{\quad ?\quad} \text{ gal. @ } \$1.458 = \$16.32$$

```
              11.19
$1,458̬)$16̬320̬00      Recall the instruction on range and approximation in Chapter 4. This
         14 58        problem (after moving decimals) divides 16 thousands by a number
          1 740       about halfway between 1 and 2 thousands. Thus, the range of the
          1 458       quotient is between 8 and 16. The approximation is 163 hundreds
            282 0    divided by 15 hundreds, or about 10. Is this an appropriate amount
            145 8    of fuel to buy for a car?
            136 20
            131 22
              4 98
```

Carry out division to two places, and round to one because pump quantities are shown to tenths of a gallon.

11.19 = 11.2 gal.

Example 3b: (metric measure)

$$\underline{\quad ?\quad} \text{ liters @ } \$38.4¢ = \$16.32$$

Find the number of liters.

38.4¢ = $.384

```
              $42.50
$̬384̬)$16̬320̬00  = 42.5 liters
       15 36
          960
          768
          192 0
          192 0
            00
```

Type II and Type III gasoline problems require division, and teachers should examine the student's work in order to diagnose deficiencies with any of the following skills.

a. Division facts.

b. Neat display of columns.

c. Subtraction facts.

85

d. Remainder as a decimal

e. Rounding to the nearest tenth.

f. Moving the decimal in the divisor and dividend.

g. Bringing down the next figure of the partial dividend.

Students will need practice with the three types of fuel problems even if they are using calculators. It is often difficult to decide when to divide or multiply. In the case of division, the student must determine which is divisor and which is dividend before entering figures on the calculator. Three aids are suggested. The first has already been discussed, the "commonsense" rule: "Does the answer seem appropriate? Does it make sense?"

A second aid is to closely examine the units. Initially, discern the unit that is desired in the answer. Then determine what operation with the given units will give the desired result. Using the previous examples:

Example 1b: Find the total cost.
Units are to be *dollars*.
Given are liters and cents per liter.

$$\text{liters} \times \frac{\text{cents}}{\text{liter}} = \frac{\text{liters}}{1} \times \frac{\text{dollars}}{\text{liter}} = \frac{\cancel{\text{liter}} \times \text{dollars}}{\cancel{\text{liter}}} = \textit{dollars}$$

Example 2b: Find the cost per liter.
Units are to be *dollars per liter* or dollars/liter.
Given are number of liters and total cost (dollars).

$$\text{dollars} \div \text{liters} = \frac{\text{dollars}}{\text{liter}} \text{ or } \textit{dollars per liter}$$

Example 3b: Find the number of *liters*.
Given are cost per liter in $\frac{\text{dollars}}{\text{liter}}$ and total cost in dollars.

$$\text{dollars} \div \frac{\text{dollars}}{\text{liter}} = \frac{\text{dollars}}{1} \times \frac{\text{liter}}{\text{dollars}} = \frac{\cancel{\text{dollars}} \times \text{liter}}{\cancel{\text{dollars}}} = \textit{liter}$$

(Explain here that the process of dividing fractions involves multiplying by the reciprocal of the divisor.)

Finally, a mnemonic device that may aid in remembering the procedure for solving gasoline purchase problems is described in Figure 5-1.

Figure 5-1
A Mnemonic Aid

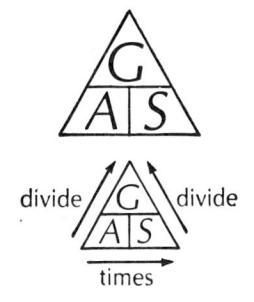

Each corner of the triangle represents one of the three variables:
G represents Gasoline or fuel expense (total)
A represents Amount of fuel purchased
S represents Selling price for each unit

The base of the triangle represents *multiplication*. The two legs of the triangle represent *division;* the dividend is always at the apex (G), and the divisor is one or the other of the base angles (A or S).

In order to calculate any one of the variables, the student should block out that variable from the triangle and perform the operation that remains.

Thus, if G is to be determined, the apex is blocked out (say, with a finger), and A times S is the solution.

If S is to be determined, G/A or G÷A is the solution.

If A is to be determined, G/S or G÷S is the solution.

Topic IV: Fuel Efficiency

Since the 1974 oil crisis, Americans have become increasingly conscious of the need to conserve fuel. The reasons are threefold: one, the supply of petroleum fuel is running out; two, the cost is rising at an alarming rate; and three, the less fuel consumed by autos, the less damage there is to the environment. The federal government now requires that all new cars sold in the United States be tested for fuel economy and that the results of the tests be posted and displayed on the automobile. The Environmental Protection Agency (EPA) is charged with the responsibility of seeing that the letter and spirit of the law are carried out. Consequently, consumers frequently hear about the "EPA mileage rating" for various cars. These mileage ratings are not always accurate, since they are calculated in the laboratory under ideal conditions. So the manufacturer adds the comment that "the mileage rating may vary" depending upon certain conditions. It is wise, therefore, for a consumer to become acquainted with the method of calculating fuel economy or fuel efficiency.

Recall the section "Odometer Readings" for the procedure to determine distance traveled. In order to determine fuel efficiency, both the distance traveled and the fuel consumed must be known. The easiest way to do this is for the consumer to record the odometer reading and the amount of gas purchased each time the tank is filled. The fuel efficiency is expressed in miles per gallon (mpg) or kilometers per liter (kpl). Since the word "per" means "for each" or "divided by," the mpg is miles divided by gallons, and the kpl is kilometers divided by liters. Hence, fuel efficiency is determined by division.

Example 4a:

Find the mpg for an auto that travels 340 miles on 8.3 gal. fuel.

Solution:

Mpg means miles divided by gallons.

```
        40.9
8.3.)340.0.0      40.9 rounded to the nearest mile is 41 mpg.
     332          (EPA mileage ratings are rounded to the nearest mile.)
      8 0
        0
      8 0 0
      7 4 7
        5 3
```

Example 4b: (metric units)

Find the kpl for an auto that travels 547 kilometers on 31.5 liters of fuel.

Solution:

kpl means kilometers divided by liters.

```
         17.3
3.5.)547.0.0     17.3 rounded to the nearest kilometer is 17 kpl.
     315
     232 0
     220 5
      11 5 0
       9 4 5
         2 0 5
```

Consumers can plan ahead for expenses when they anticipate the amount of fuel they will need. Whenever the expected number of miles traveled is known, the average miles per gallon can be used to calculate the anticipated expenses. The number of gallons of fuel needed can be determined by dividing the miles traveled by the anticipated mpg.

$$\text{mpg} \overline{\smash{\big)}\,\text{miles traveled}}^{\text{gal. of fuel used}}$$

The total fuel expense can be determined by multiplying the calculated gallons of fuel by the anticipated price per gallon.

(gal. of fuel) × (price per gal.) = total fuel expense

Example 5a:

If a traveler anticipates traveling 608 miles and has previously averaged 37 mpg, determine the amount of fuel he can expect to consume and find the cost for that fuel (assuming an average price of 153.9¢ per gal.).

Solution:

First: Divide.
$$37 \overline{)608.00} = 16.43 = 16.4 \text{ gal.}$$

Second: Multiply.   153.9¢ = $1.539

$$\begin{array}{r} \$\ 1.539 \\ \times\ 16.4 \\ \hline \$25.2396 \end{array}$$ = $25.24, anticipated fuel expense

Note again that procedure is the same with metric units. The liters of fuel can be determined by dividing the kilometers traveled by the anticipated kpl.

$$\text{kpl} \overline{)\text{kilometers traveled}}^{\text{liters of fuel used}}$$

The total fuel expense is determined by multiplying the liters by the anticipated price per liter.

(liters of fuel) × (price per liter) = total fuel expense

Example 5b: (metric units)

If a traveler anticipates traveling 977 kilometers and has previously averaged 15.7 kpl, determine the amount of fuel he can expect to consume and find the cost for that fuel (assuming an average price of 40.6¢ per liter).

Solution:

First: Divide. $15.7 \overline{)977.0.00} = 62.22 = 62.2$ liters

Second: Multiply.   40.6¢ = $.406

$$\begin{array}{r} \$\ .406 \\ \times\ 62.2 \\ \hline \$25.2532 \end{array}$$ = $25.25, anticipated fuel expense

Topic V: Renting a Car

Modern consumers are apt to have a need to rent a car at one time or another. This may occur during a vacation trip or while the owner's car is being repaired or on any number of other occasions. There are two purposes for presenting this topic here. One is to give the student experience with making consumer choices and decisions based upon objective analysis as well as subjective preference. The second is to offer a culminating experience that ties in many of the consumer math skills already presented.

When presenting this topic, students should be made conscious of the decision-making process, including goal setting, the effect of outside influence (peer group, advertisement, etc.), and the need to be realistic and objective.

LEARNING EXPERIENCES

THE DIAGNOSTIC SURVEY AND INDEPENDENCE SCALE

- Use the **Diagnostic Survey** on Reproduction Page 49 to estimate each student's degree of competence prior to instruction in the topics of this chapter.

- Upon completion of the Survey, use the **Independence Scale** on Reproduction Page 50 to identify those skills that need to be strengthened by further instruction and practice.

- Utilize the results of these two instruments to determine the appropriate assignments from the following list.

- Use the following approximate metric equivalents to convert the Survey to the metric system. The problems are changed using the appropriate factors. However, they are adjusted whenever necessary so that the conversion does not increase the difficulty of the problem.

PART I. (1-4)

1 mile = 1.61 kilometers

Multiply the mile figure by 1.61.

(1)	(2)	(3)	(4)
3734	40006	52657.1	55431.8
12144	52327	55431.8	58311.3

PART II. A. (5-8)

1 gal. = 3.79 liters (1)

Multiply the gal. figure by 3.79.

This, of course, requires that the price per unit be changed. Do this by multiplying the price per gal. by 0.264.

- 5) 40 liters @ 34¢.
- 6) 59.4 liters @ 34.3¢.
- 7) 46.7 liters @ 35.3¢.
- 8) 32.9 liters @ 38.9¢.

PART II. B. (9-12)

Change gallons to liters by multiplying gal. by 3.79; but leave the total cost the same.

- 9) 40 l.
- 10) 45 l.
- 11) 52.4 l.
- 12) 42.9 l.

TRANSPORTATION COMPUTATION

PART III. A. (17-20)

Change miles to kilometers by multiplying miles by 1.61. Change the gal. to liters by multiplying by 3.79.

(17)	(18)	(19)	(20)
240	490	280	364
60	70	45	32.7

PART III. B. (21-25)

Changes miles to kilometers as before. Change miles per gal. to kilometers per liter by multiplying the mpg by 0.425.

(21)	(22)	(23)	(24)	(25)
280	360	472	798	971
4	9	8	12	19

PART IV. (26-30)

26) 40.1 l. @ 36¢.   27) 39.5 l. @ 38.5¢.

28) 46.1 l. @ _____ = $17.78.   29) _____ l. @ 37.8¢ = $19.06.

30) 497 km. on 25.4 liters.

- Use Figure 5-2 as a general guide to convert customary units to metric for the assignments that follow. Recall, however, the admonition to adjust the figures in order to provide problems of equivalent difficulty.

Figure 5-2
Customary to Metric Conversions

From:	To:	Do this:
miles	kilometers	multiply miles by 1.61
gallons	liters	multiply gallons by 3.79
price per gallon	price per liter	multiply price per gallon by 0.264
miles per gallon (mpg)	kilometers per liter (kpl)	multiply mpg by 0.425

Topic I: Rounding and Estimating

- Refer to Chapter 1 for a review of rounding and estimating dollars-and-cents figures.
- Examples and problems of rounding and estimating appear throughout the remainder of this chapter under the other topic titles. These skills should now be treated as an integral part of all assignments and not assigned separately.

Topic II: Odometer Readings

- Ask students to locate the odometer on their family (or personal) auto and record the readings at the start and end of each day for a week.

- Have students visit several used-car lots and record the following information: year of car, odometer reading, and selling price. Have a class discussion to answer these questions: How is selling price related to odometer reading? What factors might contribute to higher or lower odometer readings? What is the average odometer reading of a one-, two-, three-year-old (etc.) car?

- Use Reproduction Page 51, Part I, for students to determine the distance traveled when given the starting and ending odometer reading.

- To give students practice with rounding off and estimating, use Reproduction Page 51, Part II.

Topic III: Fuel Purchases

- Have students survey gas stations in the area and record the cost per gallon (or liter) for these fuels: regular, no-lead, and premium gasoline; Diesel fuel; and gasohol.

- Use Reproduction Page 52, Part I, and have students calculate the total cost of fuel purchased when the cost per gallon and the number of gallons purchased are known.

- Use Reproduction Page 52, Part II, and ask students to calculate the cost per gallon when the number of gallons purchased and total cost are known.

- Use Reproduction Page 52, Part III, and have students calculate the number of gallons purchased given the cost per gallon and the total cost of purchase. Use Reproduction Page 53, Parts I and II, for a review of Type I and Type II fuel problems.

Topic IV: Fuel Efficiency

- Ask students to write to the Environmental Protection Agency for a listing of the EPA mileage ratings for all new cars. Consult the library for ratings during previous years.

- Have students survey the owners of various autos to determine their actual mileage rating. Discuss any difference between these ratings and the EPA ratings.

- Use Reproduction Page 54, Part I, and have students find the miles per gallon when given the miles traveled and gallons of fuel used.

- Use Reproduction Page 54, Parts II and III, and have students calculate the amount of fuel used and the total fuel expense.

- Have students visit various new car showrooms to examine the manufacturer's "sticker," which shows the EPA mileage rating and the estimated fuel cost for one year. Have students record these figures and in class discussion list them and ask: How did the manufacturer come up with these figures? What yearly mileage total did the manufacturer use? What cost per gallon is used?

- Conduct a class discussion about the use of mpg ratings in advertising. Use examples to illustrate how figures can be misleading. Consider the ad, for instance, that claimed that their car got more miles per *tankful* than a long list of imports. The ad stated the mpg for their car but not for that of the others. A summary of the ad appears below.

Our car gets more miles per tankful

Car A gets	363 miles per tankful.
Car B gets	337 miles per tankful.
Car C gets	263 miles per tankful.
OUR CAR gets	450 miles per tankful.

(based upon 18 mpg and our 25-gal. tank.)

Topic V: Renting a Car

- Discuss the various expenses involved with renting an automobile. Use Reproduction Page 55, and have students calculate the rental costs with the given rates.

- Conduct a class discussion on the rating scale and choices made in Part V of the above assignment. Help students clarify how they make these choices.

- Ask students to conduct a telephone survey to determine the up-to-date charges to rent a car.

ASSESSING ACHIEVEMENT OF OBJECTIVES

Ongoing Evaluation

The extent to which students have mastered the concepts covered under the five topics in this chapter can be measured by any of the activities assigned to class members individually.

Final Evaluation

For an overall evaluation of the students' mastery of the concepts in this chapter, if all topics in the chapter have been taught, a test constructed directly from the "Objectives" listed at the beginning of the chapter can be used. As an alternative, one might consider using the **Diagnostic Survey** as a final test.

RESOURCES FOR TEACHING ABOUT TRANSPORTATION COMPUTATION

Below is a selected and annotated list of resources useful for teaching the topics in this chapter, divided into audiovisual materials, games, and print materials. Addresses of publishers or distributors can be found in the alphabetic list in Appendix B.

Audiovisual Materials

A. FOR LOW ABILITY AND SPECIAL EDUCATION STUDENTS

Transportation, filmstrip and audio-cassette. Interpretive Education, 1977.

B. FOR GENERAL MATHEMATICS STUDENTS

"How's Your Driving Math?" *Consumer Math Cassettes* produced by F. Lee McFadden. The Math House, 1977.

How to Buy a Used Car, 12 minutes. General Motors Corporation, 1977.

Print

A. FOR LOW ABILITY AND SPECIAL EDUCATION STUDENTS

Working Makes Sense by Charles H. Kahn and J. Bradley Hanna. See especially pp. 31-39. Fearon-Pitman Learning, Inc., 1973.

Useful Arithmetic, Volume I, by John D. Wool and Raymond J. Bohn. See especially pp. 48-51. Frank E. Richards Publishing Company, Inc., 1972.

Useful Arithmetic, Volume II, by John D. Wool. See especially pp. 29-34. Frank E. Richards Publishing Company, Inc., 1972.

Mathematics for Today, Level Blue, by Edward Williams; Saul Katz, Ed.D.; and Patricia Klagholz. See especially pp. 129-168. Sadlier-Oxford, 1976.

B. FOR GENERAL MATHEMATICS STUDENTS

Consumer Mathematics with Calculator Applications by Alan Belstock and Gerald Smith. See especially pp. 64-67 and 248-259. McGraw-Hill Book Company, 1980.

Consumer Mathematics, Third Edition, by William E. Goe. Activities book available. See especially pp. 24-38. Harcourt Brace Jovanovich, 1979.

Mathematics for Today's Consumer by Jack Price, Olene Brown, Michael Charles, and Miriam Lien Clifford. Compiled from selections from *Mathematics for Everyday Life* and *Mathematics for the Real World.* See especially pp. 109-115 and 252-257. Charles E. Merrill Publishing Company, 1979.

Mathematics for Everyday Life by Jack Price, Olene Brown, Michael Charles, and Miriam Lien Clifford. See especially pp. 60-65 and 109-115. Charles E. Merrill Publishing Company, 1978.

Mathematics for Daily Living by Harry Lewis. See especially pp. 26-32 and 45-53. McCormick-Mathers Publishing Company, 1975.

Consumer and Career Mathematics by L. Carey Bolster, H. Douglas Woodman, and Joella H. Gipson. See especially pp. 150-159 and 167-170. Scott, Foresman and Company, 1978.

Business and Consumer Arithmetic by Milton C. Olson and A. E. McVelly. See especially pp. 180-183. Prentice-Hall, Inc., 1974.

RESOURCE UNITS, PAMPHLETS, BROCHURES, ETC.

Caring for Your Car by Clayton L. Evenden. Project Consumer, A Livonia Public Schools Project (with the Consumers' Education Office, Department of Health, Education, and Welfare, 1978).

"The Gasoline Mileage Book," *Shell Answer Book #3.* Shell Oil Company, 1976.

"Gas Mileage Guide," number 909H. "Gasoline: More Miles Per Gallon," number 513H, 1980. "How to Save Gasoline . . . And Money," number 514H, 1980. "Self-Service Gas Up and Go," number 104H, 1980. Consumer Information Center.

"Why Clark has gone to liter pricing." Clark Oil Corporation, 1981.

6

Travel and Recreation

INTRODUCTION

This chapter utilizes travel and recreation problems as a means of focusing upon necessary mathematics skills such as understanding fractions and learning concepts of time, distance, and rate of speed. In addition, the math skills associated with reading maps, using transportation timetables, determining costs, and planning for vacation trips are also emphasized. Fractions are introduced here as a necessary means to understand time, and the presentation in the "Overview" is centered upon the notion of minutes and hours. There is, for example, instruction on changing denominators from and to 60ths. This is not meant to limit the teacher. Rather, it is in keeping with the developmental process woven into the fabric of this *Guidebook*. In the same manner, multiplication and division of fractions are presented, while addition and subtraction are not. Furthermore, in keeping with the pragmatic approach, students are encouraged to change fractions to decimals so that calculators may be utilized to determine answers.

The notion of a formula is introduced in this chapter, and it should be taught as a necessary aid for memorization of a rule. The distance formula, $d = rt$, will be followed later by other formulas ($A = lw$; $i = prt$; etc.), and it is important that students become accustomed to the idea that a formula is a helpful rule. Whether using a formula or not, the significance of units is stressed, and ratio and proportion are introduced as means for converting units.

The other topics are presented primarily as interest-evoking activities that require the use of all the skills already presented, plus some new ones that can be learned quickly. Map reading has proven to be especially motivating, and the thought of traveling on a vacation keeps the interest of many students while they test their skills with money management. Much of the success will depend upon the teacher's involvement, however, because maps, the means of travel, and vacation plans ought to be geared to the needs of, and possibilities open to, the students and the particular region in which they live. Thus, local maps and maps of home states, as well as up-to-date costs for bus, train, and plane travel should be used.

> **OBJECTIVES**
>
> If all the topics in this chapter are chosen by the teacher, the student should be able to:
>
> 1. Use minutes and hours interchangeably, writing the measures in fractions of units and rounding off.
>
> 2. Demonstrate an understanding of the meaning of common fractions by being able to:
>
>   a. express common fractions in lowest terms.
>   b. express common fractions in higher terms.
>   c. change common fractions to decimal fractions.
>   d. multiply common fractions.
>   e. divide common fractions.
>   f. write mixed numbers as improper fractions.
>   g. write improper fractions as mixed numbers.
>
> 3. Change feet to miles and meters to kilometers as well as recognize fractional portions of each unit.
>
> 4. Use the concept of rate of speed in miles per hour and kilometers per hour.
>
> 5. Use the formula: distance = rate × time ($d = rt$).
>
>   a. Find d when r and t are known.
>   b. Find r when d and t are known.
>   c. Find t when r and d are known.
>
> 6. Use a roadmap successfully.
>
> 7. Use a timetable and estimate costs for travel by bus, train, or airplane.
>
> 8. Determine costs for vacation trips.

CONTENT OVERVIEW

Topic I: Fractions and Time

Expressions about time often involve fractions. The fractions of an hour are used so casually that they have become second nature. People speak of "half past the hour," "a quarter to three," and "two thirds of an hour" and are not even aware that they are using fractions. The same people who use these expressions will "avoid fractions like the plague" when it comes to math. In reality, a minute is also a fraction; it is a fraction of an hour. Because of this, fractions are here presented as they relate to time. This presentation is not meant to be inclusive but is merely an introduction to the use of fractions. Further development of the topic will continue in later chapters.

An hour is divided into 60 equal parts called minutes. A number that tells the relationship of one or more of the equal parts to the total number of equal parts is called a *fraction*. Thus, 12 minutes can be expressed as a fraction of an hour: 12/60. We normally write a fraction as a pair of numbers with a horizontal bar between them. This form is called a *common fraction*; the top or first number is the *numerator,* while the bottom or next number is called the *denominator.* A numerator may be any number, but the denominator cannot be zero. Recall the previous discussion in earlier chapters concerning the term "per." There are 60 minutes per hour; and 12 minutes is 12/60 of an hour, 12 minutes out of 60 minutes, 12 minutes per hour, or 12 per 60. The term "per" means to divide. Hence, 12/60 also shows division: 12 ÷ 60 or 60)$\overline{12}$.

All fractions signify division. By dividing the numerator by the denominator, a common fraction can be changed to a decimal fraction (or decimal). The fraction 12/60 of an hour is 0.2 of an hour because 12 ÷ 60 = 0.2. The decimal fraction, 0.2, is the same as the common fraction 2/10. Thus, 12/60 must equal 2/10. This clearly shows the interrelation between common fractions (whose denominator can be any number except zero) and decimal fractions (whose denominator is always a power of 10).

The fraction 2/10 is known as a *reduced* form of 12/60. Actually, 12/60 can be reduced even further, to 1/5. The process of reducing a fraction involves dividing both the numerator and denominator by the same number. Reducing to *lowest terms* means to divide the numerator and denominator of the fraction by the largest number that can be divided exactly into both. Thus:

$$\frac{12}{60} = \frac{12 \div 12}{60 \div 12} = \frac{1}{5}$$

Another way to reduce fractions is to *factor* the numerator and denominator and divide (or "cancel out") the like factors. Thus:

$$\frac{12}{60} = \frac{\cancel{2} \cdot \cancel{2} \cdot \cancel{3}}{\cancel{2} \cdot \cancel{2} \cdot \cancel{3} \cdot 5} = \frac{1}{5}$$

The term "cancelling out" is a poor one because it implies that the answer is zero when cancelled. In reality, the like factors divide one into the other to equal 1, the multiplicative identity. One is called the multiplicative identity because 1 times any number equals the same (identical) number. (It is said that old math teachers never die, they just reduce to lowest terms.)

When reducing a fraction to lower terms, the numbers attained are *equivalent fractions.* Thus, 12/60 = 6/30 = 3/15 = 1/5 are all equivalent. In the same manner, a fraction may be written to equivalent higher terms; for example, 1/10 = 6/60. To raise a fraction to equivalent higher terms: 1) select the higher denominator, 2) divide the new denominator by the denominator of the given fraction, and 3) multiply the numerator and denominator of the given fraction by the quotient. Thus:

$$\frac{1}{10} = \frac{?}{60}, 60 \div 10 = 6, \frac{1 \times 6}{10 \times 6} = \frac{6}{60}.$$

This method also provides a means to change a common fraction to a decimal simply by changing the denominator to 100.

The examples below show how these concepts about fractions apply to time. (The examples are taken from **Diagnostic Survey Six.**)

1) 45 minutes is what fraction of an hour?

$$45 \text{ min.} = 45 \text{ out of } 60 = \frac{45}{60} = \frac{3 \cdot \cancel{3} \cdot \cancel{5}}{2 \cdot 2 \cdot \cancel{3} \cdot \cancel{5}} = \frac{3}{4} \text{ hours}$$

$$45 \text{ min.} = \frac{45}{60} = \frac{3}{4} = \frac{3 \times 25}{4 \times 25} = \frac{75}{100} = .75 \text{ hours}$$

(or 45 ÷ 60 =
$$\begin{array}{r} .75 \\ 60\overline{)45.00} \\ \underline{42\ 0} \\ 3\ 00 \\ \underline{3\ 00} \\ 0 \end{array}$$
= .75 hours.)

2) $\frac{1}{12}$ hours is how many minutes?

$$\frac{1}{12} \text{ of an hour is } \frac{1}{12} = \frac{?}{60} \qquad \frac{1 \times 5}{12 \times 5} = \frac{5}{60} \text{ hours}$$

$\frac{5}{60}$ hour is 5 out of 60 or 5 minutes

MULTIPLICATION AND DIVISION OF FRACTIONS

Decimal fractions. Decimal fractions are easier to work with than common fractions. This is true for paper-and-pencil manipulations as well as calculator entry. For this reason, it is recommended that whenever possible common fractions be changed to decimals in order to perform the operations required. Thus, if the problem involves multiplying 3/4 miles per hour times 1/2 hour, change the problem from 3/4 X 1/2 to .75 X .5. Similarly, if the problem is 36-7/10 miles per hour times 5 hours and 3 minutes, change 36-7/10 to 36.7 and change 5 hours, 3 minutes to 5-3/60 or 5-1/20 to 5-5/100 to 5.05. Then multiply 36.7 X 5.05. Correspondingly, the same changes to decimals can be made for division.

On the other hand, some fractions do not change readily to decimals. For instance, 5 minutes is 1/12 of an hour. Changed to a decimal, 1/12 becomes $0.08\overline{3}\ldots$, a repeating decimal. Other examples are very common: $1/3 = 0.3\overline{3}\ldots$, $2/3 = 0.6\overline{6}\ldots$, $1/9 = 0.1\overline{1}\ldots$, $2/9 = 0.2\overline{2}\ldots$, and so on. Because of the nature of these decimals, when they are used in calculations, the answers will not be exact. For example:

$$\frac{1}{12} = \frac{5}{60} \text{ or 5 min.; but } 0.08\overline{3} \times 60 = 4.998 \text{ min.}$$

$$\frac{1}{3} = \frac{20}{60} \text{ or 20 min.; but } 0.3\overline{3} \times 60 = 19.8 \text{ min.}$$

Therefore, the options are to settle for an approximate answer and round off, or to learn the methods of multiplication and division of common fractions.

Common fractions. Multiplying two simple fractions (with no whole numbers) may be done by 1) multiplying the numerator of one by the numerator of the other, 2) multiplying the denominator of one by the denominator of the other, and 3) reducing the resultant fraction. Thus, 15/16 X 8/9 = 120/144. However, reducing 120/144 is rather difficult, and another method of multiplication is recommended. This method of multiplication utilizes the concept of dividing common factors before multiplication is complete and produces an already reduced answer. Thus:

$$\frac{15}{16} \times \frac{8}{9} \times \frac{15 \times 8}{16 \times 9} = \frac{\cancel{3} \times 5 \times \cancel{2} \times \cancel{2} \times \cancel{2}}{\cancel{2} \times \cancel{2} \times \cancel{2} \times 2 \times \cancel{3} \times 3} = \frac{5}{6}$$

$$\frac{9}{12} \times \frac{2}{3} = \frac{9 \times 2}{12 \times 3} = \frac{\cancel{3} \times \cancel{3} \times \cancel{1}}{\cancel{3} \times \cancel{2} \times 2 \times \cancel{3}} = \frac{1}{2}$$

(Note that in the last example, when all factors of the numerator are divided, the answer is one.)

When multiplying two *mixed numbers* (whole numbers with fractions combined), the procedure is the same except that the mixed numbers must be changed to *improper fractions*. Improper fractions are fractions whose numerator is greater than the denominator. For example, 3-5/9 becomes 32/9 because 3 = 3/1 = ?/9 = 27/9; 27/9 + 5/9 = 32/9. A shorter method is to simply multiply the whole number by the denominator and add the product to the given numerator; the sum becomes the new numerator and the given denominator is retained. When the fractions are in the appropriate form, multiplication proceeds as before.

$$3\frac{5}{9} \times 7\frac{7}{8} = \frac{32}{9} \times \frac{63}{8} = \frac{\cancel{8} \times 4 \times \cancel{9} \times 7}{\cancel{9} \times \cancel{8}} = \frac{28}{1} = 28$$

(Note above that complete factoring was not done because common factors were readily identified.)

Division of common fractions requires one preliminary step, and then it too proceeds as with multiplication. When the fractions are in the appropriate form (mixed numbers eliminated), the divisor is *inverted* (turned upside down), and then the multiplication rules are followed. Thus:

$$\frac{6}{15} \div \frac{3}{5} = \frac{6}{15} \times \frac{5}{3} = \frac{\cancel{3} \times 2 \times \cancel{5}}{\cancel{3} \times \cancel{5} \times 3} = \frac{2}{3}$$

And when a whole number is involved:

$$\frac{15}{16} \div 5 = \frac{15}{16} \div \frac{5}{1} = \frac{15}{16} \times \frac{1}{5} = \frac{\cancel{5} \times 3 \times 1}{16 \times \cancel{5}} = \frac{3}{16}$$

In the case of mixed numbers:

$$3\frac{3}{16} \div 2\frac{1}{8} = \frac{51}{16} \div \frac{17}{8} = \frac{51}{16} \times \frac{8}{17} = \frac{\cancel{17} \times 3 \times \cancel{8}}{\cancel{8} \times 2 \times \cancel{17}} = \frac{3}{2}$$

Note here that the answer is an improper fraction. Of course, an improper fraction can be changed to a mixed number. The method is to divide the numerator by the denominator and write the remainder as a fraction of the given denominator.

$$\frac{3}{2} = 2\overline{)3} \qquad = 1\frac{1}{2}$$
$$\phantom{\frac{3}{2} = }\underline{2}$$
$$\phantom{\frac{3}{2} = }1 \text{ (remainder)}$$

$$\frac{28}{5} = 5\overline{)28} \qquad = 1\frac{3}{5}$$
$$\phantom{\frac{28}{5} = }\underline{25}$$
$$\phantom{\frac{28}{5} = }3 \text{ (remainder)}$$

$$\frac{33}{15} = 15\overline{)33} = 2\frac{3}{15} = 2\frac{1}{5}$$
$$\phantom{\frac{33}{15} = 15}\underline{30}$$
$$\phantom{\frac{33}{15} = 15}3 \text{ (remainder)}$$

(Note that in the last example the answer has been reduced.)

In the final analysis, it is the degree of precision desired that will dictate the use of common fractions or decimal fractions. Generally speaking, consumers will be able to use decimal approximations of fractions more easily than common fractions (especially with calculators). In the section that follows, some examples are given for both calculations.

Topic II: Distance, Rate, and Time

Traveling great distances while on vacation can present some problems. How long will it take? How fast must you travel? How far should be traveled? What will it cost? Is the time saved by air travel worth the cost? These are all meaningful questions when planning a long trip. Each can be answered, at least in part, by the use of one simple rule: *the distance traveled is equal to the product of the rate of speed and the time traveled*. This rule can be summarized briefly in the following formula:

$$d = r \times t \quad \text{or} \quad d = rt$$

Since there are three quantities (called *variables*) in the formula, two must be known in order to calculate the third. We shall explore how the value of each variable can be determined.

FINDING THE DISTANCE

To find the distance (d), the rate (r) and the time (t) must be known. Here, the units should be considered. It is most common to use miles per hour (mph) for the rate of speed, hours (h.) for the time, and miles (mi.) for distance. In the metric system, speed is in kilometers per hour (kph), time is in hours (h.), and distance is measured in kilometers (km.). As the following examples are examined, note the significance of using proper units and how the knowledge of fractions aids in understanding the problem and arriving at the correct answer. Furthermore, also note that the rate of speed is the *average* rate. This means that the speed is not always the same; sometimes it is faster and sometimes it is slower. Indeed, there are times when the vehicle is stopped. So, the average speed must be used to calculate answers from this formula. (In each set the *b* examples show equivalent metric units.)

Example 1a:

Find the distance traveled by an auto traveling an average speed of 50 miles per hour for 6 hours.

$$d = rt \quad r = 50 \text{ mph} \quad t = 6\text{h.}$$
$$d = 50 \times 6 = 300 \text{ mi.}$$

(Units are mph $\times$ h. $= \frac{\text{mi.}}{\cancel{\text{h.}}} \times \frac{\cancel{\text{h.}}}{1} = \frac{\text{mi.}}{1} = $ miles.)

Example 1b:

Find the distance traveled by an auto traveling an average speed of 80 kilometers per hour for 6 hours.

$$d = rt \quad r = 80 \text{ kph} \quad t = 6\text{h.}$$
$$d = 80 \times 6 = 480 \text{ k.}$$

(Units are kph $\times$ h. $= \dfrac{k.}{\cancel{h.}} \times \dfrac{\cancel{h.}}{1} = \dfrac{k.}{1} =$ kilometers.)

Example 2:

$r = 40$ mph, $t = 3$ h., 30 min. Find d.

t must be expressed in hours.

$$3 \text{ h. } 30 \text{ min.} = 3\frac{30}{60} = 3\frac{1}{2} = 3.5 \text{ h.}$$
$$d = rt = 40 \text{ mph} \times 3.5 \text{ h.} = 140 \text{ mi.}$$

or 40 mph $\times 3\dfrac{1}{2}$ h. $= \dfrac{40}{1}$ mph $\times \dfrac{7}{2}$ h. $= \dfrac{20 \times \cancel{2} \text{ mi.}}{1 \; \cancel{h.}} \times \dfrac{7 \; \cancel{h.}}{\cancel{2} \; 1} = 140$ miles

Example 3a:

$r = 51$ mph, $t = 100$ min. Find d.

Express t in hours.

$$100 \text{ min.} = \frac{100}{60} = \frac{10}{6} = \frac{2 \times 5}{2 \times 3} = \frac{5}{3} \text{ h.}$$

or $\dfrac{5}{3}$ h. $= 5 \div 3 = 1.6\overline{6} \ldots = 1.7$ h. (approx.)

$$d = rt = 51 \times \frac{5}{3} = \frac{51}{1} \times \frac{5}{3} = \frac{17 \times \cancel{3}}{1} \times \frac{5}{\cancel{3}} = 85 \text{ mi.}$$

or $51 \times 1.7 = 86.7$ or 87 mi. (approx.)

Example 3b:

$r = 82$ kph, $t = 1$ h. 40 min. Find d.

Express t in hours.

$$1 \text{ h. } 40 \text{ min.} = 1\frac{40}{60} = 1\frac{4}{6} = 1\frac{2}{3} \text{ h. or } \frac{5}{3} \text{ h.}$$
$$\frac{5}{3} \text{ h.} = 5 \div 3 = 1.6\overline{6} \ldots = 1.7 \text{ h. (approx.)}$$
$$d = rt = 82 \times \frac{5}{3} = \frac{82}{1} \times \frac{5}{3} = \frac{410}{3} = 136\frac{2}{3} \text{ km.} = 137 \text{ km.}$$

or $82 \times 1.7 = 139.4$ km. $= 139$ km. (approx.)

Other calculations involving distance demonstrate the value of the metric system. They are presented here to introduce certain metric concepts that are more fully developed in another chapter. At this point, the emphasis ought to be on the use of standard units and their subunits (miles and feet for customary units; kilometers and meters in metrics) and on the relationship between fractions and decimals. The metric system uses the *meter* as the standard unit of length. In the customary system, the "standard" unit is dependent upon the length or distance being measured. Usually feet and miles are used to measure distance in customary units, and we shall explore the relationship of these units. There are 5280 feet in one mile (5280 feet per mile). Examples 4-8 are presented to show the use of fractions and decimals in customary units.

Example 4:

Change 1320 feet to miles.

Since there are 5280 feet in one mile, 1320 ft. is 1320/5280 of a mile.

Reduce as follows:

$$\frac{1320}{5280} = \frac{132 \times \cancel{10}}{528 \times \cancel{10}} = \frac{1 \times \cancel{132}}{4 \times \cancel{132}} = \frac{1}{4} \text{ mi.}$$

$$\frac{1}{4} \text{ mi.} = 1 \div 4 = 0.25 \text{ mi.}$$

Example 5:

Change 1760 feet to miles.

$$\frac{1760}{5280} = \frac{176 \times \cancel{10}}{528 \times \cancel{10}} = \frac{1 \times \cancel{176}}{3 \times \cancel{176}} = \frac{1}{3} \text{ mi.}$$

$$\frac{1}{3} \text{ mi.} = 1 \div 3 = 0.\overline{33}\ldots = 0.3 \text{ mi. (approx.)}$$

Example 6:

Change 3846 feet to miles.

$$\frac{3846}{5280} = \frac{\cancel{6} \times 641}{\cancel{6} \times 880} \text{ mi.}$$

$$\frac{641}{880} = 641 \div 880 = 0.728409\ (+) = 0.7 \text{ mi. (approx.)}$$

(Note the awkwardness of both the fraction and decimal in Example 6.)

Example 7:

Change 2/11 mile to feet.

$$5280 \text{ ft.} = 1 \text{ mi.}$$

$$\frac{2}{11} \text{ of } 1 \text{ mi.} = \frac{2}{11} \text{ of } 5280 \text{ ft.}$$

"of" means "times"

$$\frac{2}{11} \times 5280 = \frac{2}{11} \times \frac{5280}{1} = \frac{2 \times \cancel{11} \times 480}{\cancel{11} \times 1} = 960 \text{ ft.}$$

or $\frac{2}{11} = 0.\overline{18}\ldots = 0.2$ (approx.)

$0.2 \times 5280 = 1056$ ft. (approx.)

(Note the wide variance between fraction and decimal calculations in Example 7.)

Example 8a:

Change 0.8 mile to feet.

$$0.8 \times 5280 \text{ ft.} = 4224 \text{ ft.}$$

There are 1000 meters in a kilometer. (The prefix *kilo-* means 1000.) The following examples are presented to show the ease of changing units in the metric system.

Example 8b:

Change 867 meters to kilometers.

Since there are 1000 meters in 1 kilometer

$$867 \text{ m.} = \frac{867}{1000} \text{ m.}$$

Recall that dividing by multiples of ten can be accomplished by moving the decimal point to the left the same number of places as there are zeros.

$$\frac{867}{1000} = .867 = .867 \text{ km.}$$

(Note that the decimal expression is precise and there is no need to show the fraction.)

Example 9:

Change 25.36 meters to kilometers.

$$\frac{25.36}{1000} = 0.02536 \text{ km.}$$

Example 10:

Change $\frac{9}{17}$ kilometer to meters.

$$\frac{9}{17} \text{ of a km.} = \frac{9}{17} \text{ of 1000 m.}$$

$$\frac{9}{17} \times 1000 = \frac{9000}{17} = 529.41 \text{ m.}$$

$$\text{or } \frac{9}{17} = 0.529$$

$$0.529 \times 1000 = 529 \text{ m.}$$

(Recall the short methods for multiplying by powers of 10.)

(Note that even with the metric system, fractions can be somewhat awkward. Fractions are rarely, if ever, used in the metric system.)

Example 11:

Change 0.386 kilometer to meters.

$$0.386 \text{ km.} = 386 \text{ m.}$$

In summary, changing units within the metric system is much easier than changing units within the customary system. In the customary system, the quantities must be multiplied or divided by 5280. In the metric system, one needs only to move the decimal point to the right or to the left. With this evidence, it is small wonder that every nation in the world but four is already using the metric system.

The difficulty Americans have with the metric system is not actually because of the complexity of those units. It has to do with conversion from customary to metric. As pointed out before, conversion from one system to the other will not be necessary when metrics are fully adopted. Therefore, problems involving metrics should concentrate on measurements within that system, and conversion should be played down as much less important. However, the conversion factors are offered here so that one might compare the metric measures to more familiar measures:

1 mile = 1.61 kilometers

1 kilometer = 0.62 miles

To change miles to kilometers, multiply by 1.61.
To change kilometers to miles, multiply by 0.62.

FINDING THE RATE OF SPEED

Since the formula for distance has the variable for rate of speed (r), the same formula may be used to find speed when distance and time are known. The formula must be rewritten to solve for r. Thus:

$$d = rt \quad \text{or} \quad rt = d$$

Divide both sides of the equal sign by t

$$\frac{r\cancel{t}}{\cancel{t}} = \frac{d}{t}$$

$$r = \frac{d}{t}$$

$$\text{or } r = d \div t$$

The rate of speed can be determined, therefore, by dividing the distance by the time. Recall, also, that this is the average rate of speed.

Example 12a:

Find the average rate of speed that an auto must travel to go 84 miles in 2 hours and 6 minutes.

Express t *in hours.*

$$2 \text{ h. } 6 \text{ min.} = 2\frac{6}{60} = 2\frac{\cancel{6} \times 1}{\cancel{6} \times 10} = 2\frac{1}{10} = \frac{21}{10}\text{h.}$$

$$r = d \div t = 84 \div \frac{21}{10} = \frac{84}{1} \div \frac{21}{10} = \frac{84}{1} \times \frac{10}{21} = \frac{4 \times \cancel{21} \times 10}{1 \times \cancel{21}} =$$

$$\frac{40}{1} = 40 \text{ mph}$$

or: $\frac{21}{10} = 21 \div 10 = 2.1$

$84 \div 2.1 = 40$ mph

(*Units are:* $r = d \div t$ = miles ÷ hours = miles per hour.)

Example 12b:

Find the average rate of speed an auto must travel to go 135.24 kilometers in 2 hours and 6 minutes.

$$t = 2.1 \text{ h.}$$

$$r = \frac{d}{t} = \frac{135.24}{2.1} = 64.4 \text{ kph}$$

(*Units are:* $r = \frac{d}{t} = \frac{\text{kilometers}}{\text{hour}}$ = kilometers per hour.)

TRAVEL AND RECREATION

Example 13:

Find the average speed for an airplane that travels 360 mph for 10 min., 600 mph for 2 h. 22 min., and 315 mph for 16 min.

The solution is found by determining the distance traveled for each of the time intervals, adding these distances, and dividing the sum of the distances by the total time of flight.

a)   360 mph for 10 min. $t = \frac{10}{60} = \frac{1}{6}$ h.

$$d = rt = 360 \times \frac{1}{6} = \frac{360}{1} \times \frac{1}{6} = \frac{\cancel{6} \times 60}{1 \times \cancel{6}} = 60 \text{ mi.}$$

b)   600 mph for 2 h. 22 min. $t = 2\frac{22}{60} = 2\frac{11}{30}$ h.

$$600 \times 2\frac{11}{30} = \frac{600}{1} \times \frac{71}{30} = \frac{\cancel{30} \times 20 \times 71}{1 \times \cancel{30}} = 1420 \text{ mi.}$$

c)   315 mph for 16 min. $t = \frac{16}{60} = \frac{4}{15}$ h.

$$315 \times \frac{4}{15} = \frac{315}{1} \times \frac{4}{15} = \frac{\cancel{15} \times 21 \times 4}{1 \times \cancel{15}} = 84 \text{ mi.}$$

d)   60 mi.   10 min.
             1420 mi.   2 h. 22 min.
           + 84 mi.   + 16 min.
             1564 mi.   2 h. 48 min. $= 2\frac{48}{60}$ h. $= 2\frac{4}{5}$ h.

e) $r = d \div t = 1564 \div 2\frac{4}{5} = \frac{1564}{1} \div \frac{14}{5} = \frac{1564}{1} \times \frac{5}{14}$

$$\frac{1564}{1} \times \frac{5}{14} = \frac{\cancel{2} \times 782 \times 5}{1 \times \cancel{2} \times 7} = \frac{3910}{7} = 558.57 = 559 \text{ mph}$$

FINDING THE TIME

Rewriting the formula to solve for t:

$$d = rt \quad \text{or} \quad rt = d$$

Divide both sides of the equal sign by r

$$\frac{\cancel{r}t}{\cancel{r}} = \frac{d}{r}$$

$$t = \frac{d}{r} \quad \text{or} \quad t = d \div r$$

Time can be determined by dividing the distance by the average rate of speed.

Example 14:

How long will it take a traveler to go 147 miles at an average rate of 42 mph?

$$t = d \div r = 147 \div 42 = 3.5 \text{ h. or 3 h. 30 min.}$$

(*Units are:* $t = d \div r =$ miles $\div$ miles per hour $= \frac{\text{miles}}{1} \div \frac{\text{miles}}{\text{hour}}$

$\frac{\text{miles}}{1} \div \frac{\text{miles}}{\text{hour}} = \frac{\cancel{\text{miles}}}{1} \times \frac{\text{hour}}{\cancel{\text{miles}}} = $ hour.)

Example 15a:

$$d = 342, r = 54 \text{ mph. Find } t.$$

$$t = \frac{d}{r} = \frac{342}{54} = 6\frac{1}{3} \text{ h.} = 6\frac{20}{60} = 6 \text{ h. } 20 \text{ min.}$$

Example 15b:

$$d = 551 \text{ km.}, r = 87 \text{ kph. Find } t.$$

$$t = \frac{d}{r} = \frac{551}{87} = 6.3\overline{3}\ldots = 6 \text{ h. } 20 \text{ min.}$$

A MEMORY TRIANGLE

Distance, rate, or time can be determined from the formula $d = rt$. Whenever two variables are known, the third variable can be calculated. In order to facilitate this calculation so that students do not need to memorize three different forms of the same formula, a mnemonic device or memory triangle is suggested. This memory triangle is similar to the one presented in Chapter 5. A complete explanation of the memory triangle for distance, rate, and time appears in Figure 6-1.

**Figure 6-1
A Memory Triangle**

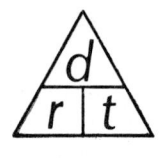

Each corner of the triangle represents one of the three variables; distance (d), rate of speed (r), and time (t).

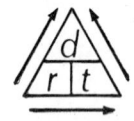

The base of the triangle represents multiplication. The two legs of the triangle represent division; the dividend is always at the apex (d), and the divisor is one or the other of the base angles (r or t).

In order to calculate any one of the variables, the student should block out that variable from the triangle and perform the operation that remains.

Thus, if d is to be determined, the apex is blocked out (say, with a finger) and $r \times t$ is the solution.

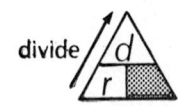

If t is to be determined, d/r or $d \div r$ is the solution.

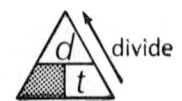

If r is to be determined, d/t or $d \div t$ is the solution.

Topic III: Map Reading

There are several advantages to presenting map reading to students at this time. Experience has shown this to be a most successful unit of study and one of the most enjoyable. It is highly motivational and calls upon several skills already presented in the *Guidebook*. The specific map-reading skills essential to successful use of maps need to be taught directly, and the sequence should show a developmental approach. These skills include using coordinates, understanding symbols, determining mileage, and following directions.

Using coordinates. Along the edges of a map there are numbers and letters marking off certain sections. Numbers usually section off the top and bottom of the map, while letters appear on the left and right edges. These are called coordinates and are related to the coordinates of a graph. The map also has a list of cities with their coordinates. Students should be instructed how to locate a city on the map when given the coordinates. The reverse task is also important. Instruction should enable students to name the coordinates of any city or place of interest whose location is known.

Understanding symbols. The legend of a map should be studied and memorized so that the symbols that appear on a map can be understood without having to refer back to the legend constantly. The most important symbols are those for type of highway (Interstate, Toll, U.S., State, Local), type of construction and pavement, and points of interest.

Determining mileage. There are four ways to determine mileage on a map: by chart, by addition of distances displayed along the highway on the map, by measurement, and by estimation. Students should become familiar with all of these methods. The use of the mileages shown along the highway is most challenging and requires some mental arithmetic. Estimation is always a part of any mileage statement owing to the specific locations desired. Of course, the same skills are necessary whether using metric units or customary units.

Following and giving directions. All of the above skills come to bear when giving or following directions. Both are important, and students should begin by learning how to follow simple directions that involve major roads only. As they progress, students should be able to follow and give lengthy directions and even draw their own maps. The teacher should rely upon maps of the local region that are familiar to the students. Thus, the imagination and creative abilities of the teacher are challenged to bring into this unit the specific materials and activities that will be of most practical value to students.

Topic IV: Bus, Train, and Plane Facts

With this topic, the teacher can relate map reading to travel by means other than car, introduce the use of a timetable, and explore the costs of various modes of travel. Students should be able to decide and plan the mode of travel for a long trip based upon interests, time, and money available. Information that is up to date and pertinent for those being taught should be utilized. Comparing costs should be modified by a determination of the length of time involved. When traveling by air, in particular, the time zones are a source of confusion, and the calculation of actual travel time can be difficult for some. In general, consumer decisions about bus, train, and plane travel involve more information gathering than calculation;

therefore, the burden of the responsibility to get the information should fall upon the student.

On the other hand, one form of calculation not already discussed that is useful in comparing costs for travel is to determine the *cost per mile*. This, of course, is obtained by determining the total cost and comparing it with the total distance traveled (i.e., dividing the total cost by the total miles traveled). This method can be deceiving, however, because long-distance travel can involve more than just fuel costs. Travel by air is fast, but travel by car is slower—1000 miles will require several days. Therefore, cost per day for travel by car should include meals and lodging.

Topic V: Vacation Trips

Most people, even if they have just returned from a vacation, are attracted by this topic. It will evoke memories of favorite places already visited and dreams of places to go. Of course, that sort of inspiration gives the teacher a great assist in teaching. In reality, this is where it all comes together: all of the skills previously learned can be applied as students examine the processes of deciding, preparing, planning, and carrying out vacation activities. There are the obvious, overall budgeting and money management activities that can be aided by the use of recordkeeping and considering the average amount spent compared with the average that is budgeted. The expenses for a trip include the initial expenses made in preparation (new clothing, new equipment, new car, maintenance of older equipment, etc.), travel expenses themselves (fuel expense, bus, train, or plane tickets), food expenses (eating out or buying and preparing groceries, as is the case with camping trips), lodging expenses (hotel, motel, or campsite), entertainment (movies, amusements, etc.), and finally the catch-all category of miscellaneous. Clearly, all basic operations with dollars and cents are involved, as well as using percents to figure sales tax and service tip; determining the three types of fuel problems; calculating mpg (or kpl); finding distance, rate, and time; using mileage tables and timetables; and more. In short, this unit may serve very well as a culmination of the *Guidebook* activities to date.

LEARNING EXPERIENCES

THE DIAGNOSTIC SURVEY AND INDEPENDENCE SCALE

- Use the **Diagnostic Survey** on Reproduction Page 56 to estimate each student's degree of competence prior to instruction in the topics of this chapter.

- Upon completion of the Survey, use the **Independence Scale** on Reproduction Page 57 to identify those skills that need to be strengthened by further instruction and practice.

- Utilize the results of these two instruments to determine the appropriate assignments from the following list.

TRAVEL AND RECREATION

Topic I: Fractions and Time

- Ask students to write minutes from 1 to 50 as fractional units of an hour.
- Use *Stein's Refresher Mathematics* (Allyn and Bacon, Inc., 1980), for practice problems with fractions, pages 74-125.

Topic II: Distance, Rate, and Time

- Use Reproduction Page 58 to provide practice problems with units of distance in the customary and metric systems.
- Use Reproduction Page 59 for exercises requiring the use of the formula, $d = rt$.

Topic III: Map Reading

For each of the assignments that follow, the teacher should obtain enough state maps so that each student may have his or her own. Maps of your own state may be obtained free from a local office of the state highway and transportation department. In addition, an outline map of your state ought to be made available to all students.

- Use the map of your own state and have students locate and list the coordinates of several cities and towns (about 20 will do). Then ask students to mark an outline map to show the approximate location of each city and town.
- To give students experience with map symbols (see the map legend or guide), have students:
  1. Identify the shape of the symbol for each type of highway (Local or County, State, U.S., and Interstate).
  2. Identify the road (by type and number) and the road classification (freeway, divided, undivided, tollway, paved, gravel, etc.) between two cities. For example, in Michigan, the road between Ann Arbor and Jackson is Interstate 94, which is a freeway.
- Use the map of your own state and have students find the distance between several pairs of cities. Select the pairs of cities so that there are several that are not on the mileage chart and students will need to add the mileage that is marked on the map along the side of the highway. Also encourage measurement and estimation of distance using the scale shown on the map.
- Use the same pairs of cities as above and have students write in words the directions to go from one city to the other, naming highway junctions and giving distances.
- Have students use the outline map to sketch the approximate location of the cities and the routes connecting them.
- Select several other pairs of cities within your state. Give students the first city of each pair and directions to follow from that starting point. Ask students to find the

second city by following your directions. Choose routes that are indirect and include several secondary roads.

- Use an outline map and ask students to sketch the approximate locations of the cities and routes described in the previous assignment.
- Ask students to find out about "road rallies" and report to the class. Investigate the possibility of having your own road rally.

Topic IV: Bus, Train, and Plane Facts

- Obtain several bus schedules for travel in your state and ask students to do the following.
    1. Determine the time and cost required to travel between several pairs of cities. Express the cost in terms of total cost and cost per mile. Also find the average rate of speed.
    2. Using a map of your own state, find the time and cost (fuel only) to travel between the same cities at an average rate of 45 mph with an auto that gets 25 mpg. Use a current fuel price.
- Obtain several train timetables for travel in your state and have students make similar comparisons as above. This time compare bus, train, and car.
- Using an airline timetable for flights from your region to distant places in the world, have students:
    1. Determine the time (be aware of time zones!) and the cost required to travel to several locations. Express the cost in terms of cost per mile. Also find the average rate of speed.
    2. Where appropriate, use bus and train schedules and compare the cost to travel to the same places by bus, train, and plane.

Topic V: Vacation Trips

- Obtain an up-to-date travel guidebook (AAA, Mobil, Rand-McNally, etc.) and have students randomly select 10 or 20 hotels and motels of various price ranges from different cities. Ask students to list the lodge and the costs for accommodations. Include prices for one person/one bed, two persons/two beds, and a family of five that includes a teenager, a six-year-old, and an infant.
- Use Reproduction Page 60 and have students determine the expenses for a three-day travel vacation trip.
- Have students determine the costs for an extended auto vacation trip that involves both camping and motel lodging. Initially, provide students with all the information they require using the forms and the directions given below. Then ask students to plan their own trip (or several trips) keeping a daily log of expenses.
- Use Reproduction Page 61 as a form for recording expenses in preparation for a vaca-

tion trip. Include the following expenses: a used van, $7895 (add sales tax, license, and title fees as necessary for your state); service to repair flat tire, get a tune-up and oil change, and general safety inspection, $57.20; install special bracket and brace so that tent can be attached to van, $25.00; groceries for camping, $65.21; new sportswear for trip, $88.42; new tent, $247.85; 3 sleeping bags @ $24.95; 3 air mattresses @ $11.10; propane stove, $39.88; camp heater, $35.47; propane fuel (6 @ $3.15); ax, $17.95; knife, $4.15; cooking supplies, $43.91; 2 cameras @ $26.49; film, 5 @ $2.57; miscellaneous, $50.00. Ask students to include all taxes as necessary for your area and record the totals on the form.

- Use multiple copies of Reproduction Page 62 and make up different expenses for each day of a simulated vacation trip. (Use Reproduction Page 60 as a prototype.) Use a different version of the same form for each day as a sort of log of activities. Let the following be a guide for planning.

    I. Auto

    A. Show consecutive odometer readings for successive days. Travel about 300 miles per day (more some days, not at all other days). Have students determine mileage each day, keep a running total, and record the average miles per day.

    B. Set up fuel problems so that each time fuel is purchased, students must calculate a Type I, II, or III fuel problem (see Chapter 5). Keep mpg about the same, say, between 19 and 25 mpg. Have students calculate mpg each day.

    II. Food
    Show cost for breakfast, lunch, and dinner and ask students to determine sales taxes, tips, and totals as needed. Some meals can be eaten at a restaurant; others can be at the campsite. Therefore, include expenses for groceries periodically.

    III. Lodging
    Lodging expenses will be motels or campsites. Use a good travel guide to determine expenses.

    IV. Entertainment and Miscellaneous
    Entertainment and miscellaneous can include amusements, movies, souvenirs, film, books, lotions, and the like.

    V. In the summary, students are to show the amount spent for that day, keep a running total on how much was spent up to date, record the average spent per day, and record the average cost per mile.

- Use Reproduction Pages 61 and 62 and have students plan their own vacation trip. They may use the same guidelines as those above plus any additional restrictions that may be appropriate. Perhaps students could also include a map showing their routes and stops along the way.

ASSESSING ACHIEVEMENT OF OBJECTIVES

Ongoing Evaluation

The extent to which students have mastered the concepts covered under the five topics in this chapter can be measured by any of the activities assigned to class members individually.

Final Evaluation

For an overall evaluation of the students' mastery of the concepts in this chapter, if all topics in the chapter have been taught, a test constructed directly from the "Objectives" listed at the beginning of the chapter can be used. As an alternative, one might consider using the **Diagnostic Survey** as a final test.

RESOURCES FOR TEACHING ABOUT TRAVEL AND RECREATION

Below is a selected and annotated list of resources useful for teaching the topics in this chapter, divided into audiovisual materials, games, and print materials. Addresses of publishers or distributors can be found in the alphabetic list in Appendix B.

Audiovisual Materials

A. FOR LOW ABILITY AND SPECIAL EDUCATION STUDENTS.

How to Read a Map, filmstrip and audio-cassette. Interpretive Education, 1977.

How to Read Schedules, filmstrip and audio-cassette. Interpretive Education, 1977.

Recreation and Leisure Time Series, 5 filmstrips, 5 audio-cassettes, and 20 student workbooks. Interpretive Education, 1977.

Transportation, filmstrip and audio-cassette. Interpretive Education, 1977.

Understanding Fractions, filmstrip, cassette, 20 student workbooks, and teacher's guide. Interpretive Education, 1977.

B. FOR GENERAL MATHEMATICS STUDENTS.

"How's Your Driving Math?" *Consumer Math Cassettes* produced by F. Lee McFadden. The Math House, 1977.

Print

A. FOR LOW ABILITY AND SPECIAL EDUCATION STUDENTS.

Useful Arithmetic, Volume I, by John D. Wool and Raymond J. Bohn. See especially pp. 48-59. Frank E. Richards Publishing Company, Inc., 1972.

Useful Arithmetic, Volume II, by John D. Wool. See especially pp. 29-33. Frank E. Richards Publishing Company, Inc., 1972.

Finding Ourselves by Dr. Eileen L. Corcoran. See entire book. Frank E. Richards Publishing Company, Inc., 1971.

Learning to Use Maps by Roger E. Franich and Jerry L. Messec. See entire book. Frank E. Richards Publishing Company, Inc., 1978.

Know the Signs—Be a Better Driver, Book 1—Driving in the Country, Book 2—Driving Interstate and Super Highways, by Eileen L. Corcoran, Ed.D., and Ambrose L. Corcoran, Ed.D. See entire books. Frank E. Richards Publishing Company, Inc., *Book 1,* 1970, and *Book 2,* 1973.

An Introduction to Everyday Skills by David H. Wiltsie. See especially pp. 17-61 and 80-89. Motivational Development, Inc., 1977.

Skills for Everyday Living, Book 1, by David H. Wiltsie. See especially pp. 2-12, 24-30, 43-49. Motivational Development, Inc., 1976.

Skills for Everyday Living, Book 2, by David H. Wiltsie. See especially pp. 2-7 and 122-126. Motivational Development, Inc., 1978.

Michigan Survival by Betty L. Hall and David Landers. Also available for all states. See especially pp. 135-151. Holt, Rinehart and Winston, 1979.

Mathematics for Today, Level Blue, by Edward Williams; Saul Katz, Ed.D., and Patricia Klagholz. See especially pp. 129-175 and 181-196. Sadlier-Oxford, 1976.

Mathematics for Today, Level Orange, by Wilmer L. Jones, Ph.D. See especially pp. 1-31. Sadlier-Oxford, 1979.

B. FOR GENERAL MATHEMATICS STUDENTS

Stein's Refresher Mathematics, Seventh Edition, by Edwin I. Stein. See especially pp. 74-126, 318, 320-322, 345-348, 614-618. Allyn and Bacon, Inc., 1980.

Trouble-Shooting Mathematics Skills, Basic Competency Edition, by Allen L. Berstein and David W. Wells. See especially pp. 195-244 and 256-259. Holt, Rinehart and Winston, 1979.

Consumer Mathematics with Calculator Applications by Alan Belstock and Gerald Smith. See especially pp. 1-15 and 260-267. McGraw-Hill Book Company, 1980.

Consumer Mathematics, Third Edition, by William E. Goe. Activities book available. See especially pp. 39-48. Harcourt Brace Jovanovich, 1979.

Mathematics for Today's Consumer by Jack Price, Olene Brown, Michael Charles, and Miriam Lien Clifford. Compiled from selections from *Mathematics for Everyday Life* and *Mathematics for the Real World*. See especially pp. 196-197, 204-206, 286-287, 292-310. Charles E. Merrill Publishing Company, 1979.

Mathematics in Life by L. Carey Bolster and H. Douglas Woodburn. See especially pp. 131-170. Scott, Foresman and Company, 1977.

Mathematics for Daily Living by Harry Lewis. See especially pp. 32-53 and 66-98. McCormick-Mathers Publishing Company, 1975.

Mathematics Plus!, Consumer, Business & Technical Applications, by Bruce R. Shaw, Richard A. Denholm and Gwendolyn H. Shelton. See especially pp. 33-64. Houghton Mifflin, 1979.

"Your Automobile Dollar," "Your Recreation Dollar," *Money Management Library*, 12 booklets. Household Finance Corporation, Money Management Institute, 1978.

Consumer and Career Mathematics by L. Carey Bolster, H. Douglas Woodman, and Joella H. Gipson. See especially pp. 174-190. Scott, Foresman and Company, 1978.

Business and Consumer Arithmetic by Milton C. Olson and A.E. McVally. See especially pp. 41-54. Prentice-Hall, Inc., 1974.

7

Getting the Best Value

INTRODUCTION

Getting the most for your money suggests a need to understand quality as well as quantity. However, quality is the more elusive concept, and this chapter deals primarily with those topics that will assist the consumer in determining quantity. Included is a look at measurement (length, area, volume, capacity, weight), unit pricing, do-it-yourself projects, and catalog buying. Fractions are again presented, this time to include the operations of adding and subtracting. The metric and customary systems of measurement are reexamined with the continuing goal of understanding the units within each system rather than conversion from one system to another. All the while that students are studying fractions and measurement, they should be made aware of how these topics are related to everyday consumer experiences. Thus, the treatment of fractions and units of measure is meant primarily to facilitate consumer awareness. For example, Topic II, "More Thought for Food," is a continuation of the presentation made in Chapter 4. This time, weights and measures are considered in budget planning and determining value.

OBJECTIVES

If all the topics in this chapter are chosen by the teacher, the student should be able to:

1. Recite the standard units of length, area, volume, capacity, and weight (mass) in both the customary and the metric systems.

2. Recognize the meaning of each unit in terms of concrete examples (e.g., a bicycle is about one meter high).

3. Convert, within the same systems, from a smaller to a larger unit and vice versa.

4. Use common and decimal fractions to express parts of whole units.

> 5. Add and subtract common fractions.
> 6. Determine and use multiples, proportions, and unit pricing to compare costs.
> 7. Determine the cost of food on a per serving basis.
> 8. Use measurement to plan and complete do-it-yourself projects.
> 9. Use weights and charge per unit of weight to determine mail order and catalog sales.

CONTENT OVERVIEW

Topic I: Measurement and Fractions

As the United States moves closer to full adoption of the metric system, there will be less and less need to use the customary units. During such a transition there is always a tendency to compare both systems, and although it has already been stated that conversion from customary to metric units is usually unnecessary, it bears repeating. What needs to be borne in mind is that only one system is used at any one time, and it is best to stick with that system when drawing comparisons of product value. Thus, it is important to understand conversion *within* the system but not necessarily conversion from one system to the other. From time to time, however, this *Guidebook* shall illustrate the inherent advantage of the metric system over the customary system. This shall be done to help motivate students and demonstrate why metric units are favored.

The relationships between the standard units of the customary system are inconsistent. The inch is 1/12 of a foot, which is 1/3 of a yard, which is 1/1760 of a mile. A weight of one ounce is 1/16 of a pound, which is 1/2000 of a ton. Incredibly, the ounce is also a measure of capacity, and it is 1/32 of a quart, which is 1/4 of a gallon. On the other hand, the standard units of the metric system are always consistent. Prefixes are used for parts of the basic unit, and these prefixes always mean the same.

kilo-	means 1000
hecto-	means 100
deka-	means 10
deci-	means $\frac{1}{10}$ or 0.1
centi-	means $\frac{1}{100}$ or 0.01
milli-	means $\frac{1}{1000}$ or 0.001

Thus, a *milli*meter is 1/1000 of a meter; a *milli*liter is 1/1000 of a liter; a *milli*gram is 1/1000 of a gram. The meter, liter, and gram are the standard units of length, capacity, and weight (mass), respectively. The advantage of the decimal fraction has already been explained. Because the customary system requires the use of common fractions, let us explore those fractions that are likely to be encountered.

GETTING THE BEST VALUE

The most common measure that uses fractions is the inch. The inch is customarily divided into 4ths, 8ths, 16ths, 32ds, and 64ths. These are, of course, the denominators of the fractions, and are all multiples of 4. (Thank goodness for some consistency.) A review of reducing fractions and changing fractions to higher terms (Chapter 5) is helpful in understanding addition and subtraction of fractions, and the teacher may wish to provide time for that review before proceeding.

ADDITION AND SUBTRACTION OF FRACTIONS

Find the sum of these measurements.

Example 1: $3\frac{3}{16}$ inches + $2\frac{5}{16}$ inches.

$3\frac{3}{16}$
$+2\frac{5}{16}$
$\overline{5\frac{8}{16}}$

Since the denominators are alike, add the numerators and write the sum over the common denominator. Reduce $\frac{8}{16} = \frac{\cancel{8} \times 1}{\cancel{8} \times 2} = \frac{1}{2}$.

$= 5\frac{1}{2}$

Example 2: $15\frac{5}{8} + 23\frac{3}{8}$.

$16\frac{5}{8}$
$+23\frac{3}{8}$
$\overline{39\frac{8}{8}}$

Add numerators and retain the common denominator $\frac{8}{8} = 1$. $39 + 1 = 40$.

$= 40$

Example 3: $13\frac{7}{16} + 2\frac{1}{2}$.

$13\frac{7}{16} = 13\frac{7}{16}$
$+2\frac{1}{2} = 2\frac{8}{16}$
$\overline{15\frac{15}{16}}$

The common denominator is 16. Change $\frac{1}{2}$ *to* $\frac{?}{16}$. $\frac{1 \times 8}{2 \times 8} = \frac{8}{16}$.

Example 4: Find the sum of $6\frac{1}{4}$, $3\frac{13}{16}$, and $7\frac{3}{8}$.

The common denominator is 16. Change $\frac{1}{4}$ and $\frac{3}{8}$ to 16ths.

$6\frac{1}{4} = 6\frac{?}{16} = 6\frac{4}{16}$
$3\frac{13}{16} = 3\frac{13}{16} = 3\frac{13}{16}$
$+7\frac{3}{8} = 7\frac{?}{16} = 7\frac{6}{16}$
$\overline{16\frac{23}{16}} \left(\frac{23}{16} = \frac{16}{16} + \frac{7}{16} = 1\frac{7}{16}\right)$
$= 16 + 1\frac{7}{16} = 17\frac{7}{16}$

For both addition and subtraction the denominators must be alike. If they are not, the first step is to determine the smallest number that is evenly divisible by all the given denominators. This is called the *lowest common denominator (LCD)*. The fractions are then changed to that denominator following the procedure described in Chapter 4 for changing to higher terms. With mixed numbers, fractions are added first, then the whole numbers are added. The resulting fraction may be reduced and simplified as required. The sum of the fractions and the whole numbers are then added.

Subtraction proceeds in a similar manner, except that regrouping is often required in order to complete the problem. When subtracting a fraction from a whole number, first regroup by taking one (1) from the whole number and changing it into a fraction whose numerator and denominator are alike and common to the given fraction. Thus, when regrouping to subtract 5 — 7/8, the "5" becomes 4-8/8. Similarly, when subtracting 3-1/8 — 1-13/16, the "3-1/8" becomes 2-8/8 + 1/8 or 2-9/8. Then, 2-9/8 must be changed to 16ths; 2-9/8 = 2-18/16. Find the difference in the following examples.

Example 5:

$$8\frac{13}{16}$$
$$-5\frac{3}{16}$$
$$\overline{3\frac{10}{16}} = 3\frac{5}{8}$$

Denominators are alike. Subtract the numerators. Reduce $\frac{10}{16} = \frac{5 \times \cancel{2}}{8 \times \cancel{2}} = \frac{5}{8}$.

Example 6:

$$12\frac{3}{8} = 11\frac{8}{8} + \frac{3}{8} = 11\frac{11}{8}$$
$$-7\frac{7}{8} = \qquad\qquad\qquad -7\frac{7}{8}$$
$$\qquad\qquad\qquad\qquad\qquad 4\frac{4}{8} = 4\frac{1}{2}$$

Example 7:

$$14\frac{1}{8} = 14\frac{1}{8} = 13\frac{8}{8} + \frac{1}{8} = 13\frac{9}{8}$$
$$-5\frac{3}{4} = -5\frac{6}{8} = -5\frac{6}{8} \qquad = -5\frac{6}{8}$$
$$\qquad\qquad\qquad\qquad\qquad\qquad\qquad 8\frac{3}{8}$$

Example 8:

$$6\frac{3}{10} = 6\frac{24}{80} = 5\frac{104}{80}$$
$$-3\frac{9}{16} = -3\frac{45}{80} = -3\frac{45}{80}$$
$$\qquad\qquad\qquad\qquad 2\frac{59}{80}$$

$\left(80 \text{ is the LCD } \frac{3}{10} = \frac{3 \times 8}{10 \times 8} = \frac{24}{80}\right.$

$\left.\frac{9}{16} = \frac{9 \times 5}{16 \times 5} = \frac{45}{80}.\right)$

GETTING THE BEST VALUE

Note that in the last example neither of the given denominators can serve as the LCD. This requires that the multiples of each fraction be examined until a common multiple is found. Usually, this will not occur if the customary fractions of an inch are followed. Nor is it likely to occur with other customary units if one is careful. For example, there are 12 inches in a foot, so that every inch represents 1/12 of a foot. There are 3 feet to a yard (or 36 inches); therefore, every foot represents 1/3 of a yard (and every inch = 1/36 of a yard). There are 16 ounces in a pound; each ounce is 1/16 pound. There are 32 ounces to a quart; each ounce represents 1/32 of a quart. Fractions are used to find the the sums and differences in the following examples.

Example 9:   Add and simplify.

$$2 \text{ ft. } 9 \text{ in.} = 2\frac{9}{12} \text{ ft.}$$
$$+ 11 \text{ ft. } 7 \text{ in.} = +11\frac{7}{12} \text{ ft.}$$
$$13\frac{16}{12} \text{ ft.} = 13 + \frac{12}{12} + \frac{4}{12} = 14\frac{4}{12} \text{ ft.} = 14 \text{ ft. } 4 \text{ in.}$$

Recall that when performing arithmetic operations with numbers denoting units of measure (called *denominate numbers*), only those numbers expressed in the same units can be added or subtracted. Thus, 7 inches cannot be added to 3 feet to obtain 10. Feet may be added only to feet, and so on. Therefore, 7 inches must be expressed as a fraction of a foot, or, 3 feet may be expressed in inches before adding or subtracting. Similar situations arise with other measures.

Example 10:   Add and simplify.

$$4 \text{ lb. } 10 \text{ oz.} = 4\frac{10}{16} \text{ lb.}$$
$$7 \text{ lb. } 9 \text{ oz.} = 7\frac{9}{16} \text{ lb.}$$
$$+ 6 \text{ lb. } 13 \text{ oz.} = 6\frac{13}{16} \text{ lb.}$$
$$17\frac{32}{16} \text{ lb.} = 17 + 2 \text{ lb.} = 19 \text{ lb.}$$

Example 11:   Add and simplify.

$$5 \text{ gal. } 3 \text{ qt.} = 5\frac{3}{4} \text{ gal.}$$
$$+ 3 \text{ gal. } 2 \text{ qt.} = 3\frac{2}{4} \text{ gal.}$$
$$8\frac{5}{4} \text{ gal.} = 9\frac{1}{4} \text{ gal.} = 9 \text{ gal. } 1 \text{ qt.}$$

Example 12:   Subtract and simplify.

$$3 \text{ qt. } 23 \text{ oz.} = 3\frac{23}{32} \text{ qt.} = 2\frac{32}{32} + \frac{23}{32} = 2\frac{55}{32} \text{ qt.}$$
$$- 1 \text{ qt. } 29 \text{ oz.} = -1\frac{29}{32} \text{ qt.} = \qquad\qquad -1\frac{29}{32} \text{ qt.}$$
$$1\frac{26}{32} \text{ qt.} = 1 \text{ qt. } 26 \text{ oz.}$$

Example 13, which follows, is completed by combining the appropriate units and then simplifying. The solution is also shown using fractions to provide a comparison.

Example 13: Subtract and simplify.

$$9 \text{ lb. } 5 \text{ oz.} = 8 \text{ lb. } 21 \text{ oz.} \qquad 8\frac{21}{16} \text{ lb.}$$

$$\text{or}$$

$$-8 \text{ lb. } 14 \text{ oz.} = -8 \text{ lb. } 14 \text{ oz.} \qquad -8\frac{14}{16} \text{ lb.}$$

$$\overline{\phantom{-8 \text{ lb. } 14 \text{ oz.} =} \quad 7 \text{ oz.}} \qquad \overline{\frac{7}{16} \text{ lb.} = 7 \text{ oz.}}$$

Common fractions are not ordinarily used with metric units. Decimal fractions are used because each subunit is expressed as 1/10, 1/100, 1/1000, and so on, of the standard unit. The advantage of using decimals has been clearly demonstrated, and common fractions can be written as decimals. However, many of the fractional expressions in the customary system are not easily converted to decimals, and some cannot be interpreted as exact decimal numbers (they form repeating decimals that are never ending).

OTHER UNITS OF MEASURE

The following table is offered as a summary of some of the most common units.

	Customary	*Metric*
Linear: (or length)	inch, foot, yard, mile	meter
Area:	square inch, square foot, square yard, square mile, acre	square meter
Capacity:	ounce, quart, gallon	liter
Volume:	cubic inch, cubic foot, cubic yard, cubic mile	cubic meter
Weight:	ounce, pound, ton	gram (mass)

The metric units actually include all of the compound units with the appropriate prefixes (kilo-, milli-, etc.). The gram is more appropriately a measure of *mass*, but in everyday use "weight" is the accepted expression.

The determination of area and volume requires that the measures be multiplied. In the case of area, two factors are multiplied. Thus, the units for area are square units (square inches, or in.^{2}; square feet, or ft.^{2}; square meters, or m.^{2}). To determine volume, three factors are multiplied. Thus, the units for volume are cubic units (cubic inches, or in.^{3}; cubic feet, or ft.^{3}; cubic meters, or m.^{3}). Linear measures are the factors that are multiplied for both area and volume. It is important to express all linear units in the same denomination when determining either the area or the volume of a geometric figure.

The number of square units contained in the surface of a geometric figure is the *area* of that surface. It is obtained by applying the appropriate formula for that geometric figure. Some common formulas for area are:

Rectangle = the length times the width: $A = lw$
Square   = the length of its side times itself: $A = s^2$
Triangle = one half of the altitude times the base: $A = 1/2ab$
Circle   = the value of pi (π) times the radius squared: $A = \pi r^2$
(π = 3.14 or 3-1/7 or 22/7)

The *volume* is the number of cubic units contained in a given space. Common formulas for determining volume are:

Rectangular solid = the length times the width times the height: $V = lwh$
Cube = the length of the edge times itself times itself:
$V = e \times e \times e$ or $V = e^3$
Cylinder = pi (π) times the square of the radius of the base times the height: $V = \pi r^2 h$ (π = 3.14 or 3-1/7 or 22/7)

Examples of practical consumer use of area measures include: painting, wallpapering; floor covering; figuring cost of cementing a sidewalk or drive; sodding, seeding, or fertilizing a lawn; and the like. Practical consumer-related use of volume measures includes: capacity of a container (e.g., refrigerator, storage box, luggage, swimming pool, etc.), amount of dirt to use for planting and landscape, room capacity, engine capacity, cylinder (piston) displacement, and the like. Further discussion of the use of linear, area, and volume measures appears in Topic III of this chapter.

ANOTHER LOOK AT METRIC MEASURE

The units of the customary system are common, and the approximate size of each is fairly well understood; most people know the size of an inch, a yard, a pound, and a quart. Consequently, when trying to describe the size of a metric unit, there is a tendency to compare it with the customary unit. This, however, often leads to confusion, and this text wishes to avoid that comparison. Instead, the following is offered as an aid in recognizing the approximate size of each metric unit.

meter	the height of a bicycle.
millimeter	the thickness of a sharpened pencil lead.
centimeter	the width of one line of loose-leaf paper.
kilometer	five city blocks.
gram	the weight of one small paper clip or one raisin.
kilogram	the weight of a one-liter carton of milk.
liter	the capacity of four regular-sized coffee mugs.
milliliter	the capacity of one eyedropper.

During the transition from customary to metric units, products sold in the United States will show both metric and customary measures. Often it appears that the metric units are "odd" or not rounded off. For example, a 7-oz. jar of peanuts is 198.4 g. It appears that the "7-oz." figure is much easier to work with. However, as metrics become more prevalent it is likely that the package size will be changed to accommodate the metric unit; perhaps the amount will be 200 g., precisely. This has already been done with soft drinks and other beverages. Formerly, they were sold by the ounce and quart; now, they are packaged in half-liter and liter bottles.

Topic II: More Thought for Food

In Chapter 4 budgeting for food was introduced, and several ideas for comparing prices were presented. These ideas can now be expanded with the use of weights and measures. Comparing prices by means of multiples, proportions, per serving and per meal costs, and unit pricing will be explored.

Recall that multiples of numbers can be used to compare values (Chapter 3). For example: What is the better buy, a package of 6 what-d'you-callits for $1.59, or a package of 10 for $2.59? The common multiple of 6 and 10 is 30; 6 X 5 = 30 and 10 X 3 = 30. Therefore:

6 for $1.59 becomes 30 for $7.95.
10 for $2.59 becomes 30 for $7.77.

Clearly, the package of 10 is the better buy, if you need 10 what-d'you-callits.

The same problem can be tackled by division to determine the price per item.

6 for $1.59 is $1.59 ÷ 6 or $.265 per item.
10 for $2.59 is $2.59 ÷ 10 or $.259 per item.

Thus, the 10 for $2.59 is again verified the better buy. This method is called unit pricing; there will be more discussion about this later on.

Still another way to compare prices was alluded to when ratios were discussed in a previous chapter. When two ratios are equal, the equality can be stated in a *proportion*. Thus, 6 for $1.59 can be stated in a ratio of amount compared with price, or 6/$1.59. To determine what 10 items would cost using the price of 6 items, the ratio amount/price = 10 items/? price is equal to 6 items/$1.59, which is a proportion.

$$\frac{10}{p} = \frac{6}{\$1.59}$$

A proportion can be solved by "cross multiplying" and then dividing.

$$\frac{10}{p} \times \frac{6}{\$1.59} \quad \text{``cross multiply''}$$

$$6p = \$15.90$$

$$\frac{\cancel{6}p}{\cancel{6}} = \frac{\$15.90}{6} \quad \text{divide both sides by 6}$$

$$p = \$2.65$$

Therefore, if 6 items cost $1.59, 10 items at the same price per item will cost $2.65. Again, the package of 10 for $2.59 is shown to be the better buy. Proportions can be used to determine prices when only one cost combination is known. The amount is always compared with the price, so that if one ratio is written with the amount (in number of items, ounces, pounds, liters, or grams) as the numerator and the cost (in dollars or cents) as the denominator, the equivalent ratio must be written in the same way. The following examples will clarify the use of the proportions to determine the better buy.

Example 14: What is the better buy, a 24-oz. package for $.98 or a 16-oz. package for $.65?

$$\frac{24 \text{ oz.}}{\$.98} = \frac{16 \text{ oz.}}{p}$$

$$24p = \$15.68$$

$$\frac{24p}{24} = \frac{\$15.68}{24}$$

$$p = \$.65\frac{1}{3}$$

Compare with a 16-oz. package for $.65.
The 16-oz. package is a slightly better buy.

Example 15: Which is the better buy, a 100-g. package for $.67 or a 75-g. package for $.51?

$$\frac{100}{\$.67} = \frac{75}{p}$$

$$100p = \$50.25$$

$$p = \$.5025$$

The 100-g. package is a slightly better buy.

UNIT PRICING

Perhaps the best way to compare prices is through the use of *unit pricing*. The unit price is simply the cost per unit of measure; it may be per item, per ounce, per gram, per liter, and the like. Several examples of finding the unit price have already been shown, although they have not been labeled as such. Division is required to determine unit price (recall that "per" means to divide), and a knowledge of units is important. Even though some stores display unit prices on their shelves, the practice is not universal, and it is to the consumer's benefit to know how to determine the unit price.

Example 16a: Find the unit price (cost per pound) for a 10-oz. jar of jelly costing 55¢.

$$55¢ = \$.55 \qquad 10 \text{ oz. is } \frac{10}{16} \text{ lb.}$$

$$\$.55 \div \frac{10}{16} = \frac{\$.55}{1} \times \frac{16}{10} = \frac{\$.11 \times \cancel{5} \times 8 \times \cancel{2}}{1 \times \cancel{5} \times \cancel{2}} = \$.88$$

or $.88 per pound

Example 16b: Find the unit price (cost per kg.) to the nearest tenth of a cent for a 283-g. jar of jelly costing 55¢.

$$55¢ = \$.55 \qquad 283 \text{ g.} = .283 \text{ kg.}$$

$$\frac{\$.55}{.283} = \$1.943 \text{ or } \$1.943 \text{ per kg.}$$

Example 17a: Which is the better laundry detergent buy: a 1-lb. 4-oz. box at 67¢ or a 5-lb. 4-oz. box at $2.59?

$$1 \text{ lb. 4 oz.} = 1\frac{4}{16} \text{ lb.} \qquad 5 \text{ lb. 4 oz.} = 5\frac{4}{16} \text{ lb.}$$

$$\$.67 \div 1\frac{4}{16} = \qquad \$2.59 \div 5\frac{4}{16} =$$

$$\frac{\$.67}{1} \div \frac{20}{16} = \qquad \frac{\$2.59}{1} \div \frac{84}{16} =$$

$$\frac{\$.67}{1} \times \frac{16}{20} = \qquad \frac{\$2.59}{1} \times \frac{16}{84} =$$

$$\frac{\$10.72}{20} = \qquad \frac{\$41.44}{84} =$$

$$\$.536 \text{ per lb.} \qquad \$.49\overline{3} \text{ per lb.}$$

The 5-lb. 4-oz. box is the better buy.

Example 17b: Which is the better laundry detergent buy: a 567-g. box at 67¢ or a 2.4-kg. box at $2.59?

$$567 \text{ g.} = .567 \text{ kg.}$$

$$\$.67 \div .567 = \$1.18 \text{ per kg.} \qquad \$2.59 \div 2.4 = \$1.079 \text{ per kg.}$$

The 2.4-kg. box is the better buy.

ESTIMATING COST PER SERVING

Cost per serving is a form of unit pricing where the "unit" is the "serving." In the previous discussion on this topic (Chapter 4), the number of servings was predetermined for the amount purchased. Indeed, servings per package now appear on many food labels. Often the number of servings is given in a cookbook or a diet meal planner on the basis of servings per pound (kilogram) or ounces (grams) per serving. In order to determine the cost per serving, the number of servings for a given purchase must be determined. This is especially necessary for homegrown foods or bulk purchases where the number of servings is not printed on the package. Also, waste must be considered, and the amount of edible food must be estimated. Consider these problems as examples of budget meal planning.

Example 18a: Duane is on a special diet prescribed by his doctor. He is to eat precise amounts of certain foods and is on a strict budget, so he plans his meals carefully. He has developed a table (see Figure 7-1) to help him plan more wisely. Complete the table by filling in the blanks.

Figure 7-1
Cost per Serving

Food	Cost per kg.	Allowable Serving	Estimated Waste per kg.	Edible Servings per kg.	Cost per Serving
ground beef	$4.36	100 g.	none	10	$.436
whole chicken	$2.10	125 g.	375 g.	5	
chicken breast	$4.15	125 g.	75 g.	7.4	
steak (w/bone)	$6.59	100 g.	100 g.		
fresh whole fish	$5.43	175 g.	250 g.		
frozen fish fillet	$6.90	175 g.	none		

GETTING THE BEST VALUE

The edible servings per kg. can be determined by subtracting the waste from 1 kg. and dividing this by the grams allowed per serving (rewritten as kilograms). Thus, for the ground beef:

$$1)\ 1\ \text{kg.} - 0\ \text{kg.} = 1\ \text{kg.} \qquad 2)\ 1\ \text{kg.} \div 0.1\ \text{kg. per serving} = 10\ \text{servings.}$$

The cost per serving is then determined by dividing the cost per kg. by the number of servings per kilogram.

$$3)\ \$4.36\ \text{per kg.} \div 10\ \text{servings per kg.} = \$.436\ \text{per serving.}$$

An examination of the units shows how they, too, are involved in the calculations.

1) Units must be alike in order to subtract. The difference carries the same unit.

2) $\text{kg.} \div \text{kg. per serving} = \text{kg.} \div \dfrac{\text{kg.}}{\text{serving}} = \dfrac{\text{kg.}}{1} \div \dfrac{\text{kg.}}{\text{serving}} = \dfrac{\cancel{\text{kg.}}}{1} \times \dfrac{\text{serving}}{\cancel{\text{kg.}}} = \text{servings.}$

3) $\text{cost per kg.} \div \text{servings per kg.} = \dfrac{\text{cost}}{\text{kg.}} \div \dfrac{\text{servings}}{\text{kg.}} = \dfrac{\text{cost}}{\cancel{\text{kg.}}} \times \dfrac{\cancel{\text{kg.}}}{\text{servings}} = \dfrac{\text{cost}}{\text{serving}} =$ cost per serving.

Similar calculations can be made with customary units, although they are more cumbersome. For example, in the case of the ground beef in Figure 7-1, 100 g. would represent 3-1/2 oz. and the edible servings per pound would be 4-4/7. Again, the advantage of metric units is clearly demonstrated.

Example 18b: Use the data from Example 18. Which is the better buy, the whole chicken or the chicken breast?

First, determine the edible servings per kg. and the cost per serving of the whole chicken.

1) 1 kg. − 0.375 kg. = 0.625 kg.

2) 0.625 kg. ÷ 0.125 kg. = 5 servings.

3) $2.10 ÷ 5 = $.42 per serving.

Second, determine the same for the chicken breast.

1) 1 kg. − 0.075 kg. = 0.925 kg.

2) 0.925 kg. ÷ 0.125 kg. = 7.4 servings.

3) $4.15 ÷ 7.4 = $.56 per serving.

The whole chicken, even with more waste, is a better value.

Example 18c: Which is the better buy, the whole fish or the fish fillet?

Whole fish:

1) 1 kg. − 0.25 kg. = 0.75 kg.

2) 0.75 kg. ÷ 0.175 kg. = about 4.3 servings.

3) $5.43 ÷ 4.3 = $1.26 per serving.

Fish fillet:

1) 1 kg. − 0 kg. = 1 kg.

2) 1 kg. ÷ .175 kg. = 5.7 servings.

3) $6.90 ÷ 5.7 = $1.21 per serving.

The frozen fish fillet is the better buy.

Estimating cost per serving requires considerable calculation, but it is more reliable in determining value because, in the final analysis the cost per unit of weight includes a cost for nonedible portions (bones, etc.). However, it is unrealistic to expect a shopper to constantly check prices in this manner. Rather, it is recommended that a periodic spot check be made, especially when extra-large purchases (as for a dinner party) are being made. Recognizing cost per serving also points out a method for curbing costs, that is, by reducing portions.

Example 19: The Hardtimers, a family of 4, have found it necessary to cut back on their budget. They normally have hamburgers three times a week, and each member has two burgers. Previously, Mrs. Hardtimer made four patties per pound of meat. Assuming that each continued to have two hamburgers three times a week, how much would the Hardtimers save in one year if they made five patties per pound? (Use $1.98 as the cost per pound.)

Solution: 4 per pound is $1.98 ÷ 4 or $.495 per burger.
5 per pound is $1.98 ÷ 5 or $.396 per burger.
4 people × 2 burgers × 3 times a week = 24 burgers.
52 weeks × 24 burgers = 1248 hamburgers per year.
1248 @ $.495 = $617.76 per year
1248 @ $.396 = $494.21
$123.55 savings per year!

If similar portion reductions are made with other high-cost foods, several hundred dollars can be saved. For every 10¢ daily savings that each person in a family of 4 makes, a $150 yearly saving can be realized.

In addition to portion reduction, other ways to reduce food costs include preshopping with newspaper ads, using coupons, raising food in a home garden, and canning and freezing foods at home. Caution should be noted, however, so that good nutrition and health are not sacrificed. Consultation with a home economist, dietician, or doctor at the local health department is a wise investment of time.

Topic III: Do-it-Yourself Projects

The most obvious use of measurement for consumers is in do-it-yourself projects. From hanging curtains and shades to painting and remodeling, from sewing a stitch to sawing a board or installing a switch, from minor repair to major construction, people are doing all kinds of projects on their own. The motive is not always to save money. Many people get involved with arts and crafts and home gardening and redecorating primarily as a hobby. It is often an expensive hobby, but it can also be very rewarding. Because of the many varieties of do-it-yourself projects, it is impossible to include them all in this *Guidebook*. However, some general guidelines are presented for consideration.

First of all, a knowledge of measurement techniques is important. Students will need to know how to use a ruler or meter stick. Fractional units are especially important when using the customary system. With the metric system, subunits must be understood. In some cases more technical measuring instruments, such as a surveyor's transit or a lathe operator's micrometer, may be needed. Often these measurements must be transferred to a drawing, usually a scale drawing. Students may wish to become skillful in making these drawings, or at least be adept at reading them.

Second, it is clear that an understanding of fractions is very important. Third, the ability to use measurements in calculations, particularly to determine area and volume, is necessary. Another calculation, not yet mentioned, is *perimeter*, the distance along the edge of an area. Perimeter measurement is required when building a fence, for example.

A fourth consideration is the cost of equipment. Sewing and making your own clothes is a good way to save money, but the initial investments must be considered, too. A sewing machine is expensive, and a great many items must be made before any savings are realized. The same is true for other projects, especially those that are short term. A table-mounted powersaw may be too large an investment if only a small remodeling project is envisioned. Perhaps a smaller, portable saw will suffice. Even when the equipment is continuously in use, another factor to consider is the cost to operate the equipment. A home freezing plan for food storage may seem to offer a grand savings, but the cost of electric power needed to run the freezer may "eat up" the savings on food.

Cost of material is yet another factor. Lumber products, cement, brick, even nails are usually purchased by construction companies, builders, and the like in very large quantities so as to realize great savings. An individual, working at home on a relatively small project, will pay substantially more for a two-by-four than will a professional builder. A wise do-it-yourselfer will plan ahead and seek out clearance sales and salvage old wood that may be otherwise thrown out. This brings to mind the cost of wasted material, especially waste as a result of inefficiency or error. A good tailor, for instance, will attempt to cut his pattern so that as much of the cloth as possible is used in the garment. Storage of material may also be a factor. Certain wood products may be subject to deterioration or rot if left in the open; seedlings may die if not transplanted quickly; paint may become useless if allowed to freeze; and so forth.

Finally, the time put into the project ought to be considered. This is especially true if time is taken off from a regular job or if a second or part-time paying job is turned down in order to complete the project. For example, a teacher who turns down a summer job of teaching driver's training in order to remodel his home may have been financially ahead if he had taken the job and hired an outside contractor. The pay received from teaching may be greater than the savings realized by doing his own work. However, as pointed out earlier, many people regard do-it-yourselfing as a hobby, and therefore it may be worth the cost to them.

In summary, the following guidelines will assure at least a modicum of success for the average home handyperson. A successful do-it-yourselfer will be able to:

1. Measure and use scale drawings or sketches accurately.

2. Understand fractions and operations with fractions.

3. Use measures to calculate with accuracy (area, perimeter, etc.).

4. Consider the cost of equipment in relation to the overall project.

5. Consider the cost of material and seek ways to salvage reusable material.

6. Factor in the investment of time compared with the results it may bring.

Topic IV: Mail-order and Catalog Buying

Some consumers often find it more convenient to shop by mail or through a department store catalog. One of the many advantages is that catalog prices are often lower than store prices. However, in order to determine the actual cost of the items purchased, the buyer must consider the cost of mailing or shipping. These "hidden" costs are not always specifically shown with the item in the catalog. There are three variables that contribute to the total cost of an item: 1) the quoted price, 2) the weight of the item, and 3) the location of the buyer. The latter two factors determine the shipping and handling charges that must be added to the listed price. The shipping weight is given in the catalog along with the price of the item. The charge for shipping can be determined by consulting another page where the buyer determines his location in relation to the shipper's address (usually designated by zone numbers). Once the zone is determined, the buyer refers to a table of charges to determine his shipping cost. Figure 7-2 shows a sample shipping table. Since the weights are given in pounds and ounces, the buyer must have a knowledge of these units.

Figure 7-2
Shipping-and-Handling Charges

	\multicolumn{8}{c}{Approximate Distance (Miles) From Distribution Center}
Zone	1 & 2	3	4	5	6	7	8
Shipping Weight	Local Zone	Less Than 150	151 to 300	301 to 600	601 to 1000	1001 to 1400	1401 to 1800	Over 1800
1 oz. to 8 oz.	$1.35	$1.37	$1.38	$1.41	$1.45	$1.49	$1.54	$1.59
9 oz. to 15 oz.	1.57	1.59	1.60	1.63	1.67	1.71	1.76	1.81
1 lb. to 2 lb.	2.06	2.16	2.20	2.31	2.42	2.48	2.59	2.76
2 lb. 1 oz. to 3 lb.	2.14	2.25	2.32	2.47	2.58	2.74	2.96	3.19
3 lb. 1 oz. to 5 lb.	2.24	2.38	2.48	2.62	2.80	3.05	3.32	3.72
5 lb. 1 oz. to 10 lb.	2.50	2.59	2.75	2.97	3.26	3.76	4.23	4.97
10 lb. 1 oz. to 15 lb.	2.75	2.92	3.18	3.55	4.03	4.53	5.47	6.66
15 lb. 1 oz. to 25 lb.	3.88	4.05	4.40	4.89	5.51	6.70	7.75	9.27
25 lb. 1 oz. to 40 lb.	4.56	4.82	5.32	6.16	7.21	8.82	10.62	12.48
40 lb. 1 oz. to 55 lb.	5.44	5.94	6.72	8.05	9.09	11.66	13.92	16.20
55 lb. 1 oz. to 70 lb.	5.45	6.83	7.84	8.92	10.68	13.65	16.52	19.04

Every catalog merchant has specific instructions for placing an order, and the appropriate catalog should be consulted for the specific directions. However, several generalizations apply to many companies. Most catalogs, for instance, can be identified by an identification number as well as by a title. Items are usually pictured, named, and described. If appropriate, the available sizes and colors are listed, and the shipping weight is indicated. The item is identified by a coded catalog number; and the price, not including shipping, handling, and tax, is displayed. The following information is usually required on an order form: account number, name, phone number, mailing address, how the order is to be delivered, page number, catalog number, quantity, color (or color code), size, price for each item, total price for quantity ordered (not including shipping, handling, and tax), and shipping weight. (If the order is phoned in rather than mailed in, the same information must be given to the phone order clerk.).

GETTING THE BEST VALUE

Ordering through a catalog is good practice for students. They learn to make discriminate decisions, practice budgeting and money management, and learn skills of recordkeeping and form completion that can be helpful job-related skills.

LEARNING EXPERIENCES

THE DIAGNOSTIC SURVEY AND INDEPENDENCE SCALE

- Use the **Diagnostic Survey** on Reproduction Page 63 to estimate each student's degree of competence prior to instruction in the topics of this chapter.

- Upon completion of the Survey, use the **Independence Scale** on Reproduction Page 64 to identify those skills that need to be strengthened by further instruction and practice.

- Utilize the results of these two instruments to determine the appropriate assignments from the following list.

Topic I: Measurement and Fractions

- Have students commit to memory the standard units of length, area, volume, capacity, and weight (or mass) for both the customary system and the metric system. Use a set of 3" × 5" cards with some cards marked inch, meter, square meter, liter, and so on. Shuffle the cards and place them face down on a table. Have students match the unit with its appropriate meaning following the rules of "memory" or "concentration." (See Chapter 1.)

- Have students measure various objects in the room so as to become familiar with the relative sizes of the following units: inch, foot, yard, meter, centimeter, millimeter, ounce, quart, gallon, milliliter, liter.

- Have students play concentration or some similar activity to memorize the meaning of the various metric prefixes.

- Use Reproduction Page 65, Part One, to give students practice in learning the relationships of units within the customary system.

- Use Reproduction Page 65, Part Two, to give students practice in learning the relationships of units within the metric system.

- Use Reproduction Page 65, Part Three, to check students' understanding of the relative sizes of various units.

- For practice with addition of fractions construct a worksheet consisting of problems similar to problems 40-43 on the Diagnostic Survey or use pages 74-91 of *Stein's Refresher Mathematics* (Allyn and Bacon, Inc., 1980).

- For practice with subtraction of fractions, construct a worksheet consisting of problems similar to problems 44-47 on the Diagnostic Survey or use pages 74-85 and 92-97 of *Stein's Refresher Mathematics.*

- Have students play concentration or some similar activity to memorize the formulas for area and volume of various geometric figures as given in the section "Other Units of Measure" of Topic I in this chapter.

- Use the assignments listed in **Topic III** of the "Learning Experiences" for practical consumer application of area and volume measurements.

Topic II: More Thought for Food

- Use Reproduction Page 66, Part One, for practice problems using multiples to determine the best buy.

- Use Reproduction Page 66, Part Two, for practice problems using proportions to determine the best buy.

- Use Reproduction Page 66, Part Three, for practice with determining the best buy by means of unit prices.

- Use **Figure 7-1** in Example 18a as an example and ask students to do some research on diet portions; estimated waste in food; and the cost of bulk, partly prepared, and prepared foods. Have students determine the cost per serving of various foods based upon a given serving portion.

- Use Example 19 as a guide and discuss how much saving can be realized when the number of servings per unit is increased.

Topic III: Do-it-Yourself Projects

- Use pages 419-432 of *Stein's Refresher Mathematics* for exercises in the practical applications with area and volume formulas.

- Use problems 62-65 of the Diagnostic Survey as examples and have students list several do-it-yourself projects they might be interested in. Make a composite list for the whole class and discuss the mathematical skills and calculations necessary to complete those projects.

- Ask students to consider redecorating the classroom. Have them measure to determine the area of the walls, floor, and ceiling (subtract from walls windows, doors, etc.). Have students consider the variables of equipment, material, and time as identified in the "Content Overview" and calculate the cost of:

    painting the walls
    wallpapering the walls
    paneling the walls
    tiling, carpeting, or covering the floor with linoleum
    covering sections of the floor with different materials

painting the ceiling
plastering the ceiling, then painting
covering the ceiling with acoustical tile
installing a baseboard and molding along the wall and floor
installing a ceiling molding
installing molding around doors and windows
building some bookcases, etc.

- Ask students to select one or more of the projects listed earlier and explore them in depth, considering the variables of equipment, material, and time as discussed in the "Content Overview." Have students simulate doing the projects they select by carrying out all calculations.

Topic IV: Mail-order and Catalog Buying

- Using Reproduction Page 67 and Figure 7-2, have students determine the cost of shipping and handling the items as presented.

- Use a current catalog and an order form similar to that shown below and refer to Chapter 3, Topic VII, and "Culminating Activities" for ideas on catalog buying. Redo the exercises, but this time include shipping and handling charges as presented in Figure 7-2, assuming different zone areas to provide variety.

Page No.	Quantity	Price (each)	Quantity Cost	Sales Tax	Shipping Weight	Shipping Cost	TOTAL COST

(Note: "Quantity Cost" refers to the cost for the amount that is purchased. "Sales Tax" should be the actual percent collected locally. If desired, "Shipping Cost" can be determined by use of the shipping table and zone map that appear in the catalog being used. Also, have students add to obtain the total shipping weight and the total shipping cost. Compare the shipping cost when each item is figured separately with the shipping cost when figured on the basis of the total weight.)

- Use a current catalog and an order form similar to that shown above and have students place simulated orders for equipment and materials needed for the do-it-yourself projects selected for Topic III. Have students use the appropriate checking account Reproduction Pages and write separate checks for each item purchased.

- Have students select items from a current catalog that can be made at home. Ask students to compare the cost of making the item and purchasing it outright. For example, select several simple clothing items (dresses, sweaters, etc.). Compare the cost of making the items from material purchased (perhaps from the same catalog) with buying them in a store. Include initial equipment investments and investigate how many clothing items must be made before real savings are realized. Other comparisons can be made with building and refinishing furniture compared with buying the finished product, constructing audio and visual equipment (radio, TV, stereo) from component parts compared with buying the assembled products, and so on.

- To provide students with practice in packaging, weighing, and shipping, obtain a balance scale that weighs items at least to 70 lb. (or about 30 kg.). (If necessary, a bathroom scale will do.) Also obtain several boxes of various sizes. Ask students to bring several items, large and small, to class and have them weigh the items. Then have students package each item separately, using newspaper as packing, and weigh them again. Discuss the time it takes to package the items, the price they would charge for "handling" (assume a minimum wage), the difference between the weight of the item and the weight of the package, and the reasonableness of "shipping-and-handling" charges.

ASSESSING ACHIEVEMENT OF OBJECTIVES

Ongoing Evaluation

The extent to which students have mastered the concepts covered under the four topics in this chapter can be measured by any of the activities assigned to class members individually.

Final Evaluation

For an overall evaluation of the students' mastery of the concepts in this chapter, if all topics in the chapter have been taught, a test constructed directly from the "Objectives" listed at the beginning of the chapter can be used. As an alternative, one might consider using the **Diagnostic Suvery** as a final test.

RESOURCES FOR TEACHING ABOUT GETTING THE BEST VALUE

Below is a selected and annotated list of resources useful for teaching the topics in this chapter, divided into audiovisual materials, games, and print materials. Addresses of publishers or distributors can be found in the alphabetic list in Appendix B.

Audiovisual Materials

A. FOR LOW ABILITY AND SPECIAL EDUCATION STUDENTS.

Mail Order Buying, filmstrip and audio-cassette. Interpretive Education, 1977.

Measurement and You, 5 filmstrips and 5 audio-cassettes. Interpretive Education, 1977.

Using Arithmetic When Shopping for Groceries, filmstrip and audio-cassette. Interpretive Education, 1977.

Understanding Fractions, filmstrip, cassette, 20 student workbooks, and teacher's guide. Interpretive Education, 1977.

B. FOR GENERAL MATHEMATICS STUDENTS

"How's Your Shopping Math?"; "How's Your Budget Math?" *Consumer Math Cassettes* produced by F. Lee McFadden. The Math House, 1977.

Budgeting, film, 11 minutes. Aetna Life and Casualty, 1975.

Managing Your Money, 4 filmstrips and 4 cassettes. Teaching Resources Films, 1974.

Consumer Education Series, 6 filmstrips and cassettes. Doubleday Multimedia, 1972.

"Food Dollars and Sense," "Money Talks," *Money Management Filmstrip Library*—kit, 4 filmstrips, 4 cassettes, workbooks, and transparencies. Household Finance Corporation, Money Management Institute, 1977.

Learning to Use Money, 10-1/2 minutes. Coronet Instructional Films, 1973.

The Consumer and the Supermarket, 15 minutes. Barr Films, 1976.

Consumer Education: Budgeting, 12 minutes. BFA Educational Media, 1968.

Consumer Education: Who Needs It, 15 minutes. Churchill Films, 1972.

Money Management and Financial Planning, 19 minutes. Aetna Life and Casualty, 1977.

Money: How Its Value Changes, 13 minutes. Coronet Instructional Films.

The Mail Order Gullibles. Pyramid Films, 1976.

Food and Money, 23 minutes. Cost of Living Council, 1973.

Games

Consumer. Western Publishing Company, 1971.

Cutting Corners developed by John Schatti. The Math Group, Inc., 1977.

"Managing Your Money." Credit Union National Association Mutual Insurance Society, 1969, 1970.

Paying the Cashier. McGraw-Hill Book Company, 1975.

Budgeting Game. EMC Publishing, 1970.

Print

A. FOR LOW ABILITY AND SPECIAL EDUCATION STUDENTS.

Using Dollars and Sense by Charles H. Kahn and J. Bradley Hanna. See especially pp. 71-112. Fearon-Pitman Learning, Inc., 1973.

Working Makes Sense by Charles H. Kahn and J. Bradley Hanna. See especially pp. 15-30 and 99-112. Fearon-Pitman Learning, Inc. 1973.

Useful Arithmetic, Volume I, by John D. Wool and Raymond J. Bohn. See especially pp. 1-11. Frank E. Richards Publishing Company, Inc., 1972.

Useful Arithmetic, Volume II, by John D. Wool. See especially pp. 18-26 and 56-66. Frank E. Richards Publishing Company, Inc., 1972.

Catalog Shopping by Martha L. Smith. See entire book. Frank E. Richards Publishing Company, Inc., 1979.

About the Home by Stephen S. Udvari and Janet Laible. See entire book. Buying Guides, Family Development Series, Steck-Vaughn Company, 1978.

An Introduction to Everyday Skills by David H. Wiltsie. See especially pp. 116-145. Motivational Development, Inc., 1977.

Skills for Everyday Living, Book 1, by David H. Wiltsie. See especially pp. 69-96 and 106-115. Motivational Development, Inc., 1976.

Skills for Everyday Living, Book 2, by David H. Wiltsie. See especially pp. 10-39 and 42-44. Motivational Development, Inc., 1978.

Using Money Series, Book III Buying Power, by John D. Wool. See especially pp. 41-62. Frank E. Richards Publishing Company, Inc., 1973.

Using Money Series, Book IV Earning Spending and Saving, by John D. Wool. See especially pp. 19-31. Frank E. Richards Publishing Company, Inc., 1973.

It's Your Money, Book 1, by Lloyd L. Feinstein and Charles H. Maley. See especially pp. 4-12 and 44-98. Steck-Vaughn Company, 1973.

It's Your Money, Book 2, by Lloyd L. Feinstein and Charles H. Maley. See especially pp. 4-13. Steck-Vaughn Company, 1973.

Math for Today and Tomorrow by Kaye A. Mach and Allan Larson. See especially pp. 15-20 and 31-48. J. Weston Walch, Publisher, 1968.

Scoring High in Survival Math by Tom Denmark. See especially pp. 20-30. Random House, Inc., 1979.

Michigan Survival by Betty L. Hall and David Landers. Also available for all states. See especially pp. 129-134. Holt, Rinehart and Winston, 1979.

Mathematics for Today, Level Red, by Saul Katz, Ed.D.; Marvin Sherman; Patricia Klagholz; and Jack Richman. See especially pp. 29-46, 61-84, 99-138. Sadlier-Oxford, 1976.

Mathematics for Today, Level Blue, by Edward Williams; Saul Katz, Ed.D.; and Patricia Klagholz. See especially pp. 169-180. Sadlier-Oxford, 1976.

Mathematics for Today, Level Orange, by Wilmer L. Jones, Ph.D. See especially pp. 74-132 and 178-191. Sadlier-Oxford, 1979.

Mathematics for Today, Level Green, by Wilmer L. Jones, Ph.D. See especially pp. 62-127 and 146-193. Sadlier-Oxford, 1979.

B. FOR GENERAL MATHEMATICS STUDENTS

Stein's Refresher Mathematics, Seventh Edition, by Edwin I. Stein. See especially pp. 74-126, 283-313, 557-567. Allyn and Bacon, Inc., 1980.

Trouble-Shooting Mathematics Skills, Basic Competency Edition, by Allen L. Bernstein and David W. Wells. See especially pp. 148-227, 291-309, 346-355. Holt, Rinehart and Winston, 1979.

Consumer Mathematics with Calculator Applications by Alan Belstock and Gerald Smith. See especially pp. 16-34, 61-74, 104-158, 300-317. McGraw-Hill Book Company, 1980.

Consumer Math, A Guide to Stretching Your Dollar, by Flora M. Locke. See especially pp. 101-122. John Wiley and Sons, Inc., 1975.

Consumer Mathematics, Third Edition, by William E. Goe. Activities book available. See especially pp. 49-118, 420-432, 441-461. Harcourt Brace Jovanovich, 1979.

Mathematics for Today's Consumer by Jack Price, Olene Brown, Michael Charles, and Miriam Lien Clifford. Compiled from selections from *Mathematics for Everyday Life* and *Mathematics for the Real World*. See especially pp. 140-175, 234-273, 278-280, 292-327, 344-402. Charles E. Merrill Publishing Company, 1979.

Mathematics for Everyday Life by Jack Price, Olene Brown, Michael Charles, and Miriam Lien Clifford. See especially pp. 42-81, 140-175, 270-791. Charles E. Merrill Publishing Company, 1978.

Mathematics for the Real World by Jack Price, Olene Brown, Michael Charles, and Miriam Lien Clifford. See especially pp. 100-119 and 136-192. Charles E. Merrill Publishing Company, 1978.

Mathematics in Life by L. Carey Bolster and H. Douglas Woodburn. See especially pp. 105-130, 151-170, 195-212, 323-360. Scott, Foresman and Company, 1977.

Mathematics for Daily Living by Harry Lewis. See especially pp. 99-144. McCormick-Mathers Publishing Company, 1975.

Mathematics Plus!, Consumer, Business & Technical Applications, by Bryce R. Shaw, Richard A. Denholm, and Gwendolyn H. Shelton. See especially pp. 33-75. Houghton Mifflin, 1979.

"Your Food Dollar," "Your Clothing Dollar," and "Your Shopping Dollar," *Money Management Library*, 12 booklets. Household Finance Corporation, Money Management Institute, 1978.

Consumer and Career Mathematics by L. Carey Bolster, H. Douglas Woodman, and Joella H. Gipson. See especially pp. 18-23, 30-33, 42-50, 324-381. Scott, Foresman and Company, 1978.

Business and Consumer Arithmetic by Milton C. Olson and A. E. McVelly. See especially pp. 41-54 and 106-111. Prentice-Hall, Inc., 1974.

Business Mathematics for the Consumer by Mearl R. Guthrie, William Selden, and Delbert Karnes. See especially pp. 245-258. Fearon-Pitman Learning, Inc., 1975.

C. RESOURCE UNITS, PAMPHLETS, BROCHURES, ETC.

U.S. Metric Board Information Kit. United States Metric Board, 1980.

"Teaching Metrication," *Educator Resource Series.* Sears, Roebuck and Company, 1976.

8

Percents Perchance

INTRODUCTION

It has been said that there are only two certainties in life: death and taxes. Perhaps another certainty can be added with regard to mathematics: that there shall be percents and that a large portion of all the people will have difficulty with percent calculation. School systems continually report that students have difficulty with this topic in spite of their academic standing. One suburban, middle- and upper-class school district reports that 38%-40% of the students completing first-year algebra do not master percent computation; and this is typical of many schools, regardless of socioeconomic position. Another certainty: the American consumer literally cannot afford this failure; percents are too prevalent in everyday life. Even in death and especially in taxes, percent calculation is everywhere.

Everyday calculations with percents include taxes (sales taxes have already been encountered), discount sales, commissions, financing, credit buying, down payment, savings, investing, borrowing, profit and loss, percent composition, cost-of-living, wage increments, census reports and polls, budgets, efficiency of machines, weather reports, statistical reporting, and on, and on. Even leisure activities ("there is a 60 percent chance of rain today"), sporting events (batting averages, won-loss percents), and school activities (percent score) do not offer escape. This chapter presents the mathematical concepts that are basic for comprehension of the common applications of percents. Beginning with ratios and proportions, and continuing with fractions, decimals, and the three basic percent calculations, this chapter provides the foundation upon which to build a practical understanding of percents. Percents cannot be left to chance. They are much too important.

> **OBJECTIVES**
>
> If all the topics in this chapter are chosen by the teacher, the student should be able to:
>
> 1. Define ratio, proportion, and percent.
> 2. Express ratios and proportions in words, using colons, as fractions, decimals, and percents.
> 3. Identify the terms of a proportion.
> 4. Solve proportions for one unknown term.
> 5. Express a percent in these terms: a) "hundredths," b) "out of 100," c) a decimal, d) a fraction.
> 6. Express each of the above terms as a percent.
> 7. Find the percent of a number.
> 8. Find what percent one number is of another number.
> 9. Find a number when a percent of it is known.
> 10. Express an everyday percent problem in terms of a soluble calculation using one or more of the aforementioned skills.
> 11. Accurately estimate solutions involving percents.

CONTENT OVERVIEW

Topic I: Ratios, Proportions, and the Meaning of Percent

Although thoroughly treated elsewhere in the text, the topic of percent is reintroduced here with special emphasis on the relationship of percent to ratio and proportion. As presented earlier, the term *percent* is derived from "per" and "centum." "Centum" means 100. Percent means "on the basis of 100," "compared with 100," or "divided by 100"; 19% means 19 divided by 100. Whenever two numbers are compared by division, the comparison is called a *ratio*. Each time the word *per* is used (miles per hour, cents per gram, etc.), a ratio is expressed. A percent is, therefore, a way of expressing a ratio where one of the numbers being compared is always 100. Other numbers, of course, can be expressed as ratios. When 24 is compared with 6, 24 is divided by 6, and it can be expressed as a division problem 24/6, which is also a fraction that can be reduced to 4/1. The ratio is 4/1, which says that 24 is 4 times as large as 6. In words, 4/1 is read, "four to one," and it may also be written 4:1. In all cases, the first number being compared is divided by the number with which it is being compared. When written as a fraction, the first number is the numerator, the second is the denominator, and the fraction is usually reduced. If, for example, 6 is compared with 24, the ratio is 6/24 or 6:24 and can be expressed as 1/4 or 1:4. The ratio can thus be written as a decimal 0.25, because 1/4 = 1 ÷ 4 = 0.25. To summarize, a ratio is a comparison by division and can be written in words, using a colon (:), as a fraction, a decimal, or a percent.

Ratios that are written with different number pairs but express the same comparison are called *equivalent ratios*. Thus, 2/4, 6/12, 9/18, 10/20, 15/30, 26/52, 47/94, and 50/100 are equivalent ratios because they also express the same ratio of 1/2. Note that the ratio 50/100 is the same as 50%. Therefore, 50% is equivalent to all of the ratios stated above. A mathematical sentence stating that two ratios are equivalent is called a *proportion*. Thus, 10/20 = 50/100 is a proportion. The same proportion can be expressed 10:20 = 50:100 or 10/20 = 50%.

This brings up an interesting and important point. It is clear that a percent is a number compared with 100, but percents are often used when the number 100 is not mentioned. For example, Matt received a score of 50% on his test in school even though there were not 100 problems on the test. If a percent is a ratio of a number always compared with 100, how can this be? Quite simply, Matt's actual score, the number correct compared with the number possible, formed a ratio that was equivalent to 50%. If there were 20 problems on the test, Matt could get 10 correct and receive a score of 50%. One may interpret from this that if there were 100 problems and Matt continued at the same rate, he would get 50 correct. Indeed, any of the ratios shown in the previous list could give him a 50% score if the second number of the ratio reflects the number of problems on the test.

Furthermore, a ratio can be written comparing an unknown number with a known number; and, if an equivalent ratio is known, a proportion can be written. For example, if it is known that another student in Matt's class obtained a score of 80%, then the number of correct answers can be determined. The known ratio is 80% or 80/100. The equivalent ratio is the number correct/the number possible or $n/20$, where n represents the unknown number of correct answers. The proportion is $80/100 = n/20$. This proportion is valuable because it can be used to determine n. In Chapter 7, proportions were used to determine the best buy. The instructions were to "cross multiply" and then divide.

$\frac{80}{100} \times \frac{n}{20}$ *Cross multiply*

$100n = 1600$

$\frac{100n}{100} = \frac{1600}{100}$ *Divide both sides by 100.*

$n = 16$ *16 correct answers out of 20 is the solution.*

Now let us examine this process more closely by analyzing the structure of proportions. Consider the equivalent ratios 6:8 and 15 to 20. The proportion is 6:8 = 15:20 or 6/8 = 15/20. Both forms may be read "6 is to 8 as 15 is to 20" or "6 compared with 8 is the same as 15 compared with 20." The four terms of the ratio are named *first, second, third*, and *fourth* in order from left to right when written 6:8 = 15:20. When written as fractions, they are named correspondingly.

$\begin{array}{c} \text{first} \rightarrow 6 \\ \text{second} \rightarrow 8 \end{array} = \begin{array}{c} 15 \leftarrow \text{third} \\ 20 \leftarrow \text{fourth} \end{array}$

In any proportion, the product of the 1st and 4th terms equals the product of the 2d and 3d terms. Thus, 6 × 20 = 120, and 8 × 15 = 120; 120 = 120. It is because of this phenomenon that the notion of "cross multiplying" is possible. Therefore, a proportion can be solved if any three of the four terms are known. Significantly, when percents are considered, one of the terms will always be 100, since all percents are numbers compared with 100. Therefore, a solution can be attained when any two of the other three variables are known.

Furthermore, there are only three basic types of percent problems, and each is based upon the general statement: *a certain percent of some number is equal to another number.* The statement can be rewritten in shorthand form as:

$$a\% \text{ of } b = c$$

This is another way of saying: *a compared with 100 is the same as c compared with b*, or

$$a : 100 = c : b$$

$$\text{or} \quad \frac{a}{100} = \frac{c}{b}$$

The final version of the proportion for percents should be memorized, because the three types of percent problems can be solved by this statement. The three types of problems are discussed more completely in later sections but are described briefly as follows.

Type One: $a\%$ of b is *what number?*

$$\frac{a}{100} = \frac{c?}{b} \qquad a \text{ and } b \text{ are known. Find } c.$$

Type Two: *What* % of b is c?

$$\frac{a?}{100} = \frac{c}{b} \qquad c \text{ and } b \text{ are known. Find } a.$$

Type Three: $a\%$ of *what number* is c?

$$\frac{a}{100} = \frac{c}{b?} \qquad a \text{ and } c \text{ are known. Find } b.$$

FURTHER DISCUSSION ON THE MEANING OF PERCENT

Percent is a way of representing hundredths. The expression "67 hundredths" can be written as 67%. Moreover, 67% can be expressed as "67 out of 100," 0.67, and 67/100. The latter two expressions, the decimal and fraction forms of percent, merit further examination.

Changing a percent to a decimal has been discussed in terms of sales tax and service tip. The method is to drop the percent sign (%) and rewrite the decimal point two places to the left of its original position (real or imaginary). To change a decimal to a percent, the opposite tack is taken: rewrite the decimal point two places to the right and insert a percent sign. Recall, also, that 1) a whole number is assumed to have a decimal point at the right of the last digit and 2) two decimal places represents hundredths. Here are several examples that involve percents and decimals.

$$67\% = .67\% = .67$$
$$57.65\% = .57.65\% = .5765$$
$$.93 = .93.\% = 93\%$$
$$.0525 = .05.25\% = 5.25\%$$

In order to change a percent to a fraction, simply drop the percent sign, write the numerical value as a numerator over 100, and then reduce the fraction. For example:

$$75\% = \frac{75}{100} = \frac{25 \times 3}{25 \times 4} = \frac{3}{4}$$

$$68\% = \frac{68}{100} = \frac{4 \times 17}{4 \times 25} = \frac{17}{25}$$

There are three ways to change a fraction to a percent: memorization, division, and proportion. Memorization is a good method for certain fractions whose decimal and percent equivalents are easily learned: 1/4 = .25 = 25%; 1/2 = .5 = 50%; 3/4 = .75 = 75%; 1/10 = .1 = 10%, etc.; 1/5 = .2 = 20%, etc.; 1/3 = .33-1/3 = 33-1/3%.

Division is always a good method and may be used to check memorization. Indeed, a fraction indicates that the numerator is divided by the denominator; the method of changing a fraction to a decimal was discussed in a previous chapter. Once a decimal numeral has been determined, it is then changed to a percent.

The use of proportions to change fractions to percents has already been alluded to, but the process is important enough to repeat. First of all, a fraction is a ratio and so is a percent. The question is: "To what percent is the fraction equivalent?" As an illustration, consider:

$$\frac{19}{32} = \text{what \%} \quad \text{or} \quad \frac{19}{32} = \frac{\text{What?}}{100} \quad \text{or} \quad \frac{19}{32} = \frac{n}{100}$$

This, being a proportion, can be solved by cross multiplying.

$$\frac{19}{32} \times \frac{n}{100} \qquad \textit{Cross multiply}$$

$$32n = 1900$$

$$\frac{32n}{32} = \frac{1900}{32} \qquad \textit{Divide both sides by 32}$$

$$n = 59.375$$

$$\frac{n}{100} = \frac{59.375}{100} = 59.375\%$$

Examples demonstrating the meaning of percent include:

Composition:

1. 100 g. of a certain food that is 76% protein and 24% fat has 76 g. of protein and 24 g. of fat.

2. 100 l. of gasohol that is 88% gasoline and 12% alcohol has 88 l. of gasoline and 12 l. of alcohol.

Discount Sales: A $100 coat selling for 25% off gives a savings of $25.

Investments: A $100 investment offering a 10.85% interest rate earns $10.85.

Cost of Living: If the rate of inflation is 12.15%, items costing $100 previously will now cost $12.15 more.

Taxes: An earning of $100 taxed at a rate of 14% means the paycheck is reduced by $14.

Budgets: A family budget plan that divides the expenses as 23% for food, 31% for home, 18% for clothing, 10% for church, 15% for automobile, and 3% for savings will divide $100 in this manner: $23 for food, $31 for home, $18 for clothing, $10 for church, $15 for automobile, and $3 for savings.

However, all of these examples show portions based upon 100 units ($, g., 1., etc.). The problems become more difficult when dealing with amounts other than 100. The following sections present methods of percent calculation that can be used for any quantity. The problems are identified in terms of the three types of percent computation.

Topic II: Type One: A Certain % of Some Number is What Number?

Recall the basic sentence for a percent problem.

$$a\% \text{ of } b = c$$

Type One problems require the determination of c *when* a *and* b *are known.* Sales tax and service tip problems are of this type, and multiplication is required (recall that "of" means "times"). Other examples of type one problems include:

1) If 24% of a certain cut of steak is fat, how much fat is in 185 g. of the steak?

2) If a store has a sale of 25% off on coats, how much can be saved on an $84 coat?

3) If the tax rate on earnings is 14%, how much tax is paid on $237?

Each of these examples can be set up in the form: $a\%$ of $b = c$, where c is unknown. These examples are treated in two ways in the sample solutions that follow. One method is to use a number sentence solution; the other utilizes proportions. When using the method of proportions, the general form for all problems is $\frac{a}{100} = \frac{c}{b}$

Example 1: If 24% of a certain cut of steak is fat, how much fat is in a 185-g. steak?

First, set up the problem in the general form.

$$24\% \text{ of } 185 \text{ g.} = c$$

$$a = 24 \quad b = 185 \quad c \text{ is unknown}$$

a) *Applying a number sentence solution:*

$$24\% \times 185 \text{ g.} = c$$

$$.24 \cdot 185 \text{ g.} = c$$

$$44.4 \text{ g.} = c$$

The steak contains 44.4 g. of fat.

b) *Using proportions:*

$$\frac{a}{100} = \frac{c}{b}$$

$$\frac{24}{100} = \frac{c}{185g.}$$

100c = 24 × 185 g. *(cross multiply)*

100c = 4440 g.

c = 44.4 g.

Example 2: If a store has a sale of 25% off on coats, how much can be saved on an $84 coat?

$$25\% \text{ of } \$84 = c$$

$$a = 25 \quad b = \$84 \quad c \text{ is unknown}$$

Using proportions:

$$\frac{a}{100} = \frac{c}{b}$$

$$\frac{25}{100} = \frac{c}{\$84}$$

100c = 25 × $84

100c = $2100

c = $21

A savings of $21 is realized.

Example 3: If the tax rate on earnings is 14%, how much tax is paid on $237?

$$14\% \text{ of } \$237 = c$$

$$a = 14 \quad b = \$237 \quad c \text{ is unknown}$$

Applying a number sentence solution:

.14 · $237 = c

$33.18 = c

The tax is $33.18.

Often problems may be a bit more complex and require more calculation, as the case would be in Example 2 if the sale price of the coat were to be determined. In that case, the $21 savings must be subtracted from the $84 original cost to obtain the sale price of $63. In other situations, several percent determinations might be needed followed by addition and subtraction. In Example 3, for instance, several taxes might be deducted from a salary. Such taxes as a 14% federal tax, a 3.75% state tax, a 1-1/2% city tax, and a 6.35% social security tax are not at all uncommon. In order to determine the actual take-home pay after taxes, the following calculations must be made (note the alternate use of number sentence and proportion solutions).

federal tax: (already determined at 14%) is $33.18

state tax: 3.75% of $237 = c

$$\frac{3.75}{100} = \frac{c}{\$237}$$

$$100c = \$888.75$$

$$c = \$8.89 \text{ state tax}$$

city tax: 1.5% of $237 = c

$$.015 \cdot \$237 = c$$

$$\$3.555 = c \text{ or } \$3.56 \text{ city tax}$$

social security tax: 6.35% of $237 = c

$$\frac{6.35}{100} = \frac{c}{\$237}$$

$$100c = \$1504.95$$

$$c = \$15.0495 = \$15.05 \text{ social security tax}$$

total taxes:

$33.18	*federal*
8.89	*state*
3.56	*city*
+ 15.05	*social security*
$60.68	

total take-home pay:

$237.00
− 60.68
───────
$176.32

In earlier chapters much was made of "common cents" decisions and the ability to tell at a glance if an answer is apparently correct. "Common cents" is invaluable for percent calculations, too. The continuation of Example 3 is a case in point. Consider the reasonableness of the answers. Any solution is subject to error, particularly when there are multiple calculations. A most common error is to misplace the decimal point, and a "common cents" inspection will likely discover it. Looking over the solutions to Example 3 (as extended), are the answers reasonable? A misplaced decimal point for the federal tax could yield a figure of $3.31 or $331.80. There are several ways to spot these as unreasonable answers. First, is it reasonable for a payroll tax to be higher than the wage? Second, 10% of any number is 0.1 of that number, and it can be obtained by moving the decimal point to the left one place. The federal tax in this case is slightly more than 10% but much less than 20%. Does the solution comply? Third, the other tax rates are less than the federal tax, and the answers should reflect this. Furthermore, the state tax should be about 3 times the city tax. (These relative measures apply only to this particular problem and should not be construed as applicable for all payroll tax computations.) In a similar manner, this kind of reasoning can be applied to estimate answers prior to a full-scale calculation. (See **Topic V: "Mentally Estimating Percents."**)

Topic III: Type Two: What % of a Certain Number is Another Number?

If a dress that regularly sells for $97.50 now sells for $17.55 less, what is the rate (percent) of reduction?

When problems of this type are encountered, they can be put into the following form.

$$a\% \text{ of } b = c \quad \text{or} \quad \frac{a}{100} = \frac{c}{b}$$

where b and c are known and a is to be determined

Approaching the problem with a number sentence solution yields:

$$a\% \text{ of } \$97.50 = \$17.55$$

$$a\% \cdot \$97.50 = \$17.55$$

$$\text{or } \$97.50 \cdot \frac{a}{100} = \$17.55$$

$$\frac{\cancel{\$97.50}}{\cancel{\$97.50}} \cdot \frac{a}{100} = \frac{\$17.55}{97.50}$$

$$\frac{a}{100} = .18$$

$$\cancel{100} \times \frac{a}{\cancel{100}} = .18 \times 100$$

$$a = 18 \text{ or } 18\%$$

(Note that the decimal form must be written as a percent.)

Applying the same data to a proportion results in a much simpler solution.

$$\frac{a}{100} = \frac{\$17.55}{\$97.50}$$

$$\$97.50 \, a = \$1755$$

$$a = 18 \text{ or } 18\%$$

A careful reading is necessary for all percent problems and is especially important in order that the proper figures are used in calculations. As an illustration, suppose the initial example was worded this way:

> "If a dress that regularly sells for $97.50 now sells for $79.95, what is the rate (percent) of reduction?"

Here the new selling price is given rather than the price reduction. If this figure is used as *c* in the calculations, an incorrect answer (82%) would have been obtained. The error can be avoided by attentive examination of the general forms and the given information. The general form is : *a*% of *b* = *c*. The problem asks "what is the . . . (percent) reduction?" Therefore, *a% reduction* of *b* = *c*. Further, the data clearly require that the original price ($97.50) be represented by *b*. Consequently, *c* must represent the *amount of reduction*, not the new selling price. For it makes sense to say, "*a% reduction of the original price (b) = the amount of reduction (c)*." The calculation is completed, therefore, by first subtracting $79.95 from $97.50 to determine *c*, then continuing the computation to find *a*.

Additional verification can be made by applying the "common cents" rule: Does it seem reasonable that the reduction would be 18%?

Example 4: Amanda answered 23 questions correctly and missed 7 questions. What percent of the questions did she answer correctly?

$$a\% \text{ of } b = c$$

$$a\% \text{ of } 23 \text{ is } 7$$

a is unknown, b is 23, c is 7

Solution by word statement.

$$a\% \cdot 23 = 7$$

$$23 \cdot \frac{a}{100} = 7$$

(Note: this step is a proportion)
$$\frac{a}{100} = \frac{7}{23}$$

$$\frac{a}{100} = 0.304$$

$$a = 30.4 = 30.4\% \text{ correct}$$

Does it seem reasonable that Amanda would receive a score of 30.4%? No! Redo the problem.

a% *correct* of the *number possible* (b) is the *number correct* (c).

the number possible (*b*) is 23 + 7 = 30

$$a\% \cdot 30 = 23$$

$$30 \cdot \frac{a}{100} = 23$$

$$\frac{a}{100} = \frac{23}{30}$$

$$\frac{a}{100} = 0.7\overline{66}$$

$$a = 76.\overline{6} = 76.\overline{6}\% \text{ or } 76.7\% \text{ correct}$$

Example 5: A chair that normally sells for $225 is on sale for $185. What is the rate (percent) of reduction?

$$a\% \text{ of } b = c$$

a% reduction of the regular price = the amount of reduction

$$a\% \text{ of } \$225 = \$40$$

Solving by proportion:

$$\frac{a}{100} = \frac{\$40}{\$225}$$

Note that the ratios are equivalent because each is comparing a reduction to a base number. On the left the base number is 100; on the right the base number is the original price. The answer for a tells what the reduction would be if the original price was $100. Corresponding terms of a proportion must match. Continuing the solution:

PERCENTS PERCHANCE

$$\$225a = \$4000$$

$$a = \frac{\$4000}{\$225}$$

$$a = 17.\overline{7} = 17.\overline{7}\% = 17.8\% \text{ reduction}$$

Example 6: Gail bought a dress for $60.42. The salesclerk said that the price represented a savings of $19.08. Find the percent discount from the original price.

$$a\% \text{ of } b = c$$

$a\%$ *reduction* of the *original price* (b) = the *amount saved* (c).

Determine the original price: $60.42 + $19.08 = $79.50

a is unknown, $b = \$79.50$, $c = \$19.08$

$a\%$ of $79.50 = $19.08

$$\frac{a}{100} = \frac{\$19.08}{\$79.50}$$

$$\$79.50a = \$1908$$

$$a = 24 = 24\% \text{ discount}$$

Example 7: Donna was undecided about buying a stereo that was marked $179.49. It was not on sale, but the salesman said that there would soon be a price increase to offset a recent wage-and-benefit adjustment that cost the company 8%. The new price would be $199.99. If this is true, is the percent increase in price equivalent to the wage-and-benefit adjustment?

$$a\% \text{ of } b = c$$

$a\%$ *increase* of the *original price* (b) = the *amount of increase* (c).

The amount of increase is $199.99 − $179.49 = $20.50

a is unknown, $b = \$179.49$, $c = \$20.50$

$a\%$ of $179.49 = $20.50

$$\frac{a}{100} = \frac{\$20.50}{\$179.49}$$

$$\$179.49a = \$2050$$

$$a = \frac{\$2050}{\$179.49}$$

$$a = 11.42 = 11.42\%$$

The price increase far exceeds the wage-and-benefit increase.

(A consumer advocate might be interested.)

Topic IV: Type Three: A Certain % of What Number is Another Number?

Example 8: A store advertised a 20% reduction on all basketball equipment. Tom bought a goal-and-backboard set and was told that he saved $9.50. What were the original and the sale prices?

$$a\% \text{ of } b = c$$

$a\%$ reduction of the original price (b) = the savings (c).

$a = 20\%$, b is unknown, $c = \$9.50$

20% of b = $9.50

A problem of this sort, where b is to be determined, represents the third type of percent problem. Again, language is important. The data must be interpreted correctly and identified in terms of the general statement using a, b, and c. Continuing with Example 8 and applying a word statement solution:

$$.20 \cdot b = \$9.50$$

$$b = \frac{\$9.50}{.20}$$

$$b = \$47.50, \text{ the original cost}$$

Using a proportion to solve the problem yields:

$$\frac{20}{100} = \frac{\$9.50}{b}$$

The proportion may be read: A savings of $20 on a $100 purchase is the same as a savings of $9.50 on what purchase?

$$20b = \$950$$

$$b = \frac{\$950}{20}$$

$$b = \$47.50, \text{ the original cost}$$

Example 9: Brenda is on a special diet and is restricted to drinking only skimmed milk for 10 days. She is required to have a minimum of 65 g. of protein per day. If skimmed milk contains 3.6% protein, how many grams of milk must she consume in order to obtain the required protein?

$$a\% \text{ of } b = c$$

3.6% of how much milk = 65 g. of protein?

$$3.6\% \cdot b = 65 \text{ g.}$$

$$.036b = 65 \text{ g.}$$

$$b = \frac{65 \text{ g.}}{.036}$$

$$b = 1805.\overline{55} \text{ g. of milk}$$

(If one liter of milk weighs about 1 kg. (1000 g.), $1805.\overline{55}$ is nearly 2 liters of milk.)

Example 10: A store recently completed a sale where several items were marked down by 20%. After the sale, a store clerk was to mark each item back to its original price. The clerk discovered one item marked only with the sale price of $13.80. Help that clerk out and determine the original price.

This problem is somewhat different from the others because only one of the three variables is known, and simple subtraction or addition does not seem to yield another. In the general statement, $a\%$ of $b = c$, a is 20%, but b and c are unknown. The figure $13.80 represents the reduced price and, at first glance, does not seem to be workable with the other data. However, the general form $a\%$ of $b = c$ may be interpreted:

$a\%$ of the original price = the reduced price, where $13.80 is the reduced price (c).

In this manner a is not 20% but is 100% − 20% or 80%. This is derived from the notion that 100% of the original price minus 20% of the original price equals 80% of the original price, which equals the reduced price. Therefore:

$$80\% \text{ of } b = \$13.80$$

$$.8 \cdot b = \$13.80$$

$$b = \frac{\$13.80}{.8}$$

PERCENTS PERCHANCE

$b = \$17.25$, *the original price*

Example 11: Find the original price of a clotheswasher that had a sale price of $204.60 when a 33-1/3% reduction was allowed.

$$a\% \text{ of } b = c$$

$$c = \$204.60, b \text{ is unknown}$$

$$a\% = 100\% - 33\tfrac{1}{3}\% = 66\tfrac{2}{3}\%$$

$66\tfrac{2}{3}\%$ of the original price is the reduced price.

$$66\tfrac{2}{3}\% \text{ of } b = \$204.60$$

$$.66\tfrac{2}{3}b = \$204.60$$

$$\text{or } \tfrac{2}{3}b = \$204.60 \quad \left(\text{note that } .66\tfrac{2}{3} = \tfrac{2}{3}\right)$$

$$b = \$204.60 \div \tfrac{2}{3} = \tfrac{204.60}{1} \times \tfrac{3}{2}$$

$$b = \$306.90, \textit{the original price}$$

Example 12: Find the regular price of a chair that has been priced at $166.60 after being reduced by 15%.

$$a\% = 100\% - 15\% = 85\%$$

$$85\% \text{ of } b = \$166.60$$

$$\frac{85}{100} = \frac{\$166.60}{b}$$

$$85b = \$16660$$

$$b = \frac{\$16660}{85}$$

$$b = \$196, \text{ the regular price}$$

Example 13: John was shopping for a car and received the following prices for the same car from two different dealers: $8599 from dealer 1 and $8942 from dealer 2. However, the second dealer included the 4% state sales tax on the auto. What was the price of the second car without tax?

Solution: The total price of the car is the original price plus the sales tax.

If n is the original price,

$$100\%n + 4\%n = \text{the total price (\$8942)}$$

$$1.00n + .04n = \$8942$$

$$1.04n = \$8942$$

$$n = \frac{\$8942}{1.04}$$

$$n = \$8598.08 \text{ from dealer 2}$$

Clearly, it is not possible to present all of the varieties of percent problems that one is likely to encounter. The generalizations made here, however, ought to be helpful in attacking any problem. Most important, the data must be scrutinized so that corresponding (and necessary) terms are known and matched (e.g., % reduction matched with amount reduced, not with new price, and the problem may be identified as one of the three types of percent problems presented in this chapter. Even after the type of problem is determined, the data must be applied correctly. Indeed, some calculations may require several computations and might involve all three types of problems. These possibilities are made exceedingly clear with a closer look at payrolls, taxes, credit and financing, saving and investing, housing, car buying, and other topics in the remaining chapters of this *Guidebook*.

Topic V: Mentally Estimating Percents

Estimating has been included repeatedly in several topics of the *Guidebook*, and many of the previous skills (such as rounding off) come to bear when estimating percents. There are several reasons to encourage the refinement of estimating skills. An estimate may be reckoned prior to calculation to anticipate or predict a solution. It may be done after computation to check and verify an answer. Or, perhaps there is no time for a thorough accounting of the problem and an estimate that is rendered quickly will suffice. Recall the discussion of taxes in Example 3. There, estimating skills were used to verify the results. In a similar situation an employee might wish to estimate his payroll taxes prior to receiving his check so as to anticipate how much he will have available to spend. Other consumer examples include mentally calculating sales tax to help decide on a purchase, mentally determining down payment and finance charges, or mentally reckoning which food item to buy based upon nutritional value. Here are a few suggestions to aid the consumer in making accurate estimates. They are not necessarily given in any order, for there are many factors that determine which methods are best to apply.

1) Round off all numbers that are given.

2) Recognize the decimal quality of the percent.

Whenever finding the percent of a number (Type One), mentally move the decimal point of the number as follows.

2 places to the left for 1% to 7% or 8%

1 place to the left for 8% or 9% to 80% or 90%

no movement for 90% or above

If the percent can be rounded to 1%, the estimate is attained when the decimal point is moved 2 places to the left. If the percent can be rounded to 10%, the estimate is attained when the decimal point is moved one place to the left. If the percent can be rounded to 100%, nothing needs to be done.

For example:

1.3% of $38.90 becomes 1% of $40.00 = $.40

8.85% of $258.82 becomes 10% of $250.00 = $25.00

94.35% of $44.51 becomes 100% of $40.00 = $40.00

3. **Utilize multiples of 10.** After moving the decimal point one place to the left, multiply the rounded-off number by the appropriate multiplier.

   22% of $392.41 becomes 20% of $400.00

   10% of $400.00 = $40.00

   $40.00 X 2 = $80.00 *(because 2 X 10% = 20%)*

   76% of $561.68 becomes 80% of $550.00

   10% of $550.00 = $55.00

   $55.00 X 8 = $440.00 *(because 8 X 10% = 80%)*

4. **Recognize the ratio of the percent figure compared with 100.** Every percent can be written as the numeral of the percent over 100; for example, 39% = 39/100, and so on. When trying to determine what percent one number is of another (Type Two), set up the ratio and try to manipulate the numbers so that the denominator becomes 100. Mentally set up a proportion; c is what % of b becomes:

$$\frac{c}{b} = \frac{What?}{100}$$

   To be specific, consider:

   32 is what percent of 37?

$$\frac{32}{37} = \frac{What?}{100}$$

   37 X 3 is greater than 100

   and 32 X 3 is 96   so

$$\frac{32}{37} \text{ is less than } \frac{96}{100}$$

   An estimate of 90 (90%) might be appropriate. (The actual answer is 86.4869.)

5. **Use "common cents"** (or sense, if money is not involved). Ask yourself: "Is it reasonable for a percent (less than 100%) of a number to be *greater* than the number itself?" Of course not! Redo the problem, if this is the case, and check the decimal point.

6. **Utilize known fractions for percents.** Some commonly used fraction-percent equivalents can be very helpful in computations.

   27% of $238.71 becomes 25% of $240.00

   25% is $\frac{1}{4}$

   $\frac{1}{4}$ of $240.00 = $60.00 (The actual answer is $64.45!)

Continuing with another example:

69% of $547.91 becomes

$\frac{2}{3}$ of $540.00 or 2 times ($\frac{1}{3}$ of 540.00)

2 × $180 = $360.00 (The actual answer is $378.06!)

7. **Compare with answers already verified.** If you know the answer to 19% of $73, then 21% of the same number should be a little more, and 9% should be somewhat less than one half of the known answer.

To summarize, one can readily see that mental estimation is not a trick or game that leaves the answer up to guesswork or chance. Nor is it a shortcut to learning about percents. The more that consumers understand about percents, the more expert they are at estimation. In fact, none of the suggestions listed here can be used without a basic knowledge of percent computation. One might say that if left to chance instead of diligent study, success with percents might perchance to be, oh, somewhere around 0.000321%.

LEARNING EXPERIENCES

THE DIAGNOSTIC SURVEY AND INDEPENDENCE SCALE

- Use Reproduction Page 68, Part Three, and have students solve the given proportions. degree of competence prior to instruction in the topics of this chapter.

- Upon completion of the Survey, use the **Independence Scale** on Reproduction Page 69 to identify those skills that need to be strengthened by further instruction and practice.

- Utilize the results of these two instruments to determine the appropriate assignments from the following list.

Topic I: Ratios, Proportions, and the Meaning of Percent

- Use problems 1, 2, and 3 of the **Diagnostic Survey** as examples and have the class discuss the meaning of and the ways to express various examples of ratios (include miles per gallon, price per unit, servings per unit, etc.).

- Use Reproduction Page 70, Part One, and have students practice writing ratios.

- Use Reproduction Page 70, Part Two, and have students determine equivalent ratios.

- Use problems 5 and 6 of the **Diagnostic Survey** as examples and have the class discuss the meaning of and ways to express proportions.

- Use Reproduction Page 70, Part Three, and have students solve the given proportion.

- Use problems 7, 8, 9, 10, and 11 of the **Diagnostic Survey** and have students discuss the meaning of and ways to express percents.

- For additional practice with ratios and proportions, assign problems on pages 507-512 of *Stein's Refresher Mathematics* (Allyn and Bacon, Inc., 1980).

- Use *Stein's Refresher Mathematics*, pages 206-217, for additional practice with the meaning of percents and ways to express percents.

Topic II: Type One: A Certain % of Some Number is What Number?

- For drill and practice with Type One percent problems, use *Stein's Refresher Mathematics*, pages 218-222 and 232-233.

- To provide experience with translating practical situations to mathematical statements and proportions for Type One percent problems, use Reproduction Page 71.

Topic III: Type Two: What % of a Certain Number is Another Number?

- For drill and practice with Type Two percent problems, use *Stein's Refresher Mathematics*, pages 223-227 and 232-233.

- Use Reproduction Page 72 to provide experience with translating practical situations to mathematical statements and proportions for Type Two percent problems.

Topic IV: Type Three: A Certain % of What Number is Another Number?

- Use *Stein's Refresher Mathematics*, pages 228-233, for drill and practice with Type Three percent problems.

- For experience with translating practical situations to mathematical statements and proportions for Type Three percent problems, use Reproduction Page 73.

Topic V: Mentally Estimating Percents

- Use any of the previous percent assignments and have students either determine estimated values prior to calculation or estimate to verify already calculated answers or both.

- Use newspapers and catalogs and have students estimate sales taxes, discounts, interest payments, and the like for advertised items.

- Use newspapers and catalogs and have students estimate possible price increases for a given inflation rate (cost-of-living increases).

> ## ASSESSING ACHIEVEMENT OF OBJECTIVES
>
> **Ongoing Evaluation**
>
> The extent to which students have mastered the concepts covered under the five topics in this chapter can be measured by any of the activities assigned to class members individually.
>
> **Final Evaluation**
>
> For an overall evaluation of the students' mastery of the concepts in this chapter, if all topics in the chapter have been taught, a test constructed directly from the "Objectives" listed at the beginning of the chapter can be used. As an alternative, one might consider using the **Diagnostic Survey** as a final test or use Reproduction Page 74 to test competency with all three types of percent problems.

RESOURCES FOR TEACHING ABOUT PERCENTS PERCHANCE

Below is a selected and annotated list of resources useful for teaching the topics in this chapter, divided into audiovisual materials, games, and print materials. Addresses of publishers or distributors can be found in the alphabetic list in Appendix B.

Audiovisual Materials

A. FOR LOW ABILITY AND SPECIAL EDUCATION STUDENTS.

Understanding Percents, filmstrip, audio-cassette, and 20 student workbooks. Interpretive Education, 1977.

Games

Big Deal! Creative Teaching Associates, 1976.

Consumer. Western Publishing Company, 1971.

Cutting Corners developed by John Schatti. The Math Group, Inc., 1971.

Print

A. FOR LOW ABILITY AND SPECIAL EDUCATION STUDENTS.

Working Makes Sense by Charles H. Kahn and J. Bradley Hanna. See especially pp. 49-58, 79-82, 108. Fearon-Pitman Learning, Inc., 1973.

Useful Arithmetic, Volume II, by John D. Wool. See especially pp. 45-72. Frank E. Richards Publishing Company, Inc., 1972.

An Introduction to Everyday Skills by David H. Wiltsie. See especially pp. 131-135. Motivational Development, Inc., 1977.

Skills for Everyday Living, Book 1, by David H. Wiltsie. See especially pp. 97-105. Motivational Development, Inc., 1976.

Mathematics for Today, Level Green, by Wilmer L. Jones, Ph.D. See especially pp. 128-167. Sadlier-Oxford, 1979.

B. FOR GENERAL MATHEMATICS STUDENTS.

Stein's Refresher Mathematics, Seventh Edition, by Edwin I. Stein. See especially pp. 206-238. Allyn and Bacon, Inc., 1980.

Trouble-Shooting Mathematics Skills, Basic Competency Edition, by Allen L. Bernstein and David W. Wells. See especially pp. 310-331. Holt, Rinehart and Winston, 1979.

Activities Handbook for Teaching with the Hand-Held Calculator by Gary G. Bitter and Jerald L. Mikesell. See especially pp. 191-251 and 301-306. Allyn and Bacon, Inc., 1980.

Consumer Mathematics with Calculator Applications by Alan Belstock and Gerald Smith. See especially pp. 16-34. McGraw-Hill Book Company, 1980.

Consumer Mathematics, Third Edition, by William E. Goe. Activities book available. See especially pp. 441-456. Harcourt Brace Jovanovich, 1979.

Mathematics in Life by L. Carey Bolster and H. Douglas Woodburn. See especially pp. 213-232. Scott, Foresman and Company, 1977.

Mathematics for Daily Living by Harry Lewis. See especially pp. 514-524. McCormick-Mathers Publishing Company, 1975.

Mathematics Plus!, Consumer, Business & Technical Applications, by Bryce R. Shaw, Richard A. Denholm, and Gwendolyn H. Shelton. See especially pp. 76-100. Houghton Mifflin, 1979.

Consumer and Career Mathematics by L. Carey Bolster, H. Douglas Woodman, and Joella H. Gipson. See especially pp. 24-41. Scott, Foresman and Company, 1978.

Business and Consumer Arithmetic by Milton C. Olson and A. E. McVelly. See especially pp. 77-90. Prentice-Hall, Inc., 1974.

Business Mathematics for the Consumer by Mearl R. Guthrie, William Selden, and Delbert Karnes. See especially pp. 103-120. Fearon-Pitman Learning, Inc., 1975.

9

Paychecks, Deductions, and Taxes

INTRODUCTION

Most people like surprises. But one surprise that is not at all appreciated is the one that a new wage earner gets when he is "surprised" by a paycheck that is much smaller than he anticipated. The reason for the reduced paycheck is the reason for this chapter; that is, several items are subtracted from the paycheck automatically. These subtractions are called *deductions*; the amount of pay before subtraction is called the *gross pay*; and the amount of spendable income is called *take-home pay*. This chapter includes the topics of finding a job, understanding payroll deductions, determining pay, comparing incomes, and filing annual income tax reports.

It is in the area of income that consumer and job-related skills and attitudes come together. Ironically, the motive of "getting the most" out of what is invested often results in a clash that is difficult to resolve. This is evident when one considers the wage-price relationship to inflation. Consumers want low prices, but they want high wages too. Of course, there are other reasons to hold a particular job besides the paycheck. Many people choose a job because of personal pride, interest, social responsibility, and other motivations. Nevertheless, even people fortunate enough to be comfortable and happy with their job are still faced with the possibility of an ever shrinking paycheck.

OBJECTIVES

If all the topics in this chapter are chosen by the teacher, the students should be able to:

1. Define the terms for payroll deductions and income taxes as identified in this chapter.

2. Determine or recognize how to determine the various payroll deductions.

3. Determine and use the effective tax and deduction rate on income.

> 4. Understand the various methods of determining pay.
>
> 5. Compare incomes that may be reported in different ways (e.g., annual salary compared with hourly rate, etc.).
>
> 6. Recognize the use and purpose of different tax tables, forms, and schedules.
>
> 7. Describe the difference between a regressive and a progressive income tax.
>
> 8. File a simple annual income tax return (federal, state, and city).

CONTENT OVERVIEW

Topic I: Paychecks and Deductions

One of the first things that confronts a new job seeker is the myriad of forms and papers to be filled out. There are application forms and usually employment tests from each prospective employer. Of course, the first thing that is required is a social security card. This is a simple verification that the job hunter is registered according to the Federal Insurance Contribution Acts (FICA) through the Social Security Administration. This registration may be started at any local social security office and is completed upon receipt of the card. The applicant is assigned a number that remains with him all his life. The FICA requirement is that nearly everyone (some professional people are exempt) who earns a wage will pay a certain percent of their income to the federal government for the purpose of establishing a fund that is used to pay out benefits to retirees, widows and dependent survivors, and disabled workers. A social security card is usually a mandatory requirement for employment.

Another form required before an applicant may begin work is the *W-4 federal withholding tax form.* This is a simple statement of registration that directs the employer to take taxes out of each paycheck so that a large lump-sum payment at the end of the year is avoided. The important information includes the name, address, marital status, and social security number of the applicant and the number of "allowances" claimed. The term "allowances" has to do with the amount of tax the worker is required to pay. Both social security (FICA) and withholding contributions are only part of the many deductions that help make that first paycheck a "surprise." Although these deductions were alluded to in Chapter 8, they are explained in greater detail here.

PAYROLL DEDUCTIONS
A *payroll deduction* is an amount subtracted from a paycheck. There are several different payroll deductions that can be categorized as either taxes or nontaxes. Two terms that are constantly used when discussing payroll deductions are:

> *gross pay*—the amount of pay before any deductions are subtracted.
>
> *net pay*—the amount of spendable income after deductions are made (also known as "take-home pay").

Taxes. There are several income taxes that might be deducted from a paycheck, depending upon where the worker lives. They include: federal tax, state tax, and city tax. Everyone does not pay the same income tax even if their income is the same. This is because people who have certain expenses can claim "allowances," so that their tax is lowered. Allowances are claimed on the W-2 form for the federal government. Although the federal tax is subject to the withholding statement of allowances on the W-2 form, the other taxes may or may not be. This is because the federal tax rules allow a reduction of tax for several types of expenses, whereas the city and state taxes may not. One type of allowance usually granted on all income taxes is the *dependent* allowance. One allowance may be claimed for each member of the household who is dependent upon the income of the wage earner (thus, the term often used is *dependents*). Dependents may include the wage earner himself or herself, a spouse, children, and parents, or other persons who are members of the household and receive their support from the wage earner. Additional allowances (called exemptions on the annual tax return) are given for blindness and for age (sixty-five or older). Federal tax rules enable taxpayers to claim other allowances for expenses that are permitted as deductions on the annual tax return. The number of allowances is determined by the expected total expenses that can be subtracted from the annual income and is presented on a table that is attached to the W-4 form. Considering all sources, dependents, exemptions, and allowances for tax-reducing expenses, the total number of allowances is entered on the W-4 form. This number, along with the gross pay, consequently determines the amount of federal tax to be withheld from the paycheck. As already stated, this allowance may also be considered for the state and city taxes. However, some states and cities allow a reduction in taxes for dependents and other exemptions only, and do not consider the itemized expenses that are allowed by the federal government.

Furthermore, the federal tax rate is a progressive tax, because it increases as the income increases. When taxpayers refer to "raising their tax bracket," they are referring to the fact that the tax rate remains the same within certain limits; as soon as those limits are exceeded, the tax rate increases. For example, if the taxable pay for a married worker is $125, the federal tax rate is 15%. However, if the taxable pay for the same worker is $200, the first $127 is taxed at a rate of 15%, but the remaining $73 is taxed at 18%. And a worker earning over $556 will pay $119.35 plus 37% on the excess over $556. The increment continues to climb to a maximum of 70%. The actual tax rate also depends upon the number of allowances claimed by the earner. The particular amount for each wage may be computed by percentage calculation or by using a tax table. Tables showing current percent rates and taxes for various incomes and allowances may be obtained from the Internal Revenue Service (IRS), and from your state and local taxation departments. Figures 9-1 and 9-2 are shown here only as examples and ought not to be considered as truly representing any current tax structure. When using the percentage calculation of Figure 9-1, the *taxable pay* is determined by subtracting the product of the number of allowances claimed times $19.23 from the gross pay. When using the tax tables, the full gross pay is used and the tax appears in the column that corresponds to the number of allowances claimed.

Figure 9-1
Percentage Method of Withholding

WEEKLY Payroll Period

(a) SINGLE person—including head of household:

If the amount of wages is: The amount of income tax to be withheld shall be:

Not over $27 0

Over—	But not over—		of excess over—
$27	—$75	.....14%	—$27
$75	—$100	....$6.72 plus 16%	—$75
$100	—$181	....$10.72 plus 19%	—$100
$181	—$269	....$26.11 plus 24%	—$181
$269	—$331	....$47.23 plus 29%	—$269
$331	—$433	....$65.21 plus 32%	—$331
$433		..........$97.85 plus 37%	—$433

(b) MARRIED person—

If the amount of wages is: The amount of income tax to be withheld shall be:

Not over $46 0

Over—	But not over—		of excess over—
$46	—$147	....14%	—$46
$147	—$210	....$14.14 plus 16%	—$147
$210	—$296	....$24.22 plus 20%	—$210
$296	—$447	....$41.42 plus 25%	—$296
$447	—$556	....$79.17 plus 31%	—$447
$556	—$658	....$112.96 plus 34%	—$556
$658		..........$147.64 plus 37%	—$658

Source: Adapted from Department of the Treasury, IRS Publication 15, October 1981.

Upon examination of Figures 9-1 and 9-2, it becomes evident that the effective tax rate is somewhat less than what is apparent. Take, for example, the married worker who earns a wage of $535 in one week. If he claims 5 allowances, his taxable income computed by Figure 9-1 is:

$$\$535 - (5 \times \$19.23) = \$535 - \$96.15 = \$438.85$$

The tax is:

$$\$62.91 + 28\% \text{ of } (\$438.85 - \$369) =$$

$$\$62.91 + (.28 \times \$69.85) = \$62.91 + \$19.56 = \$82.47$$

The effective tax rate is determined by answering the question: "$82.47 is what % of $535.00?" Rewriting in the form used in Chapter 8:

$$a? \% \text{ of } \$535.00 \text{ is } \$82.47$$

$$\frac{a}{100} = \frac{\$82.47}{\$535.00}$$

$$\$535a = \$8247$$

$$a = \frac{\$8247}{\$535} = 15.41\%$$

The effective rate appears to be much lower than the rate shown on the table because the entire gross pay is considered, and the effective rate is reduced by averaging the 15% to 28% that is applied at various levels. Indeed, the effective rate will be even lower when more allowances are considered.

Similar computations are made to determine city and state taxes. However, these taxes are not usually progressive; they are, instead, flat-rate taxes that remain the same percent regardless of income. They are easier to compute because, once the exemption allowance is subtracted from the gross pay, the same percent rate is applied. For example, a certain state has a 4.6% tax rate and permits a $23.08 weekly allowance for each exemption ($1200 per exemption per year). If a worker earns $280 in one week and claims 2 exemptions, $280 − (2 × $23.08) or $233.84 is the taxable income. The tax is computed as:

$$\$233.84 \times 4.6\% = \$233.84 \times .046 = \$10.76$$

PAYCHECKS, DEDUCTIONS, AND TAXES 161

Figure 9-2
Federal Withholding—Married Persons—Weekly Payroll Period

And the wages are—		And the number of withholding allowances claimed is—										
At least	But less than	0	1	2	3	4	5	6	7	8	9	10
		The amount of income tax to be withheld shall be—										
$310	$320	$46.20	$41.40	$37.50	$33.70	$29.80	$26.00	$22.50	$19.50	$16.40	$13.40	$10.70
320	330	48.70	43.80	39.50	35.70	31.80	28.00	24.10	21.10	18.00	14.90	12.10
330	340	51.20	46.30	41.50	37.70	33.80	30.00	26.10	22.70	19.60	16.50	13.50
340	350	53.70	48.80	44.00	39.70	35.80	32.00	28.10	24.30	21.20	18.10	15.00
350	360	56.20	51.30	46.50	41.70	37.80	34.00	30.10	26.30	22.80	19.70	16.60
360	370	58.70	53.80	49.00	44.20	39.80	36.00	32.10	28.30	24.40	21.30	18.20
370	380	61.20	56.30	51.50	46.70	41.90	38.00	34.10	30.30	26.40	22.90	19.80
380	390	63.70	58.80	54.00	49.20	44.40	40.00	36.10	32.30	28.40	24.60	21.40
390	400	66.20	61.30	56.50	51.70	46.90	42.10	38.10	34.30	30.40	26.60	23.00
400	410	68.70	63.80	59.00	54.20	49.40	44.60	40.10	36.30	32.40	28.60	24.80
410	420	71.20	66.30	61.50	56.70	51.90	47.10	42.30	38.30	34.40	30.60	26.80
420	430	73.70	68.80	64.00	59.20	54.40	49.60	44.80	40.30	36.40	32.60	28.80
430	440	76.20	71.30	66.50	61.70	56.90	52.10	47.30	42.50	38.40	34.60	30.80
440	450	78.70	73.80	69.00	64.20	59.40	54.60	49.80	45.00	40.40	36.60	32.80
450	460	81.60	76.30	71.50	66.70	61.90	57.10	52.30	47.50	42.70	38.60	34.80
460	470	84.70	78.80	74.00	69.20	64.40	59.60	54.80	50.00	45.20	40.60	36.80
470	480	87.80	81.90	76.50	71.70	66.90	62.10	57.30	52.50	47.70	42.90	38.80
480	490	90.90	85.00	79.00	74.20	69.40	64.60	59.80	55.00	50.20	45.40	40.80
490	500	94.00	88.10	82.10	76.70	71.90	67.10	62.30	57.50	52.70	47.90	43.10
500	510	97.10	91.20	85.20	79.20	74.40	69.60	64.80	60.00	55.20	50.40	45.60
510	520	100.20	94.30	88.30	82.30	76.90	72.10	67.30	62.50	57.70	52.90	48.10
520	530	103.30	97.40	91.40	85.40	79.50	74.60	69.80	65.00	60.20	55.40	50.60
530	540	106.40	100.50	94.50	88.50	82.60	77.10	72.30	67.50	62.70	57.90	53.10
540	550	109.50	103.60	97.60	91.60	85.70	79.70	74.80	70.00	65.20	60.40	55.60
550	560	112.60	106.70	100.70	94.70	88.80	82.80	77.30	72.50	67.70	62.90	58.10
560	570	116.00	109.80	103.80	97.80	91.90	85.90	80.00	75.00	70.20	65.40	60.60
570	580	119.40	112.90	106.90	100.90	95.00	89.00	83.10	77.50	72.70	67.90	63.10
580	590	122.80	116.30	110.00	104.00	98.10	92.10	86.20	80.20	75.20	70.40	65.60
590	600	126.20	119.70	113.10	107.10	101.20	95.20	89.30	83.30	77.70	72.90	68.10
600	610	129.60	123.10	116.50	110.20	104.30	98.30	92.40	86.40	80.40	75.40	70.60
610	620	133.00	126.50	119.90	113.40	107.40	101.40	95.50	89.50	83.50	77.90	73.10
620	630	136.40	129.90	123.30	116.80	110.50	104.50	98.60	92.60	86.60	80.70	75.60
630	640	139.80	133.30	126.70	120.20	113.70	107.60	101.70	95.70	89.70	83.80	78.10
640	650	143.20	136.70	130.10	123.60	117.10	110.70	104.80	98.80	92.80	86.90	80.90
650	660	146.60	140.10	133.50	127.00	120.50	113.90	107.90	101.90	95.90	90.00	84.00
660	670	150.20	143.50	136.90	130.40	123.90	117.30	111.00	105.00	99.00	93.10	87.10
670	680	153.90	146.90	140.30	133.80	127.30	120.70	114.20	108.10	102.10	96.20	90.20
680	690	157.60	150.50	143.70	137.20	130.70	124.10	117.60	111.20	105.20	99.30	93.30
690	700	161.30	154.20	147.10	140.60	134.10	127.50	121.00	114.40	108.30	102.40	96.40
700	710	165.00	157.90	150.80	144.00	137.50	130.90	124.40	117.80	111.40	105.50	99.50
710	720	168.70	161.60	154.50	147.40	140.90	134.30	127.80	121.20	114.70	108.60	102.60
720	730	172.40	165.30	158.20	151.10	144.30	137.70	131.20	124.60	118.10	111.70	105.70
730	740	176.10	169.00	161.90	154.80	147.70	141.10	134.60	128.00	121.50	115.00	108.80
740	750	179.80	172.70	165.60	158.50	151.40	144.50	138.00	131.40	124.90	118.40	111.90
750	760	183.50	176.40	169.30	162.20	155.10	147.90	141.40	134.80	128.30	121.80	115.20
760	770	187.20	180.10	173.00	165.90	158.80	151.60	144.80	138.20	131.70	125.20	118.60
770	780	190.90	183.80	176.70	169.60	162.50	155.30	148.20	141.60	135.10	128.60	122.00
780	790	194.60	187.50	180.40	173.30	166.20	159.00	151.90	145.00	138.50	132.00	125.40
790	800	198.30	191.20	184.10	177.00	169.90	162.70	155.60	148.50	141.90	135.40	128.80
800	810	202.00	194.90	187.80	180.70	173.60	166.40	159.30	152.20	145.30	138.80	132.20
810	820	205.70	198.60	191.50	184.40	177.30	170.10	163.00	155.90	148.80	142.20	135.60
820	830	209.40	202.30	195.20	188.10	181.00	173.80	166.70	159.60	152.50	145.60	139.00
830	840	213.10	206.00	198.90	191.80	184.70	177.50	170.40	163.30	156.20	149.10	142.40
840	850	216.80	209.70	202.60	195.50	188.40	181.20	174.10	167.00	159.90	152.80	145.80
850	860	220.50	213.40	206.30	199.20	192.10	184.90	177.80	170.70	163.60	156.50	149.40
		37 percent of the excess over $860 plus—										
$860 and over		222.40	215.30	208.10	201.00	193.90	186.80	179.70	172.60	165.50	158.30	151.20

Source: Adapted from Department of the Treasury, IRS Publication 15, October 1981.

Figure 9-3
Flat Rate Income Tax Withholding Tables

WEEKLY payroll period

And the wages are —		And the number of withholding exemptions claimed is —										
At least	But less than	0	1	2	3	4	5	6	7	8	9	10
		The amount of Massachusetts income tax to be withheld shall be —										
$220	$230	$11.35	$ 9.29	$ 8.56	$ 7.84	$ 7.11	$ 6.39	$ 5.67	$ 4.94	$ 4.22	$ 3.50	$ 2.77
230	240	11.86	9.79	9.07	8.34	7.62	6.90	6.17	5.45	4.72	4.00	3.28
240	250	12.36	10.29	9.57	8.85	8.12	7.40	6.68	5.95	5.23	4.51	3.78
250	260	12.87	10.80	10.08	9.35	8.63	7.90	7.18	6.46	5.73	5.01	4.29
260	270	13.37	11.30	10.58	9.86	9.13	8.41	7.69	6.96	6.24	5.51	4.79
270	280	13.88	11.81	11.08	10.36	9.64	8.91	8.19	7.47	6.74	6.02	5.30
280	290	14.38	12.31	11.59	10.87	10.14	9.42	8.69	7.97	7.25	6.52	5.80
290	300	14.88	12.82	12.09	11.37	10.65	9.92	9.20	8.48	7.75	7.03	6.30
300	310	15.39	13.32	12.60	11.87	11.15	10.43	9.70	8.98	8.26	7.53	6.81
310	320	15.89	13.83	13.10	12.38	11.66	10.93	10.21	9.48	8.76	8.04	7.31
320	330	16.40	14.33	13.61	12.88	12.16	11.44	10.71	9.99	9.27	8.54	7.82
330	340	16.90	14.84	14.11	13.39	12.66	11.94	11.22	10.49	9.77	9.05	8.32
340	350	17.41	15.34	14.62	13.89	13.17	12.45	11.72	11.00	10.27	9.55	8.83
350	360	17.91	15.84	15.12	14.40	13.67	12.95	12.23	11.50	10.78	10.06	9.33
360	370	18.42	16.35	15.63	14.90	14.18	13.45	12.73	12.01	11.28	10.56	9.84
370	380	18.92	16.85	16.13	15.41	14.68	13.96	13.24	12.51	11.79	11.06	10.34
380	390	19.43	17.36	16.63	15.91	15.19	14.46	13.74	13.02	12.29	11.57	10.85
390	400	19.93	17.86	17.14	16.42	15.69	14.97	14.24	13.52	12.80	12.07	11.35
400	410	20.43	18.37	17.64	16.92	16.20	15.47	14.75	14.03	13.30	12.58	11.85
410	420	20.94	18.87	18.15	17.42	16.70	15.98	15.25	14.53	13.81	13.08	12.36
420	430	21.44	19.38	18.65	17.93	17.21	16.48	15.76	15.03	14.31	13.59	12.86
430	440	21.95	19.88	19.16	18.43	17.71	16.99	16.26	15.54	14.82	14.09	13.37
440	450	22.45	20.39	19.66	18.94	18.21	17.49	16.77	16.04	15.32	14.60	13.87
450	460	22.96	20.89	20.17	19.44	18.72	18.00	17.27	16.55	15.82	15.10	14.38
460	470	23.46	21.39	20.67	19.95	19.22	18.50	17.78	17.05	16.33	15.61	14.88
470	480	23.97	21.90	21.18	20.45	19.73	19.00	18.28	17.56	16.83	16.11	15.39
480	490	24.47	22.40	21.68	20.96	20.23	19.51	18.79	18.06	17.34	16.61	15.89
490	500	24.98	22.91	22.18	21.46	20.74	20.01	19.29	18.57	17.84	17.12	16.40
500	510	25.50	23.44	22.71	21.99	21.26	20.54	19.82	19.09	18.37	17.65	16.92
510	520	26.04	23.97	23.25	22.53	21.80	21.08	20.36	19.63	18.91	18.18	17.46
520	530	26.58	24.51	23.79	23.06	22.34	21.62	20.89	20.17	19.45	18.72	18.00
530	540	27.12	25.05	24.32	23.60	22.88	22.15	21.43	20.71	19.98	19.26	18.54
540	550	27.65	25.59	24.86	24.14	23.41	22.69	21.97	21.24	20.52	19.80	19.07
550	560	28.19	26.12	25.40	24.68	23.95	23.23	22.51	21.78	21.06	20.33	19.61
560	570	28.73	26.66	25.94	25.21	24.49	23.77	23.04	22.32	21.60	20.87	20.15
570	580	29.27	27.20	26.47	25.75	25.03	24.30	23.58	22.86	22.13	21.41	20.69
580	590	29.80	27.74	27.01	26.29	25.56	24.84	24.12	23.39	22.67	21.95	21.22
590	600	30.34	28.27	27.55	26.83	26.10	25.38	24.66	23.93	23.21	22.48	21.76
600	610	30.88	28.81	28.09	27.36	26.64	25.92	25.19	24.47	23.75	23.02	22.30
610	620	31.42	29.35	28.62	27.90	27.18	26.45	25.73	25.01	24.28	23.56	22.84
620	630	31.95	29.89	29.16	28.44	27.71	26.99	26.27	25.54	24.82	24.10	23.37
630	640	32.49	30.42	29.70	28.98	28.25	27.53	26.81	26.08	25.36	24.63	23.91
640	650	33.03	30.96	30.24	29.51	28.79	28.07	27.34	26.62	25.90	25.17	24.45
650	660	33.57	31.50	30.77	30.05	29.33	28.60	27.88	27.16	26.43	25.71	24.99
660	670	34.10	32.04	31.31	30.59	29.86	29.14	28.42	27.69	26.97	26.25	25.52

Source: Table adapted from Massachusetts Department of Revenue, Income Tax Withholding Tables, 1979.

PAYCHECKS, DEDUCTIONS, AND TAXES

Similar calculations are made for other wages and exemptions. The percent rate is always 4.6%. Figure 9-3 is representative of a tax table for a state or city flat-rate tax that is not progressive.

Social security (FICA) deductions are a form of flat-rate taxation. With few exceptions, nearly everyone has an FICA deduction on every paycheck. This tax, however, is *regressive*; that is, the tax is lower as income increases. In fact, the tax becomes zero after a certain income is reached. The FICA rate for 1981 was 6.65% on all income up to $29,700. There is no income reduction allowance, but the $29,700 was the maximum income subject to FICA tax. (See Figure 9-4.) After that amount is attained within one calendar year, no more social security taxes are collected from that wage earner on that particular job. However, if the worker has more than one job, he will continue to pay even if his combined income exceeds $29,700. Fortunately, this excess social security payment is available as a refund when the annual tax report is filed, but it must be specifically requested. The amount actually paid into the social security fund is double the amount the wage earner pays, because the employer must match the payment of the employee. In the case of self-employed workers, a different, higher percent is paid.

Other Deductions. Insurance premiums (usually health and life), union and professional dues, savings, loan payments, United Fund contributions, and retirement and investment payments make up the bulk of the other deductions that may appear on a paycheck. Some of them are voluntary, and others are mandatory. Usually they do not represent a percentage rate of the income but are flat amounts based upon the item and the circumstances (or choices) of the wage earner. Often they reflect a contribution to a benefit supported by the employer. Such may be the case when an employee pays a certain amount for added insurance coverage that is over and above the benefit paid for by the employer. Although these deductions are not taxes, they do affect the net take-home pay and, therefore, must be considered when planning effective money management. The effective deduction rate for miscellaneous deductions varies with each individual but may range anywhere from 1% to 20%.

As an illustration of effective deduction rate, consider a paycheck with a gross pay of $618.21 and the following miscellaneous deductions: insurances, $12.51; union dues, $6.30; credit union, $53.00; United Fund, $1.75; annuity investments, $25.00. The sum of all these deductions is $98.56. The effective rate of reduction is:

$$a\% \text{ of } \$618.21 \text{ is } \$98.56$$

$$\frac{a}{100} = \frac{\$98.56}{\$618.21}$$

$$a = \frac{9856}{618.21}$$

$a = 15.9$ or 15.9% of this gross pay is deducted for miscellaneous deductions.

There are two other items that reduce net pay, although they are not considered real deductions. One is *dock pay* (not to be confused with health insurance). Dock pay is the amount subtracted from the gross pay because of absence, tardiness, or the result of disciplinary actions. Employers will dock a certain amount from a worker's pay if the rules of employment have been violated. The other reduction is not caused by the worker, at least not directly. *Inflation* is sometimes called the invisible deduction because it is not readily apparent. Inflation causes the net pay to be reduced in terms of buying power. It is invisible and silent, but it is real. The relationship among wages, prices, and inflation is a complicated one, and economists are not in agreement as to what the cause is or what the effect is.

Topic II: Methods of Determining Pay

Paychecks may be received weekly, biweekly, semimonthly, monthly, or at some other regular interval. The methods used to determine the amount of pay are also numerous. Many professional and semiprofessional persons receive an annual salary. This salary is not usually paid in one lump sum but is paid in equal installments during the year or working period. Teachers, professional athletes, management personnel, salespersons, and others are often paid in this way. However, most workers are paid in either one or a combination of several of the following ways: hourly, piecework, commission, fees, and tips. Other forms of compensation that may also make up a person's income are bonuses, holiday and vacation pay, and sick pay.

HOURLY PAY
Individuals paid by the hour usually must punch a clock to record their arrival and departure. Aside from the obvious need to be at work on time, employees who wish to keep track of the pay they have earned will also need to be able to determine the hours and fractions of hours worked. A review of this topic from "Fractions and Time" in Chapter 4 might be warranted. Sample *time cards*, showing time in and time out, should be examined and translated to hours and fractions of an hour. Pay is determined by an *hourly rate*. The weekly wage is found by multiplying the hours worked during the week by the hourly rate. If, for example, Neal N. Hammer is paid $6.45 per hour and he works 40 hours in one week, his gross pay is 40 X $6.45 or $258. On the other hand, if he works more than 40 hours in one week (or more than 8 hours in one day) it is customary to pay the excess hours at an *overtime rate*, usually 1-1/2 times the regular rate (called "time and a half"). Overtime is normally determined by a company or union rule and may call for a premium of from 1-1/2 to 4 times the regular rate. Normally a regular week is considered as 5 days and 40 hours, a regular day is 8 hours, and Sundays and holidays are nonworking days. Any work beyond these "normal" definitions is considered overtime. Consequently, if Mr. Hammer works 40 regular hours and 4-1/2 overtime hours (at time and a half), his gross pay is:

$258.00 for regular hours

$(1\frac{1}{2} \times \$6.45)$ or $\$9.675 \times 4\frac{1}{2} = \underline{\$\ 45.54}$ for overtime

$303.54 Gross Pay

What would his pay have been if he worked the same hours at regular pay?

In addition to overtime, hourly workers may receive an incentive for working during a difficult shift (late evenings, perhaps), weekends, or perhaps under hazardous circumstances (e.g., utility workers during a storm). This is often called *shift incentive* or *shift differential*. Employers often provide an extra 5% to 20% hourly incentive to those who will work these "odd" times. This increase in hourly rate is not dependent upon overtime; it simply establishes a new hourly rate. Overtime can still be accumulated. Suppose Neal N. Hammer worked

44-1/2 hours during the midnight shift and was paid a 5% hourly incentive. How much did he make during that week?

$$\$6.45 + (5\% \text{ of } \$6.45) = \$6.45 + \$.32 = \$6.77 \text{ per hour}$$

$$\$6.77 \times 40 = \$270.80 \text{ for 40 hours work}$$

$$\$6.77 \times 1\text{-}1/2 = \$10.16 \text{ per hour for overtime}$$

$$\$10.16 \times 4\text{-}1/2 = \$45.72 \text{ for overtime pay}$$

$$\$270.80 + \$45.72 = \$316.52 \text{ Net Pay}$$

All that happens to hourly workers is not good, especially if they break the rules of their employer. Hourly workers who are late to work may find they have a penalty to pay (or dock pay). An employer may require, for example, a penalty for tardiness beyond 10 minutes. The rule might be that the loss of salary begins after the first 10 minutes late. Beyond that point, one fourth of an hour's pay for each 10-minute period (or fraction of 10 minutes) is subtracted from the weekly wage. Oftentimes, the tardiness time is accumulative for the pay period, so that 6 minutes late one day is added to 9 minutes late of another day, and so on. Suppose, for instance, an employee arrives at work 9 minutes late nearly every day, smugly thinking he has beaten the "10-minute" rule, only to find that at the end of the week his 36 minutes of tardiness result in some dock pay. Ten minutes are subtracted from the 36, according to the rule, leaving 26 minutes or 2-6/10 ten-minute periods. If his hourly pay is $4.76, he will be docked 1/4 of $4.76 for each of 3 periods. In other words, he will lose 3/4 of $4.76, or $3.57. At this rate, he may find that his weekly pay turns out to be weakly paid.

PIECE WORK

Some employees in certain fields are paid by the number of items they produce. As more articles are produced, the employee is entitled to more pay. For example, if a certain worker is paid 57¢ for each item he produces and he is able to produce 372 items in one week, his gross pay will be:

$$372 \times \$.57 = \$212.04$$

Sometimes the company will encourage greater production by offering a graduated piecework scale that increases the per item pay as more items per week are produced. For example, consider a scale that pays:

47¢ for the first 150 items

57¢ for the next 100 items

69¢ for the next 50 items

84¢ for the next 50 items

$1.04 for all items thereafter

Using this scale, the worker making 372 items will earn:

$$\$.47 \times 150 = \$70.50$$
$$.57 \times 100 = 57.00$$
$$.69 \times 50 = 34.50$$

$$.84 \times 50 = 42.00$$
$$\$1.04 \times 22 = \underline{\$\ 22.88}$$
$$\$226.88$$

The first method of piecework pay is called *straight piecework*; the second is called *differential piecework*. One of the drawbacks associated with piecework pay is that the worker may sacrifice quality for quantity, and this form of pay is much less common than hourly pay.

COMMISSIONS

Salespersons are often paid on a *commission* basis. This means that the employee receives a certain percent of the sales he or she has made during the pay period. The examples that follow illustrate several ways that commission pay may be determined.

Example 1: Estelle is a real estate salesperson, and she earns 3-1/2% of the sale on homes listed by her company and 1-3/4% of the sale on homes listed by other companies. During some months she has hardly any sales but is able to make up for this by making several good sales during other months. What is her total commission if she sells two homes, one listed by her company for $79,500 and another listed by another company for $85,400?

$$\$79{,}500 \times 3\tfrac{1}{2}\% = \$79{,}500 \times .035 = \$2782.50$$

$$\$85{,}400 \times 1\tfrac{3}{4}\% = \$85{,}400 \times .0175 = \underline{\$1494.50}$$

Total Commission Received $4277.00

Example 2: Manuel sells home insulation. He is paid 17% of all sales above $800. How much does he earn if he sells insulation contracts totaling $690, $975, $1450, and $1810?

$$\$690 + \$975 + \$1450 + \$1810 = \$4925$$

$$\$4925 - \$800 = \$4125$$

$$\$4125 \times 17\% = \$4125 \times .17 = \$701.25$$

Example 3: Sophie is a salesperson for a furniture store that pays a commission of a graduated scale of 7% of the first $900, 9% of the next $700, 12.6% of the next $500, and 16-1/2% of the sales over $2100. How much does Sophie make if her total sales are $3298?

$$7\% \text{ of } \$900 = .07 \times \$900\quad = \$\ 63.00$$
$$9\% \text{ of } \$700 = .09 \times \$700\quad = \ \ \ 63.00$$
$$12.6\% \text{ of } \$500 = .126 \times \$500\ = \ \ \ 63.00$$
$$16\tfrac{1}{2}\% \text{ of } \$1298 = .165 \times \$1198 = \underline{\ \ 197.67}$$

Total Commission = $386.67

OTHER MEANS OF EARNING PAY

Earnings may be determined in several other ways. Some professional people charge a specified fee for a service that is performed (doctor's fee, attorney's fee, speaker's fee, etc.). Waiters and waitresses earn tips (a type of commission) based upon the meals they serve (see Chapter 4). Some wage earners receive a bonus at the end of the year or at the culmination of a successful project. Although a bonus can be a substantial amount, it is usually paid in addition to regular earnings paid out in one of the manners described earlier. Indeed, many people are paid by a combination of two or more of the ways described in this section.

PAYCHECKS, DEDUCTIONS, AND TAXES

A teacher on a yearly salary may be paid hourly for additional work or receive a fee for assuming a special responsibility. Hourly workers may also receive piecework incentives. Waiters usually receive an hourly wage as well as tips. Because of the variety of ways to determine pay, it is often difficult to compare one job with another. The next section explores this idea.

COMPARING INCOMES

When searching for a job one might encounter ads that make it difficult to compare employment possibilities and thus determine which job is best. For the sake of clarity, four areas of consideration are suggested: advertised periods of wage payments, method of determining wages, fringe benefits, and job-related expenses.

Advertised periods of wage payments. Consider this dilemma: Gail Gemini was at a loss as to which job to choose. She had the following choices.

- A. Office aide, $8000 per year.
- B. Clerk, $650.00 per month.
- C. Salesgirl, $310.00 biweekly.
- D. Office assistant, $320.00 twice a month.
- E. Cashier, $155.60 weekly.

Assuming that all are full-time jobs, which offers the best pay?

The simplest way to compare is to convert them all to an annual income. This may be done in several ways, including using a proportion, multiplying, or dividing. Job A is already expressed in terms of annual income. Job B can be converted by multiplying by 12. Thus:

$$\$650 \times 12 = \$7800 \text{ annually for job B}$$

Job C is expressed as a biweekly pay, which is every other week. This results in 26 pays per year and an annual salary of:

$$\$310 \times 26 = \$8060 \text{ annually for job C}$$

Job D, paid twice a month, appears to offer more than job C, but it does not. This job has only 24 pays per year and provides a yearly wage of:

$$\$320 \times 24 = \$7680 \text{ annually for job D}$$

Finally, the last job on the list, paid weekly, amounts to:

$$\$155.60 \times 52 = \$8091.20 \text{ annually for job E}$$

If Gail is insistent upon making a minimum of $8200 per year, she can determine the minimum weekly wage by the proportion: $8200/1 year = ?$/1 week. However, corresponding terms of a proportion must match. Therefore, 1 week equals 1/52 of a year, and the proportion becomes:

$$\frac{\$8200}{1 \text{ year}} = \frac{\$n}{1/52 \text{ year}}$$

or $\quad \$n = \dfrac{1}{52} \times \$8200 \quad$ *by cross multiplying the minimum weekly wage that Gail will accept.*

$\$n = \$157.69,$

Similar proportions can be used for other wage payment periods.

Method of wage determination. Some wages are easy to compare. For instance, an hourly rate can easily be converted to a weekly, monthly, or yearly wage simply by dividing or multiplying. However, jobs that offer a piecework wage or a commission, or include a combination of an hourly scale with either of these, are more difficult to compare. Indeed, wages based upon piecework, commissions, fees, tips, and bonuses are all highly speculative and nearly impossible to determine with certainty. People who take these jobs are accepting a certain risk that they will be satisfied with their earnings. When exploring the possibility of employment, it is wise to estimate the potential income based upon what may be reasonably anticipated.

Example 4: Determine the job that is likely to pay the most based upon the given wage structures and the anticipated production.

A) Waitress: $2.75 per hour plus tips. There are 3 other waitresses, and the restaurant averages $968 in sales per day during the hours the waitress will work.

B) A widget-gizmo maker: $3.50 per hour plus 63¢ per widget-gizmo made. It is anticipated that 9 items can be made in two hours.

C) Salesperson: $4.28 per hour plus 9-3/4% of sales. It is anticipated that $950 will be the average sales per week.

Job A.

$2.75 *per hour* × 40 *hours per week* = $110.00 *wages per week*

$968 *per day* ÷ 4 *waitresses* = $242

$242 × 12% *anticipated average tip* = $29.04 *per day*

$29.04 × 5 = $145.20

$110 (*wages*) + $145.20 (*Anticipated tips*) = $255.20 *anticipated income per week.*

Job B.

$3.50 *per hour* × 40 *hours per week* = $140 *wages per week*

4-1/2 *items in one hour* × 40 = 180 widget-gizmos

180 × $.63 = $113.40

$140 (*wages*) + $113.40 (*piecework earnings*) = $253.40 *per week.*

Job C.

$4.28 *per hour* × 40 *hours per week* = $171.20 *wages per week*

$950 *in sales* × 9-3/4% *commission rate* = $92.63 *commission*

$171.20 (*wages*) + $92.63 (*commission*) = $263.83 *per week.*

Fringe benefits and job expenses. On the surface it appears that the salesperson job (C) is the best. However, there are other factors that may be considered, especially when the anticipated incomes are so similar. For example, it is estimated that employers spend anywhere from 30% to 60% of their employee benefit expenses (including wages) in providing those things called *fringe benefits*. This means that, on the average, each worker receives benefits worth about one half of his pay in addition to his regular pay. These fringe benefits may include insurance (health, life, unemployment, workmen's compensation, etc.), vacation pay, sick pay, holiday pay, and retirement. Included as a benefit is the employer's share of social security contributions which matches the amount paid in by the employee. Incidental benefits might include such things as working conditions, coffee breaks, availability of dining and recreation facilities, and the like. A restaurant waiter or waitress may be entitled to a lunch or dinner allowance, and this could add up to quite an advantage. The cost of eating away from home, in fact, is a major consideration for job expenses. Other job-related expenditures may include the cost of travel to and from work; the costs of uniforms, special clothing, tools, equipment; and the costs that are the result of on-the-job health hazards.

Topic III: Annual Income Tax Reports

Each year, by April 15, every wage earner must file an income tax report with each of the governmental bodies that collects an income tax. Most people file a federal and state income tax return, and an increasing number of people are also required to report a city income tax. The reports are to be made even if no taxes were withheld. Taxpayers must become familiar with several terms, forms, schedules, and tables in order to carry out this responsibility properly. Some of them are listed below.

gross income—the total of all income.

deductions—certain expenditures that may be subtracted from the gross income.

exclusions—certain income that is not taxable.

adjusted gross income—the total of all income minus the allowable deductions and exclusions.

W-2 form—a report from the wage earner's employer stating the gross income and the taxes and FICA payments withheld.

1040—the standard federal income tax form. Also called the *long form*.

1040A or *short form*—a shorter version of 1040.

schedules—specialized forms for reporting income, expenses, and deductions. There are more than a dozen different schedules. Some schedules show tax rates for tax computation.

tables—data related to reporting and computing taxes. Separate tables are provided for specific purposes (e.g., state sales tax, earned income credit, and general tax). The general tax tables show listings of wages and exemptions and the taxes corresponding to these or directions on how to calculate taxes.

zero bracket amount—that part of income not subject to tax. The amount varies according to filing status. The amounts are built into tax rate schedules and tax tables.

filing status—identification of taxpayer for taxing purposes as 1) married, filing jointly, or a qualifying widow or widower, 2) single, 3) married, filing separately, and 4) head of household.

tax liability—the amount of income tax owed.

FEDERAL INCOME TAX

The federal income tax is a progressive tax because the rate increases with increased income. This has been pointed out in the weekly withholding table shown in Figure 9-1. (To obtain the annual rate for the filing status shown in Figure 9-1, simply multiply each figure by 52.) The correct amount of tax to pay may be different from the amount withheld, and the annual return is necessary to determine a refund or balance due. The precise amount of tax depends on several factors, including filing status, exemptions, and deductions.

Short Form 1040A. This form should be used only when a) the only sources of income are wages, salaries, tips, or unemployment compensation and less than $400 in interest or dividends; b) deductions are not itemized; and c) no excess FICA payments are claimed. Persons who meet all of the above criteria usually need to be concerned only with which schedule or table should be used to determine their tax liability. The taxpayer's filing status determines which table or schedule to use, and each status carries a different minimum taxable income and different tax rates. Therefore, it is essential to use the correct table or schedule.

Example 5: I. M. Broke is married and earned $5214 in wages as well as $146 from a savings account. His employer withheld $140.40 from his weekly wages. He has no children and his wife did not work.

Mr. Broke may use form 1040A; his filing status is married, filing a joint return; and he must use Tax Table B. On 1040A he will report his total income and the amount withheld. Upon consulting the table, he finds that his gross income of $5360 gives him no tax liability, and he therefore enters "-0-" on the appropriate line. Consequently, he is entitled to a refund of $140.40, the entire amount of federal tax withheld from his wages.

Example 6: Tex Weary is single, and he too earned $5214 in wages and $146 from a savings account. He had $298.40 withheld from his wages.

Mr. Weary may use form 1040A, his filing status is single, and he must use Tax Table A. On 1040A he will report his total income and the amount withheld. Upon consulting the table, he finds that his gross income of $5360 gives him a tax liability of $310. He therefore enters $310 on the appropriate line, subtracts the $298.40 already withheld, and enters $11.60 on the proper line as a balance due. He must include a check for $11.60 with his return.

Example 7: Another taxpayer earning the same amount who is married but filing a separate return uses Tax Table C and has a tax liability of $419.

Other factors that may affect the tax liability include a dividend exclusion of $100 ($200 if a joint return), contributions to candidates for public office, and earned income credit. The rules for these and other tax questions are spelled out in detail in several publications available free from the Department of the Treasury, IRS.

Long Form 1040. Any taxpayer who cannot use 1040A must use 1040. These taxpayers come under two categories: those that itemize deductions and those that do not.

Example 8: Steven and Susan Long are married and have three children. Steven earned $31,305 as an auto mechanic and night watchman. Susan's only income was a $2500 prize she won in a raffle. They had a savings account that earned $160, and their stock paid a dividend of $293. Steven's two jobs resulted in an overpayment of $159.68 to FICA (social security). His employers withheld $5057 for federal tax.

Because of the income from the raffle and the excess FICA payment, the long form 1040 must be used. They file a joint return and claim 5 exemptions. The entire income from wages and interest is taxable. They are entitled to a $200 exclusion on dividend income, and therefore only $93 is taxable. If this was all they had to report, the short form 1040A could have been

PAYCHECKS, DEDUCTIONS, AND TAXES   171

used. However, the $2500 raffle prize, and the $159.68 FICA overpayment can be reported only on the long form. The winnings are reported as "other income" on the front side of 1040, and the overpayment on social security is reported under "payments" on the line entitled "excess FICA and RRTA tax withheld" on the back of the 1040. The Longs are not itemizing deductions and, therefore, determine their tax on $34,058 directly from Tax Table B, which already includes a consideration for a standard deduction of $3400. (If the Longs had expenses for allowable deductions that totaled more than $3400, they would have benefited by itemizing deductions.) They find (on Tax Table B) that their tax is $5937. However, their "payments" include $5057 (federal tax withheld) and $159.68 excess FICA. The excess FICA may be applied to the federal tax already withheld, and the Longs report (and pay) a balance due of $720.32.

Example 9: Phyllis and Bert Knut each worked during the previous year. Phyllis earned $22,355 and Bert earned $7847. They have two children and a niece (all under twenty-one) and Bert's mother (who has no income) living with them. Bert is legally blind. They examined their expenses and found they had more than $5000 in deductions. These included $275 for medical care insurance, $350 for medicine and drugs, $859 for doctor's fees, $27.50 for hospital costs, $120 for transporation to receive medical treatment, $967 for state and city income taxes, $1027.40 for real estate taxes, $254 for sales tax, $1989.07 for home mortgage interest, $27.65 for credit-card interest, $675 for church, $165 for union dues, $15 for a safe-deposit box, and $35 for tools required on the job. Phyllis had $3322 withheld from her pay, and Bert's employer withheld $346 from his earnings.

Form 1040 and Schedule A are required. Schedules A and B usually appear on the same paper, A on one side, B on the other. Schedule A is used to itemize deductions. The deductions appear as categories: "Medical and Dental Expenses," "Taxes," "Interest Expenses," "Contributions," "Casualty or Theft Loss," and "Miscellaneous Deductions."

The medical expenses allowed to be claimed are subject to some restrictions. Only one half (but not more than $150) of the medical insurance is deductible. Therefore, the Knuts can claim $137.50. Medicine and drugs expense is limited to any amount over 1% of the adjusted gross income. In the Knuts' case, 1% of $30,202 is $302, and $350 − $302 is $48. The $48 is entered on the schedule and is incorporated with other factors and further restrictions before becoming a bona fide deduction. The balance of the medical insurance premium is added to the $48 ($137.50 + $48.00). To this is added the $859 for doctors, $27.50 for hospitals, and $120 for transportation. The total of $1192 must be greater than 3% of the adjusted gross income, and only the excess is allowed. Thus:

$$3\% \text{ of } \$30,202 \text{ is } \$906$$

$$\$1192 - \$906 = \$286$$

Added to the $137.50 allowed for insurance premiums, this gives a total medical expense of $423.50. (Note: the claim for one half of the medical insurance premium is allowed even if no other medical expenses qualified.)

Taxes amount to $967 + $1027.40 + $254 or $2248.40.

Interest is $1989.07 + $27.65 or $2016.72.

Contributions are $675.

There is no casualty or theft loss.

And miscellaneous deductions are $165 + $15 + $35 or $215.

The total of all allowable deductions is $423.50 + $2248.40 + $2016.72 + $675 + $215 or $5578.62. This amount is then subject to another restriction based upon the filing status of the taxpayer. For the Knuts, they must subtract $3400 (the standard deductible amount already figured into the tables) from $5578.62. This results in an allowable deduction of $2178.62.

This reduces their adjusted gross income to $28,023.38. When Phyllis and Bert determine their tax on Tax Table B, they may claim 7 exemptions (one each for the dependents and one extra exemption for Bert's blindness). Their tax (from Tax Table B) is $3504. Since a total of $3668 was withheld from their wages, they are entitled to a refund of $164. (Had they not itemized deductions, their tax would have been $4120, requiring a balance due payment of $452!)

Example 10: Rob Morebanks is a financial consultant and earned $47,291. His wife, Minnie, worked part time and earned $6427 as a store clerk. They have no children or other dependents, and their allowable deductions total $8796. Rob paid $12,219 in withholding taxes, and Minnie paid $668.

Filing a joint return, the Morebankses may subtract $5396 (the result of $8796 − $3400) from their adjusted gross income. Thus:

$$\$47{,}291 + \$6427 = \$53{,}718$$
$$\$53{,}718 - \$5396 = \$48{,}322$$

Since the taxable income is greater than $40,000 ($20,000 is the limit for single taxpayers), they must compute their tax on Schedule TC and Schedule Y. On Schedule TC, the Morebankses are allowed to reduce their taxable income by $1000 for each exemption, since the exemptions are not included on Schedule Y. This means that they compute their tax on $46,322. The portion of Schedule Y pertaining to the Morebankses is:

Income is over—	But less than—	Tax is—
$45,800	$60,000	$12,720 + 49% of excess over $45,800

Observing this, the Morebankses realize that they are just above a lower tax bracket, and they reexamine their expenses to see if they have overlooked $522 in allowable deductions. They discover that since they purchased a new car last year they are entitled to deduct the sales tax paid on the car over and above the sales tax deduction allowed on the sales tax tables. They purchased a car for $9878 and paid 5-3/4% sales tax on that amount. The $567 additional deduction now puts their taxable income at $45,755 and allows them to drop to a lower tax bracket on Schedule Y:

Income is over—	But less than-	Tax is—
$35,200	$45,800	$8,162 + 43% of excess over $35,200

The Morebankses tax is calculated as follows:

$$\$45{,}755 - \$35{,}200 = \$10{,}555$$
$$43\% \text{ of } \$10{,}555 = .43 \times \$10{,}555 = \$4538.65$$
$$\$8162 + \$4536.65 = \$12{,}698.65$$

The total withholding taxes paid amounted to $12,219 + $668 or $12,887. Therefore, the Morebankses will receive a refund of:

$$\$12{,}887 - \$12{,}698.65 = \$188.35$$

Had they not found the additional deduction, their tax would have been $12,975.78 and they would have owed $88.78 on the balance due.

Several other factors to consider on the long form include other sources of income, adjustments to income, credits, other taxes, and other payments. The area of particular concern to the average taxpayer is the listing of itemized deductions, the categories of which have already been enumerated. In all cases the rules governing these aspects of the tax return are available free from the IRS. However, there are many complicated procedures for reporting certain gains and losses, and oftentimes a tax consultant is needed. The IRS provides free consultation, and there are many private tax advisory businesses.

FICA OR SOCIAL SECURITY

The social security tax was discussed briefly in the section on withholding taxes. A summary of the FICA contributions for 1981 and succeeding years is shown in Figure 9-4. Recall that excess FICA taxes paid through payroll withholding may be claimed on the income tax return (see Example 8). The effect of limiting the income subject to taxation is that persons who earn more money will pay at a lower rate. To illustrate, consider a wage earner whose income is $49,700. He will pay 6.65% of $29,700 toward social security. This amounts to $1975.05. However, since his income is actually $49,700, the amount paid to social security is $1975.05/$49,700.00 or only 3.97% of his income, which leaves 96+% for other expenses. The taxpayer who earns $29,700 pays the same dollar amount, but he has only 93.35% of his pay remaining. For this reason, the maximum income subject to tax has been increased in recent years and will continue to be increased (see Figure 9-4).

Figure 9-4
FICA Contributions
1981-1987

Year	Maximum Income Subject To FICA Tax	Rate
1981	$29,700	6.65%
1982	$31,800	6.70%
1983	$33,900	
1984	$36,000	
1985	$38,100	7.05%
1986	$40,200	7.15%
1987	$42,300	

By 1990 the FICA tax rate will be 7.65%.
(All figures are anticipated and subject to change.)

STATE AND CITY INCOME TAXES

Some of the data used for the federal tax establish the basis for taxation by a city or state. Primarily, the adjusted gross income and the number of exemptions reported on the federal return are also used on the other tax returns. Most state and city income taxes are flat-rate taxes applied to the income after an allowance for exemptions. Some, however, are progressive graduated taxes like the federal income tax. In either case, refunds and balances due are handled in a manner similar to those on the federal returns.

Example 11: Ida Hoe and her husband, Garfield Daniel (known to his friends as Gar Dan), had a combined income of $28,291. They have four children and were allowed 6 exemptions on the federal return. Their state income tax rate is 4.7%, but they are allowed $1250 per exemption. Their city income tax rate is 1-1/2%, and the city allows $1500 per exemption. Find the Hoes' state and city income taxes.

The state tax is:

$1250 × 6 = $7500 exemption allowance

$28,291 − $7500 = $20,791 taxable income

4.7% of $20,791 = .047 × $20,791 = $977.18.

The city tax is:

$$\$1500 \times 6 = \$9000 \text{ exemption allowance}$$
$$\$28{,}291 - \$9000 = \$19{,}291$$
$$1\text{-}1/2\% \text{ of } \$19{,}291 = .015 \times \$19{,}291 = \$289.37.$$

Example 12: Orvile and Wilma Wrong have no children and live in a state that allows a reduction of the tax (*after* it is computed) for filing status and dependents. Their income is $26,403; the tax rate is 5.37%; and they are allowed $185 for being married and $97 per dependent. Find their tax liability and determine their effective tax rate.

$$5.37\% \text{ of } \$26{,}403 = .0537 \times \$26{,}403 = \$1417.84$$
$$\$185 + (\$97 \times 2) = \$185 + \$194 = \$379$$
$$\$1417.84 - \$379 = \$1038.84 \text{ state tax liability.}$$

The effective tax rate is:

$$a\% \text{ of } \$26{,}403 = \$1038.84$$
$$\frac{a}{100} = \frac{\$1038.84}{\$26{,}403}$$
$$a = \frac{103{,}884}{26{,}403}$$
$$a = 3.93 \text{ or } 3.93\% \text{ is the effective tax rate.}$$

Example 13: Harry and Mary Goround have a yearly income of $32,490. They claimed 7 exemptions on their federal tax return and reduced their taxable income by $1782 through itemized deductions. Their state allows a deduction of $1150 for each exemption and a deduction for expenses that is 50% of their itemized federal deductions. Find the state tax liability for the Gorounds if the tax rate is a progressive graduated tax of:

> 3% on first $1000
> plus 3-1/2% on next $1000
> plus 4-1/2% on next $1000
> plus 5% on next $1000
> plus 6% on next $2000
> plus 7-1/4% on next $2000
> plus 8-3/4% on next $2000
> plus 10% on any amount over $10,000

The Gorounds' taxable income is reduced by:

$$(7 \times \$1150) + (50\% \text{ of } \$1782)$$
$$\$8050 + \$891 = \$8941$$

This means that their taxable income is:

$$\$32{,}490 - \$8941 = \$23{,}549$$

And their tax liability is:

```
       .03  × $1000                                  $  30.00
    +  .035 × $1000                    =                35.00
    +  .0425 × $1000                   =                42.50
    +  .05  × $1000                    =                50.00
    +  .06  × $2000                    =               120.00
    +  .0725 × $2000                   =               145.00
    +  .0875 × $2000                   =               175.00
    +  .10  × ($23,549 − 10,000) = .10 × $13,549 = $1354.90
                                                    $1952.40
```

PAYCHECKS, DEDUCTIONS, AND TAXES

The effective state income tax rate is:

$a\%$ of $\$32,490$ is $\$1952.40$

$$\frac{a}{100} = \frac{\$1952.40}{\$32,490}$$

Multiply both sides of equation by 100 to remove decimal point:

$$a = \frac{195,240}{32,490}$$

$a = 6$ or 6% is the effective state tax rate.

Some states and cities have a tax rate schedule similar to Schedule Y of the federal tax; and of course there may be some tax methods not shown here. Although it is valuable for students to have experience with various types of calculations and tax tables, it is most important that they become familiar with, and learn about, their own local and state taxation systems. Therefore, tax forms should be obtained from local government, and attention given to meeting the responsibilities for the area in which the students live.

LEARNING EXPERIENCES

THE DIAGNOSTIC SURVEY AND INDEPENDENCE SCALE

- Use the **Diagnostic Survey** on Reproduction Page 75 to estimate each student's degree of competence prior to instruction in the topics of this chapter.

- Upon completion of the Survey, use the **Independence Scale** on Reproduction Page 76 to identify those skills that need to be strengthened by further instruction and practice.

- Utilize the results of these two instruments to determine the appropriate assignments from the following list.

(Note: Figures 9-1—9-3 are required for student use while doing the **Diagnostic Survey**.)

Topic I: Paychecks and Deductions

- Obtain information and literature from your local social security office and have students discuss the purpose, benefits, value, and financing of the social security system.

- Obtain applications for social security cards and have students go through the application process. If students already have a card, have them complete a lost-card report and reapplication.

- Inquire at local stores, shops, factories, and employment agencies and ask for samples of their application forms and employment tests. Distribute these among students for practice exercises.

- Obtain a current W-4 form from an employment office (your school personnel office will have them). Have students examine them and provide the information requested. Be sure to go over the explanation and directions for determining allowances, particularly dependents, exemptions, and deductions.

- Discuss the following terms with the students, and have them learn definitions of each: *payroll check, payroll deductions, withholding taxes, FICA, gross pay, net pay, dependents, exemptions, allowances, progressive tax, regressive tax.*

- Have students learn the various payroll deductions they are likely to confront: *city, state,* and *federal taxes; social security; insurances; dues; savings and loans; United Fund; retirement and investments; dock pay.*

- Have students discuss inflation as "the invisible deduction." Ask them to investigate the consumer price index (include CPI, CPI-U, and CPI-W) and the average wage index (AWI) and discuss how these reflect the status of inflation and its effect on the average consumer.

- Use Reproduction Page 77, Part I, and have students determine the sum of all deductions and find the net pay.

- Discuss the meaning of "effective tax rate" and "effective deduction rate." Use Reproduction Page 77, Part II, A and B, and have students determine the effective rates for each deduction. Use the statement $a\%$ of $b = c$ and the proportion $a/100 = c/b$.

- Use Reproduction Page 78, Part One, for students to determine payroll deductions using the effective rates that are given.

- Use Reproduction Page 78, Part Two, for students to determine payroll deductions using the tax table for federal tax (Figure 9-2) and state tax (Figure 9-3). Use the current social security tax rate and an effective rate of 1-1/2% for city tax and 8% for miscellaneous deductions. Be sure students use the *allowances* figure for the federal taxes and the *exemptions* figure for the state tax.

- Have students calculate the federal withholding taxes for Reproduction Page 77, Part Three, using the schedule in Figure 9-1.

Topic II: Methods of Determining Pay

- Obtain sample time cards used for hourly employees and distribute them within the class. Discuss the various aspects of time card usage including using a time clock, interpretation of hours worked, hourly rate of pay, overtime, shift incentive (or differential), and dock pay.

- Use Reproduction Page 79 and ask students to determine weekly pay from hourly pay.

- Discuss piecework pay, examining the pros and cons as compared with hourly pay.

- Use Reproduction Page 80, Part I, and have students determine the weekly pay from the data given.

PAYCHECKS, DEDUCTIONS, AND TAXES

- Discuss *commission* pay, examining the pros and cons as compared with simple hourly pay.

- Use Reproduction Page 80, Part II, and have students determine the weekly pay based on the sales and commissions given.

- Use Reproduction Page 81, Part I, and have students determine the weekly pay for various combinations of pay scales.

- Discuss fringe benefits and job expenses with the class.

- Use Reproduction Page 81, Part II, and have students determine the best-paying jobs.

- Use the effective tax rates given on Reproduction Page 78 and determine the payroll deductions and net pay for the weekly wages determined in Reproduction Pages 79, 80, and 81.

- Have students use the tax table for federal tax (Figure 9-2) and state tax (Figure 9-3), the current FICA tax rate, and effective rate of 1-1/2% for city tax, and 8% for miscellaneous deductions to determine the payroll deductions and net pay for each weekly wage found in Reproduction Pages 79, 80, and 81.

- Use the tax schedule in Figure 9-1 and recalculate the federal withholding taxes found in each of the above assignments (which used the tax table) and note the differences.

Topic III: Annual Income Tax Reports

- Write to your local and state governments to obtain several tax forms and instructions for each student. Federal income tax teaching materials can be obtained from the Department of the Treasury, Internal Revenue Service, Washington, D.C. Ask for the *Understanding Taxes* unit and be sure to obtain Publications 17, 19, 21 (and 22 for farm areas) as well as additional Forms 1040, 1040A, and Schedules A and B. Use these materials along with the assignments and Reproduction Pages of this *Guidebook*.

- Discuss and have students define the terms identified in the "Content Overview" for this topic.

FEDERAL INCOME TAX
- Use Form 1040A (short form) and Publication 17, *Your Federal Income Tax*, from the federal IRS and have students complete several returns based upon imaginary incomes and the following guidelines.

  Use Table A: 1. Single, all taxes withheld are refundable because minimum income not attained.

  2. Single, partial refund.

  3. Single, balance due. (See Example 6.)

  Use Table B: 1. Married, joint return with and without other dependents, partial refund. (See Example 5.)

2. Married, joint return with and without other dependents, balance due.

3. Married, joint return, with and without other dependents, with dividend and interest income and contributions to candidates for public office.

4. Do each of above with one income and also with two incomes.

Use Table C: Married, filing separately, with and without dependents. Set up situations that would compare separate and joint returns for households with two incomes. (See Example 7.)

Use Table D: Single, Head of Household, claiming several dependents.

- Use IRS Publication 21, *Understanding Taxes*, pages 10-19, for further practice with Form 1040A.

- Use Form 1040 (long form) and IRS Publication 17, *Your Federal Income Tax*, and any other appropriate form from the IRS and have students complete several returns based upon imaginary incomes and the following guidelines.

Use Table A:
1. Single, more than $400 in interest and dividend income.
2. Single, income from contest award.
3. Single, two jobs, excess FICA payments.
4. Single, itemized deductions (without medical claims).
5. Single, itemized deductions (with medical claims).
6. Single (any combination of above).
7. Single (see items 4-8 for married couples, below).

Use Table B:
1. Married, joint return, dependent children, income from contest award, excess FICA payments. (See Example 8.)

2. Married, joint return, dependent children, itemized deductions (without medical claims).

3. Married, joint return, dependent children and additional exemptions (foster-care persons, age, blindness, etc.), itemized deductions including medical. (See Example 9.)

4. Married, joint return with various combinations of above and/or other income factors that tend to complicate the return (e.g., dividend income, state and local income tax refunds, alimony income, business income, etc.).

5. Married, joint return with other complicating factors affecting adjusted gross income (e.g., moving expenses, employee business expense, Keogh payments, interest penalties, alimony paid, etc.).

6. Married, filing joint return with credits that will reduce the tax liability directly (contribution to political candidates, credit for elderly, for child and dependent care, for investments, for residential energy, etc.).

PAYCHECKS, DEDUCTIONS, AND TAXES

7. Married, filing joint return with other taxes to pay (e.g., self-employment tax, FICA tax on tips not already reported, etc.).

8. Married, filing joint return with other payments (e.g., estimated tax payments and credit from previous year, Regulated Investment Company credit, etc.).

Use Table C: Married, filing separately. Set up situations that would compare joint and separate returns for households with two incomes and circumstances similar to those described in the previous paragraphs.

Use Table D: Single, head of household with circumstances similar to those outlined in the previous paragraphs.

Use Schedule X, Y, Z: See Example 10 and IRS Publication 17 for rules governing the use of these tables.

SOCIAL SECURITY

- Have students examine the data in Figure 9-4 concerning social security rates and maximum incomes subject to tax. Ask them to consider that from 1937 to the early 1950s only 1% of the first $3000 income was subject to FICA taxation. This meant that only $30 per year (maximum) was deducted from incomes. Also, ask students to examine the effective rates of contribution by wage earners whose incomes far exceed the maximum taxable income. Ask that they determine the effective rate (in terms of gross income and actual payment) for these payments and incomes.

    1. $2130.60 contribution (based upon the 1982 rate of 6.70% on a maximum income of $31,800) for these actual incomes.

        a) $40,200   b) $41,800   c) $51,800   d) $63,600   e) $81,800

    2. $2874.30 contribution (based upon the 1986 rate and maximum) for each of the incomes shown in item 1.

State and City Income Tax

- Use the tax forms obtained from your local and state governments and have students complete several returns that parallel the federal tax returns created in the previous sections.

- To give students practice with various types of tax computation, consider creating assignments that require state and city tax computation with the following guidelines.

    1. Flat-rate taxes on the amount of income remaining after an allowance for dependents and other exemptions is subtracted from the gross income. (See Example 11).

    2. Flat rate on gross income, subtractions made from resultant tax based upon filing status, dependents, and other exemptions. (See Example 12.)

    3. Progressive graduated tax:
        a. with allowance for dependents only.
        b. with allowance for exemptions as per federal tax.
        c. with allowance for itemized deductions as allowed on federal return. (See Example 13.)

4. Progressive graduated tax that requires the use of a computation system similar to Schedule Y of the federal income tax. (See also Figure 9-1.)

ASSESSING ACHIEVEMENT OF OBJECTIVES

Ongoing Evaluation

The extent to which students have mastered the concepts covered under the three topics in this chapter can be measured by any of the activities assigned to class members individually.

Final Evaluation

For an overall evaluation of the students' mastery of the concepts in this chapter, if all topics in the chapter have been taught, a test constructed directly from the "Objectives" listed at the beginning of the chapter can be used. As an alternative, one might consider using the **Diagnostic Survey** as a final test.

RESOURCES FOR TEACHING ABOUT PAYCHECKS, DEDUCTIONS, AND TAXES

Below is a selected and annotated list of resources useful for teaching the topics in this chapter, divided into audiovisual materials, games, and print materials. Addresses of publishers or distributors can be found in the alphabetic list in Appendix B.

Audiovisual Materials

A. FOR LOW ABILITY AND SPECIAL EDUCATION STUDENTS

Income Tax Series, 5 filmstrips, 5 audio-cassettes, and 20 student workbooks. Interpretive Education, 1977.

What Are Company Benefits?, filmstrip and audio-cassette. Interpretive Education, 1977.

B. FOR GENERAL MATHEMATICS STUDENTS

"How's Your Paycheck Math?"; "How's Your Tax Math?" *Consumer Math Cassettes* produced by F. Lee McFadden. The Math House, 1977.

"Money Talks," 26 minutes. A brief history of taxation. And "What Happened to My Paycheck?" 16-1/2 minutes. Discusses payroll deductions, tax responsibilities, and processing tax returns. Department of the Treasury, 1977.

Check Your Paycheck, 13-minute slide presentation with worksheets. Chamber of Commerce of the United States, 1977.

Games

Big Deal! Creative Teaching Associates, 1976.

Cutting Corners developed by John Schatti. The Math Group, Inc., 1977.

Print

A. FOR LOW ABILITY AND SPECIAL EDUCATION STUDENTS

Working Makes Sense by Charles H. Kahn and J. Bradley Hanna. See especially pp. 3-28, 41-42, 79-98. Fearon-Pitman Learning, Inc., 1973.

Useful Arithmetic, Volume I, by John D. Wool and Raymond J. Bohn. See especially pp. 33-39, 42-48, 57-60. Frank E. Richards Publishing Company, Inc., 1972.

Useful Arithmetic, Volume II, by John D. Wool. See especially pp. 38-45. Frank E. Richards Publishing Company, Inc., 1972.

Banking, Budgeting, and Employment by Art Lennox. See especially pp. 33-35 and 44-75. Frank E. Richards Publishing Company, Inc., 1979.

An Introduction to Everyday Skills by David H. Wiltsie. See especially pp. 22-38 and 146-157. Motivational Development, Inc., 1977.

Skills for Everyday Living, Book 1, by David H. Wiltsie. See especially pp. 8-12 and 130-136. Motivational Development, Inc., 1976.

Skills for Everyday Living, Book 2, by David H. Wiltsie. See especially pp. 98-101 and 134-160. Motivational Development, Inc., 1978.

Using Money Series, Book IV Earning Spending and Saving, by John D. Wool. See especially pp. 1-19. Frank E. Richards Publishing Company, Inc., 1973.

It's Your Money, Book 1, by Lloyd L. Feinstein and Charles H. Maley. See especially pp. 18-28. Steck-Vaughn Publishing Company, 1973.

Math for Today and Tomorrow by Kaye A. Mach and Allan Larson. See especially pp. 49-86. J. Weston Walch, Publisher, 1968.

Scoring High in Survival Math by Tom Denmark. See especially pp. 2-5 and 12-14. Random House, Inc., 1979.

Michigan Survival by Betty L. Hall and David Landers. Also available for all states. See especially pp. 18-65 and 121-128. Holt, Rinehart and Winston, 1979.

Mathematics for Today, Level Red, by Saul Katz, Ed.D.; Marvin Sherman; Patricia Klagholz; and Jack Richman. See especially pp. 1-6 and 139-158. Sadlier-Oxford, 1976.

Mathematics for Today, Level Blue, by Edward Williams; Saul Katz, Ed.D.; and Patricia Klagholz. See especially pp. 5-26. Sadlier-Oxford, 1976.

B. FOR GENERAL MATHEMATICS STUDENTS

Stein's Refresher Mathematics, Seventh Edition, by Edwin I. Stein. See especially pp. 541-556 and 607-613. Allyn and Bacon, Inc., 1980.

Trouble-Shooting Mathematics Skills, Basic Competency Edition, by Allen L. Bernstein and David W. Wells. See especially pp. 332-345 and 380-384. Holt, Rinehart and Winston, 1979.

Consumer Mathematics with Calculator Applications by Alan Belstock and Gerald Smith. See especially pp. 75-103. McGraw-Hill Book Company, 1980.

Consumer Math, A Guide to Stretching Your Dollar, by Flora M. Locke. See especially pp. 1-35 and 272-304. John Wiley and Sons, Inc., 1975.

Consumer Mathematics, Third Edition, by William E. Goe. See especially pp. 159-240. Activities book available. Harcourt Brace Jovanovich, 1979.

Mathematics for Today's Consumer by Jack Price, Olene Brown, Michael Charles, and Miriam Lien Clifford. Compiled from selections from *Mathematics for Everyday Life* and *Mathematics for the Real World*. See especially pp. 4-33 and 86-99. Charles E. Merrill Publishing Company, 1979.

Mathematics for the Real World by Jack Price, Olene Brown, Michael Charles, and Miriam Lien Clifford. See especially pp. 4-33 and 86-99. Charles E. Merrill Publishing Company, 1978.

Mathematics for Daily Living by Harry Lewis. See especially pp. 145-206. McCormick-Mathers Publishing Company, 1975.

Mathematics Plus!, Consumer, Business & Technical Applications, by Bryce R. Shaw, Richard A. Denholm, and Gwendolyn H. Shelton. See especially pp. 101-132. Houghton Mifflin, 1979.

"Your Financial Plan," *Money Management Library*, 12 booklets. Household Finance Corporation, Money Management Institute, 1978.

Consumer and Career Mathematics by L. Carey Bolster, H. Douglas Woodman, and Joella H. Gipson. See especially pp. 68-87. Scott, Foresman and Company, 1978.

Business and Consumer Arithmetic by Milton C. Olson and A. E. McVelly. See especially pp. 255-278 and 297-310. Prentice-Hall, Inc., 1974.

Business Mathematics for the Consumer by Mearl R. Guthrie, William Selden, and Delbert Karnes. See especially pp. 223-244. Fearon-Pitman Learning, Inc., 1975.

C. RESOURCE UNITS, PAMPHLETS, BROCHURES, ETC.

Business Mathematics by Sally A. Loughrin, 1979. *Money* by Pamella Pruett. See especially pp. 54-60, 1979. *Applications of Algebra to Consumer Problems* by Edward Segowski and Roger Strong, 1979. Project Consumer, A Livonia Public Schools Project (with the Consumers' Education Office, Department of Health, Education, and Welfare, 1978).

Understanding Taxes, Publication 21, Publication 22, Publication 19, Publication 17. Department of the Treasury, current year.

The Tax Jungle. Oregon Department of Revenue, 1976.

10

Spending and Saving

INTRODUCTION

It seems fitting that the final chapter of a text concerned with money management be entitled "Spending and Saving," since the purpose, after all, of managing money is either to save it or to spend it. This chapter deals with the biggest single expenses any individual or family is likely to have during a lifetime: buying a home and buying an automobile. The topics covered here are probably the most comprehensive of all subjects studied in the *Guidebook*, and they require more sophisticated knowledge to be fully grasped by the consumer. Much of the task of the student is information gathering, and the purpose here is to provide direction as to what information is necessary to make an intelligent decision. Although students may be able to assume a role of real home buyer, they may not, in fact, be ready for that kind of decision. Caution should be exercised, therefore, not to overburden the student with more than he or she needs to know. It is better that the student feel comfortable and knowledgeable about the task of deciding so that when the time comes to really buy a home, for example, he or she will be confident that it is not an overwhelming task. Therefore, the topics are presented so as to provide the student with the basic direction for asking the right questions to get the answers most important in making appropriate decisions. The subjects are dealt with in a general way so that students may provide the specifics as they pertain to their own needs. In addition to home and car buying, this chapter explores obtaining credit and loans, securing savings, making investments, buying insurance, and planning retirement. At the end of the unit, a suggested culminating activity is to have students simulate an experience of receiving a large sum of money and, by careful spending (following specified directions), spend all their money in such a way (including several investments) that they will actually *increase* their net worth.

OBJECTIVES

If all the topics in this chapter are chosen by the teacher, the students should be able to:

1. Determine the dollar cost of credit or loan transactions (i.e., the amount of interest paid), given the amount of loan, monthly payments, and number of payments.

2. Determine the monthly payments on a loan, given the amount of the loan, amount of interest, and number of payments.

3. Determine the number of payments on a loan, given the amount of the loan, amount of interest, and monthly payments.

4. Calculate simple interest from the formula $i = prt$.

5. Use the formula $i = prt$ to calculate any one of the variables when the other three are known.

6. Determine and compare the effective rate of interest for single-payment, lump-sum loans and loans paid in monthly installments.

7. Determine the "true" interest rate using the constant ratio formula.

$$r = \frac{2mi}{p(1 + n)}$$

8. Understand and describe the difference between discounted loans and add-on-interest loans.

9. Use and understand the term *amortize*.

10. Understand three methods for determining finance charges for credit cards and charge accounts: a) adjusted balance b) previous balance, and c) average balance methods.

11. Compare the cost of credit or loans from banks, savings and loan associations, credit unions, finance corporations, pawnshops, and debt consolidation firms.

12. Determine and compare simple interest on savings paid annually, semiannually, quarterly, and daily.

13. Determine compound interest by table and by formula.

$$a = p(1 + r)^n$$

14. Understand the effective rate of compound interest.

15. Understand the salient features of higher-risk savings and income sources such as stocks and bonds.

16. Understand the salient features of insurances for life, home, health, and auto, including such terms as *premium* and *rates*, *face value* and *coverage*, *deductible*, *cash value*, and *dividends*.

17. Recognize various ways to plan for retirement security.

18. Understand the comparative cost and net value of renting or buying a home.

19. Determine cost of a home mortgage.

20. Determine cost of utilities.

21. Determine the costs of buying and maintaining an automobile.

CONTENT OVERVIEW

Topic I: Credit and Loan

"Enjoy now and pay later" is a familiar theme expressed by merchants eager to have people buy on *credit*. Credit is a form of a loan; it is the promise to pay later for something that is received immediately. However, there is a cost; it is called *interest*. Interest is rent paid for money that is used to pay the merchant who sold the goods and services on credit. The money is, in effect, loaned to the buyer so that he or she may obtain the purchase and the merchant may receive the money due. Various types of financial institutions handle this transaction (even the merchant might "loan" money to buy his or her goods and thus receive profit on the sale and interest on the credit). Because of the myriad of calculations, consumers are easily confused by credit plans. Irreputable firms will take advantage of the ill-prepared buyer, and the ancient motto, "Let the buyer be aware," is especially true for credit and loans. Although much of the language and computation may seem very confusing, it is not necessary for the consumer to go through complicated calculations to determine his financial obligations. Some simple calculation is often enough to allow the average buyer to "shop for credit" and make a wise selection of a loan or payment plan that is suitable and economical. This section will present some guidelines to help simplify the process. Information is also given for those who wish to have a greater understanding of how interest is determined, but this is optional. In addition, several consumer references, aids, and legal protections are identified as sources of support and assistance.

DOLLAR COST OF CREDIT AND LOANS

The simplest way to determine what the cost of credit is, is to carefully read the loan or credit contract. The federal Truth in Lending legislation requires that several things be disclosed to the buyer when making a contract. Although many items appear on the contract, there are only four that are necessary to determine the consumer cost: 1) *the amount of loan or amount of purchase financed*, 2) *the charge for credit (interest charge or finance charge)*, 3) *the amount of monthly payments, and* 4) *the number of monthly payments*. Thus, if a buyer is comparing the cost of credit from one company to another, these four items should be considered. Notice that the *interest rate* is not on the list. This is because the rate, depending upon how it is applied, can be confusing. Given the same purchase price and the same number of months to pay, it is the total *amount* of interest charged that is important. Of course, given this information, the total amount to pay back can be determined by adding:

the original purchase price (with tax) +
the finance charge = *total amount to pay.*

If the finance charge is not known, it can be determined from the other data. Similarly, if any one of the four factors is unknown, it can be determined from the other three.

Example 1: A purchase was made for $1872, tax included. A down payment of $372 was made resulting in monthly payments of $75.44 for 24 months. What was the finance charge?

Solution:

$1872 (*original price*) − $372 (*down*) = $1572 (*financed*)

$75.44 for 24 months = $75.44 × 24 = $1810.56 total paid

$1810.56 − $1572 = $238.56 finance charge

Example 2: Find the amount of monthly payments if the finance charge on a $900 loan is $108 and there are to be 15 monthly payments.

Solution:

$900 + $108 = $1008 total amount to pay back

$$15 \overline{)\$1008.00} = \$67.20 \quad \text{monthly payments are \$67.20}$$

Example 3: Find the number of payments if $500 is put down on a purchase of $8742, the monthly payments are $371.12, and the finance charge is $5118.32.

Solution:

Need to know the total amount to pay back.

$8742 − $500 = $8242 *amount financed*

$8242 + $5118.28 = $13,360.32 *amount to pay back*

$$\$371_{\text{x}}12_{\wedge} \overline{)\$13,360_{\text{x}}32_{\wedge}} = 36 \text{ } payments$$

Calculating Simple Interest

The basic formula for simple interest is:

$$i = prt$$

where i = total amount of interest, p = principal or amount that is financed, r = rate of interest, t = time to complete payments. This is the single most important interest formula, but in terms of determining payments it is not the whole story. There may be other factors to consider, and there are other ways to calculate interest. However, whenever the expression "simple interest" is used, the formula $i = prt$ is used.

Example 4: Find the simple interest for these.

    a) $2800 for 1 yr. at 9.25%
    b) $5400 for 3 yr. at 17-1/2%
    c) $450 for 7 mo. at 10%
    d) $2120 for 3 yr., 8 mo. at 12.7%

Solution:

$$i = prt$$

a) i = $2800 × 9.25% × 1 = $2800 × .0925 × 1 = $252
b) i = $5400 × 17-1/2% × 3 = $5400 × .175 × 3 = $2835
c) i = $450 × 10% × 7 mo. = $450 × .10 × 7/12 = $26.25
d) i = $2120 × 12.7% × 3 yr. 9 mo. = $2120 × .127 × 3-3/4 = $1009.65

The simple interest formula may be used to calculate any one of the four variables if the other three are known. (See Figure 10-1.) Thus, from $i = prt$ we can obtain:

$$prt = i$$

$$\frac{p\cancel{r}\cancel{t}}{\cancel{r}t} = \frac{i}{rt}$$

$$p = \frac{i}{rt} \quad \text{to calculate the principal}$$

$$\frac{\cancel{p}r\cancel{t}}{\cancel{p}t} = \frac{i}{pt}$$

$$r = \frac{i}{pt} \quad \text{to calculate rate of interest}$$

$$\frac{\cancel{p}\cancel{r}t}{\cancel{p}\cancel{r}} = \frac{i}{pr}$$

$$t = \frac{i}{pr} \quad \text{to calculate time}$$

Figure 10-1
A Memory Diamond

The diamond represents the formula

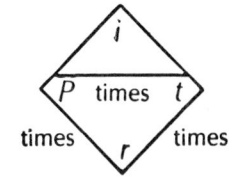

$$i = prt$$

Where i = interest, p = principal, r = rate, and t = time.

Note: The line under the i in the diamond means "divided by."

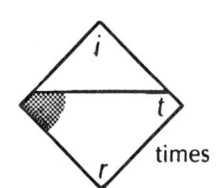

If p is to be determined, simply cover the corner at p and read the remainder of the diamond.

$$p = \frac{i}{rt}$$

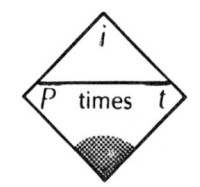

If r is to be determined, simply cover the corner r and read the remainder of the diamond.

$$r = \frac{i}{pt}$$

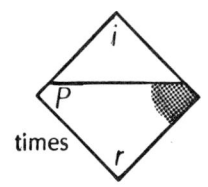

If t is to be determined, simply cover the corner t and read the remainder of the diamond.

$$t = \frac{i}{pr}$$

Ways to use simple interest. Discounted loans and add-on-interest loans, credit cards, charge accounts, auto loans, and home mortgages use $i = prt$ to calculate loan payments, but not all of them will have the same payment for the same amount borrowed. Some loan or credit contracts will reflect this and report an effective annual percentage rate.

First, let us look at a loan of $900 paid in two different ways. If a person borrows $900 and is charged $81 interest for one year, he is paying $81 rent to use the money for one year. At the end of one year, if he pays $981 in one lump sum, he has had use of all the money for one year, and his rate of interest can be determined as:

$$r = \frac{i}{pt}$$

$$r = \frac{81}{900 \times 1} = \frac{81}{900} = .09 \text{ or } 9\%$$

On the other hand, if the same person begins to pay back the $981 only 1 month after receiving it and continues to pay each month for one year, he does not have use of the full $900 for the entire year. Therefore, if he pays the same *amount* of interest, the effective *rate* of interest (on the money he has had use of) is greater. For example, the full $900 is available only for the first month; then the second month, $825 of the loan is useful because of payment of $75 (plus interest payment). The third month $750 is the amount of principal owed, then $675, $600, $525, $450, $375, $300, $225, $150, and finally $75 for the last month. The *effective annual rate*, r, is:

$$r = \frac{i}{pt}$$

where i = the actual interest paid ($81), p = the *average* amount of money (or credit) actually used during the length of time of the loan (t). The average amount of credit or money actually used is the average principal owed per month. Note that a short method for determining the average of any series which decreases at the same rate is to add the first and the last numbers in the series and divide by 2. Thus, the average principal here is ($900 + $75)/2 or $487.50. The effective rate, therefore, is:

$$r = \frac{i}{pt} = \frac{81}{\$487.50 \times 1} = 0.166 \text{ or } 16.6\% \text{ (approx.)}$$

The effective annual percentage rate is about 16.6%.

Parenthetically, it is interesting to note that whenever a loan payment schedule is made so that the amount of interest and amount of principal is the same for every payment, another formula may be used to calculate the effective annual percentage rate. Because the ratio of interest to principal is constant, this formula is called the constant ratio formula and is stated:

$$r = \frac{2mi}{p(1+n)}$$

where p, i, and r represent the same values as previously stated, m is the number of payment periods in one year (usually one per month or 12), and n is the number of payments (altogether).

Another way to demonstrate why there is so much confusion with credit financing is to compare the *discount* loan and the *add-on-interest* loan. When a discount loan is made, the lender takes off all of the interest from the principal immediately right off the top, and the borrower receives only the difference. However, the borrower must pay back the entire principal. If someone needs $800, therefore, he must borrow $800 *plus* the amount of interest he will be charged in order that he may receive $800 cash (e.g. if interest is $144, the borrower must borrow $944 in order to receive $800).

An *add-on-interest* loan provides the borrower with the full principal at the onset of the loan. The interest is added to the principal, and the borrower must repay the entire principal plus interest.

SPENDING AND SAVING

Example 5: Val Kainoh blew her top when she heard about the monthly payments required to secure a $900 add-on-interest loan at 9% for one year. She was much more pleased with the payments for a $900 discounted loan at 9% for one year. Did the discount loan really cost her less?

Solution:

Add-on-Interest Loan		Discount Loan	
$900	principal and amount received	$900	principal
+ 81	interest	− 81	interest
$981	to be repaid	$819	amount received

$$\frac{\$81.75}{12\,)\overline{\$981.00}} \text{ monthly payment} \qquad \frac{\$75}{12\,)\overline{\$900}} \text{ monthly payment}$$

Both loans cost the same, but one provides more money to be used each month until the loan is paid. (Just for fun, calculate the effective rate of interest for each.)

Credit cards and charge accounts offer still another maze of options that could catch the unwary, unaware. There are essentially three ways used to calculate interest for charge customers. Interest is calculated monthly as a certain percent of the:

1. adjusted balance

2. previous balance

or 3. average balance

Example 6: Teddy Ruffrider made several purchases on his San Juan Charge Card. The total came to $600. He also charged an identical amount for the identical period to two other charge accounts: Missda Charge and Battree Charge. Each reported a 1.7% interest rate per month. The San Juan Charge used the adjusted balance method, Missda Charge used the previous balance, and Battree Charge used the average daily balance. Teddy made a $500 payment to each on the fifteenth day of the paying period after his first billing. He compared his second billing from each and found the following:

San Juan Charge (adjusted balance) $1.70 interest fee

Missda Charge (previous balance) $10.20 interest fee

Battree Charge (average daily balance) $5.95 interest fee

This is how it works. The adjusted balance method calculates interest on the basis of 1.7% of $100, since the balance was adjusted for the $500 payment ($600 − $500 = $100). The previous balance method calculates on the basis of 1.7% of $600, the previous balance before he made a payment. The average daily balance method calculates on the basis of 1.7% of $350, the average amount still owed each day of that month (15 days @ $600 and 15 days @ $100). (At this time one might wonder why it's called simple interest.)

Mortgages and other high-dollar loans are computed by simple interest, too. However, they are calculated so that each monthly payment consists of a reduction in the principal and the interest on the new value of the principal as reduced each month. Interest is charged on the unpaid balance. Thus, each month's equal payment includes an increasing amount that goes toward the principal and a decreasing amount that goes to pay the interest. This is sometimes called an *amortized* loan. The borrower does not pay interest on the full amount of the loan for the entire period of the loan as he does with the discount or add-on-interest loans. Tables such as that shown in Figure 10-2 are usually used to determine monthly payments for amortized loans. The precise amount of each payment that goes toward interest and principal varies; an example is shown in Figure 10-3. This breakdown of interest and principal is called an amortization schedule and can be obtained from a local bank or mortgage institution for a nominal fee. (See also Topic IV.)

Figure 10-2
Monthly Amortized Loan Payments
at 12%

Amount	Term 10 Years	15 Years	20 Years	25 Years	30 Years	35 Years	40 Years
5,000	71.74	60.01	55.06	52.67	51.44	50.78	50.43
10,000	143.48	120.02	110.11	105.33	102.87	101.56	100.85
15,000	215.21	180.03	165.17	157.99	154.30	152.34	151.28
20,000	286.95	240.04	220.22	210.65	205.73	203.11	201.70
25,000	358.68	300.05	275.28	263.31	257.16	253.89	252.13
30,000	430.42	360.06	330.33	315.97	308.59	304.67	302.55
35,000	502.15	420.06	385.39	368.63	360.02	355.45	352.98
40,000	573.89	480.07	440.44	421.29	411.45	406.22	403.40
45,000	645.62	540.08	495.49	473.96	462.88	457.00	453.83
50,000	717.36	600.09	550.55	526.62	514.31	507.78	504.25
60,000	860.83	720.11	660.66	631.94	617.17	609.33	605.10
75,000	1076.04	900.13	825.82	789.92	771.46	761.67	756.38

Figure 10-3
Interest-Principal Breakdown of Monthly Payments
for 25-Year, $30,000 Loan at 8%

Payment	Amount	Interest	Principal	Balance
1	231.55	200.00	31.55	29,968.45
24	231.55	194.79	36.76	29,181.80
48	231.55	188.44	43.11	28,222.17
72	231.55	180.98	50.57	27,096.63
96	231.55	172.24	59.31	25,776.45
120	231.55	161.98	69.57	24,228.04
144	231.55	149.96	81.59	22,411.93
168	231.55	135.85	95.70	20,281.84
192	231.55	119.31	112.24	17,783.51
216	231.55	99.90	131.65	14,853.22
240	231.55	77.14	154.41	11,416.32
264	231.55	50.44	181.11	7,385.24
288	231.55	19.13	212.42	2,657.21
300	226.52	1.50	225.02	.00

Finance companies, pawnshops, and debt consolidation firms are other sources of credit. The consumer is wise if he or she is careful about obtaining credit or loans from any source; the primary question ought to be: "What will it cost me overall when I've made my last payment?" Several laws have been enacted to help the buyer, and there are agencies at the federal, state, and local levels that are eager to help. Some of the legislation includes the Equal Credit Opportunity Act, the Fair Credit Reporting Act, the Fair Credit Billing Act, the Fair Debt Collection Practices Act, among others.

SPENDING AND SAVING

Additional credit-related topics that can be explored are: shortcuts in calculation and other computations (e.g., Rule of 78s), secured loans, life insurance loans, how to file a credit complaint, and more.

Topic II: Savings and Investing

"How much of my income should I save?"
"What sort of gain can I expect from my savings?"
"What's the difference between simple and compound interest?"
"What savings institution is best for me?"
"Are stocks and bonds worth the risk?"

These are the kinds of questions often asked by consumers. The answers depend on many variables, and although this section will not provide all the solutions, many of the variables will be explored. The topics are presented in a general way so that individuals may continue to investigate their interests through the many resources cited.

SIMPLE INTEREST

We have already examined the simple interest formula, $I = prt$, as it pertains to loans and credits. This same formula is used to determine interest for many savings accounts. The amount of interest actually accumulated is, however, dependent upon when the interest is paid. Three methods of interest payment will be explored: the semiannual and quarterly balance and withdrawal methods and the daily balance method.

Example 7: Consider the savings account that pays interest semiannually at 6%. If the account is opened with $1000 on January 2, is followed by another deposit of $375 on March 1 and a withdrawal of $117 on April 1, how much interest is earned for the first half year on June 30? What is the account balance on June 30?

Solution: The rules governing semiannual and quarterly balance and withdrawal accounts are such that interest is paid on money deposited during an interest period from the first of the month if the deposit is made before the tenth of the month. Each month represents 1/12 of a year. Also, money that is withdrawn during an interest period does not earn any interest. Therefore, in this example:

a) *Interest is paid for 6 months on the $1000 deposited on January 1.*

$$I = 100 \times .06 \times \frac{1}{2} = \$60 \times \frac{1}{2} = \$30.00$$

b) *$375 was deposited on March 1 and $117 was withdrawn on April 1. Therefore $375 − $117 or $258 was on deposit for 4 months or 1/3 of a year.*

$$I = \$258 \times .06 \times \frac{1}{3} = \$15.48 \times \frac{1}{3} = \$5.16$$

c) *The total interest is:*

$$\$30.00 + \$5.16 = \$35.16$$

d) *The account balance on June 30 is:*

$1000.00
+ 375.00   *deposit*
$1375.00
− 117.00   *withdrawal*
$1258.00
+ 35.16   *interest June 30*
$1293.16   *new balance*

Similar calculations are done on quarterly paid interest.

Example 8: Find the interest and new balance on the following account that pays at a rate of 6% using the daily balance method:

January 2	open account with	$850
February 14	withdraw	$798
March 1	deposit	$ 64
March 31	interest paid	?

Solution:

Daily balance accounts are governed by these procedures.
If the interest is paid on March 31, the quarterly rate must be divided by 4. Therefore:

$$6\% \div 4 = 1.5\% \text{ or } .015$$

The quarterly rate is then divided by the number of days in the quarter, in this case 90. Thus:

$$.015 \div 90 = .0001666$$

The interest is determined by this factor times the amount of money on deposit times the number of days the money was held. Consequently:

$850 was on deposit for 45 days.
$850 × .0001666 × 45 − $6.37 *interest to February 14.*
$850 − $798 = $52 was on deposit from February 14 through March 1 or 15 days.
$52 × .0001666 × 15 = $.13 *interest from February 14 through March 1.*
$52 + $64 = $116 was on deposit from March 1 through March 31 or 31 days.
$116 × 0001666 × 31 = $.60 *interest from March 1 through March 31.*

The total interest is:
$6.37
+ $ .13
+ $ .60
$7.10

The new balance is:
$850.00
− $798.00   *withdrawal*
$ 52.00
+ $ 64.00   *deposit*
$116.00
+ $ 7.10 *interest*
$123.10   *new balance*

COMPOUND INTEREST

When interest is added, a new balance is created from which to calculate the interest for the next pay period. Interest is, therefore, earned on the interest as well as on the principal. Whenever interest is paid on interest, the interest is said to be *compounded*. Compound interest, of course, is greater the more often it is compounded. Usually compound interest is determined by using tables such as that shown in Figure 10-4. However, it can also be calculated by the formula:

COMPOUND INTEREST TABLE
Showing How Much $1 Will Amount to at Various Rates

Periods	½%	1%	1¼%	1½%	2%	2½%
1	1.005000	1.010000	1.012500	1.015000	1.020000	1.025000
2	1.010025	1.020100	1.025156	1.030225	1.040400	1.050625
3	1.015075	1.030301	1.037971	1.045678	1.061208	1.076891
4	1.020151	1.040604	1.050945	1.061364	1.082432	1.103813
5	1.025251	1.051010	1.064082	1.077284	1.104081	1.131408
6	1.030378	1.061520	1.077383	1.093443	1.126162	1.159693
7	1.035529	1.072135	1.090851	1.109845	1.148686	1.188686
8	1.040707	1.082857	1.104486	1.126493	1.171659	1.218403
9	1.045911	1.093685	1.118292	1.143390	1.195093	1.248863
10	1.051140	1.104622	1.132271	1.160541	1.218994	1.280085
11	1.056396	1.115668	1.146424	1.177949	1.243374	1.312087
12	1.061678	1.126825	1.160755	1.195618	1.268242	1.344889
13	1.066986	1.138093	1.175264	1.213552	1.293607	1.378511
14	1.072321	1.149474	1.189955	1.231756	1.319479	1.412974
15	1.077683	1.160969	1.204829	1.250232	1.345868	1.448298
16	1.083071	1.172579	1.219890	1.268986	1.372786	1.484506
17	1.088487	1.184304	1.235138	1.288020	1.400241	1.521618
18	1.093929	1.196147	1.250577	1.307341	1.428246	1.559659
19	1.099399	1.208109	1.266210	1.326951	1.456811	1.598650
20	1.104896	1.220190	1.282037	1.346855	1.485947	1.638616

Periods	3%	4%	4½%	5%	5½%	6%
1	1.030000	1.040000	1.045000	1.050000	1.055000	1.060000
2	1.060900	1.081600	1.092025	1.102500	1.113025	1.123600
3	1.092727	1.124864	1.141166	1.157625	1.174241	1.191016
4	1.125509	1.169859	1.192519	1.215506	1.238825	1.262477
5	1.159274	1.216653	1.246182	1.276282	1.306960	1.338226
6	1.194052	1.265319	1.302260	1.340096	1.378843	1.418519
7	1.229874	1.315932	1.360862	1.407100	1.454679	1.503630
8	1.266770	1.368569	1.422101	1.477455	1.534687	1.593848
9	1.304773	1.423312	1.486095	1.551328	1.619094	1.689479
10	1.343916	1.480244	1.552969	1.628895	1.708144	1.790848
11	1.384234	1.539454	1.622853	1.710339	1.802092	1.898299
12	1.424561	1.601032	1.695881	1.795856	1.901207	2.012196
13	1.468534	1.665074	1.772196	1.885649	2.005774	2.132928
14	1.512590	1.731676	1.851945	1.979932	2.116091	2.260904
15	1.557967	1.800944	1.935282	2.078928	2.232476	2.396558
16	1.604706	1.872981	2.022370	2.182875	2.355263	2.540352
17	1.652848	1.947901	2.113377	2.292018	2.484802	2.692773
18	1.702433	2.025817	2.208479	2.406619	2.621466	2.854339
19	1.753506	2.106849	2.307860	2.526950	2.765647	3.025600
20	1.806111	2.191123	2.411714	2.653298	2.917757	3.207135

Source: Table from *Stein's Refresher Mathematics* (Allyn and Bacon, Inc., 1980), page 595.

$$b = p(r + 1)^n$$

where b is the new balance after interest, p is the principal, r is the rate of interest, and n is the number of interest periods. The expression $(r + 1)^n$ means that $(r + 1)$ is used as a factor n times. The exponent, n, tells how many times $(r + 1)$ will "multiply itself." Thus, $(r + 1)^2 = (r + 1)(r + 1)$ and $(r + 1)^3 = (r + 1)(r + 1)(r + 1)$, and so on.

Example 9: What is the new balance after 2 years of a $1000 savings certificate that earns 6% interest compounded quarterly?

Solution:

$$b = p(r + 1)^n$$

$p = \$1000, r = 6\% \div 4$ or $1.5\%, n = 2$ yr. $\times 4 = 8$

$b = \$1000 (.015 + 1)^8$
$= \$1000 (1.015)^8$
$= \$1000 (1.015)(1.015)(1.015)(1.015)(1.015)(1.015)(1.015)(1.015)$
$= \$1000 (1.126493)$
$= \$1126.49$ New balance

Note that the result of $(1.015)^8$ is the same as the figure given on the table in Figure 10-4 for period 8 and 1-1/2% interest. The table is simply the calculation of $(r + 1)^n$. The new balance is obtained by multiplying the principal by the factor shown on the table.

A comparison of the various methods for calculating interest will clearly show that for a specific annual interest rate, interest compounded more frequently (i.e., has more interest-paying periods, n) will result in more interest accumulated. For this reason, savings institutions will advertise their annual rates and their effective rates. If bank A has the same annual rate as bank B but bank A has a higher effective rate, it means that bank A compounds its interest more frequently. Consequently, it pays to investigate the number of payment periods as well as the interest rate when deciding upon where to save. This option is available for small deposits in regular savings accounts as well as for larger deposits such as saving certificates. Savings certificates usually pay a higher rate of interest than regular savings but must be kept in the account longer to earn the higher amount.

Savings institutions include commercial banks, mutual savings banks, savings and loan associations, and credit unions. Each type offers its own specialized services but operates in a similar manner. All require some form of deposit or withdrawal slip (similar to checking accounts), and each pays interest on the money deposited in savings. Credit unions differ somewhat in that they refer to savings as *shares*, since the effect is to invest in the credit union corporation. Earnings on the shares are also called *dividends* to carry on the ownership connotation. All four institutions offer insured savings through some agency of the federal government. These insuring agencies are the Federal Deposit Insurance Corporation (FDIC), the Federal Savings and Loan Insurance Corporation (FSLIC), and the Federal Share Insurance Agency (FSIA).

SECURITY WITH RISK

Savings accounts, particularly because of the insurance agencies mentioned, offer a great deal of security without a great deal of risk. However, the yield or return on the investment is small compared with the yield potential in other investment areas, particularly stocks. On the other hand, savings accounts do not carry the risks of stocks; and just as there is a potential of high yield in other investments, there is also the potential of complete loss. What follows are brief, generalized descriptions of some of the terms and processes involved with investment securities such as stocks and bonds.

Stocks are issued by a company and sold to investors. Ownership of stock means part ownership of the company, and therefore the value of the stock fluctuates as does the profitability of the company. *Bonds* are promissory notes (sort of IOUs) issued by a business or government to raise cash. They are contracts promising that the money will be paid back by a certain date with interest. Certain federal government bonds are called Treasury notes. Bonds are less risky than stocks. *Mutual funds* are investment companies that pool the resources of large numbers of people and invest the money in stocks and bonds. Individuals who buy into a mutual fund are indirectly investing in stocks but are actually investing in a company, which in turn invests their money along with that of several thousand others. *Annuities* are similar to mutual funds but are less risky. They provide a guaranteed income after a long period of investment and are used primarily as retirement income sources. Because of the potential risk involved, it is wise to consult an investment counselor or broker before investing in securities. Brokers, however, charge a fee based upon the buying and selling price of the stock. The fee is similar to a salesman's commission because it is paid as a percent of the total sale.

Gain or loss. There are several factors that affect the amount of profit (or loss) experienced from securities investment.

1. The broker's fee or commission.
2. In the case of bonds, the amount of simple interest to be expected, how often it is to be paid (similar to the considerations for savings accounts), and the date of maturity.
3. Initial cost of the stock or bond.
4. Dividends likely to be declared (based upon previous years).
5. Annual earnings of the company. A "price-to-earnings" ratio is often used to compare various companies.
6. The market value of the stock or bond. Securities may be sold at any time after purchase. If they are in demand, a price much higher than the purchase price may be offered, providing a huge profit. Or, a stock or bond may lose appeal and sell for less than the price at which it was purchased. The key questions in the securities market are when to buy and when to sell.

In order to determine buying and selling prices, one must be acquainted with the methods of quoting prices for various stocks and bonds. Stock prices are quoted in dollars and eighths of a dollar. Thus, a stock that sells for 66-1/8 costs $66.125 per share. These prices are listed (or "quoted") at broker's offices and in the newspapers. Usually current prices (which fluctuate regularly), opening and closing prices for the day, the net change, the highs and lows for the year, and the previously declared dividend are listed with the stock. Fractions are used extensively.

Bonds, on the other hand, are listed to report their guaranteed interest rate (simple) per year; their current actual interest rate; the number of trades made; the highest, lowest, and closing price of the day; and the net change. Fractions and percents are used in bond quotations. (The high, low, close, and net change are all reported as percents of the face value.)

Other investments. In addition to seeking profit from stocks and bonds, other sources of investment include buying real estate (both land and building), going into business for oneself, and education. Indeed, education represents an investment of and in the most important person in your life—yourself.

Topic III: Insurance and Retirement

Insurance is a form of saving, but its primary prupose is to provide financial protection for family and business survivors. It is a cooperative method of sharing risks. The value of insurance can be easily seen when a disaster strikes. If an uninsured home burns to the ground, it must be rebuilt on funds that are borrowed at high interest rates. However, every home does not burn down. Most people live long enough to see their family grow up, and the majority of people do not have major auto accidents. The value of insurance cannot be measured by the monetary returns actually received. It must be measured by the amount of protection or potential return in case of need. On the other hand, there are cost-and-return comparisons that can be made in terms of premiums, protection, cash value, and other factors.

This section provides a broad overview and gives references for such insurance and retirement topics as life, health, and home insurance; social security; annuities; and private pension plans. (Auto insurance is covered in Topic V.) In some cases, all or part of the cost of these plans is paid by an employer as a fringe benefit. However, the real question to the consumer is, "What does it cost me?" The purpose of this section is to shed some light on the matter so that buyers can shop and compare prices more effectively.

LIFE INSURANCE
Life insurance is primarily a financial protection for the survivors (called *beneficiaries*) upon the death of the insured. There are four basic types of life insurance: *term, ordinary, limited payment,* and *endowment.* All require a certain payment, called a *premium*, based upon the amount of protection, called a *face of policy*, and other factors. In each type of insurance, the beneficiary receives the face amount in case of death of the insured. When comparing prices, it is most important that these "other factors" be considered, because they may contribute as much as any factor to the value of the policy.

Term life. A policy that provides protection only for a specified period of time (or term) is a term life policy. Premium rates and protection may vary with the age of the insured. There is no *cash value* whatsoever to the insured. The only people to benefit from a term policy are the beneficiaries in case of death to the insured.

Ordinary life. "Straight life" or "whole life" or "permanent life" are other names for ordinary life policies. These policies provide protection until the death of the insured. An important difference between ordinary and term life is that ordinary life does provide some cash benefit to the insured, while term life does not. First, ordinary life provides a buildup of cash value for the policy that continues to grow as long as the insured lives. The cash value is dependent upon the premiums and face value and the length of time the insurance is in force. This cash value means that the insured may stop payment on his policy and 1) cash it in to receive the cash instead of the protection, or 2) receive paid-up insurance protection based upon the cash value. The cash value is never nearly as great as the face value of the policy. The insured may also use the cash value to 3) borrow from the insurance company at a specified low rate of interest. The entire cash value may be borrowed, and only the interest must be paid for each year of the loan as long as the premiums are continued to be paid. In the event of death before the loan is paid off, the amount of the loan is subtracted from the face value. These low-interest loans are an important consideration for consumers. For example, compare $500 borrowed from an insurance policy at 5% to $500 borrowed from a bank at 14%.

A second cash benefit available as an option to the insured is the opportunity to earn dividends. This is called *participating* insurance. It is a form of investment in the insurance company, and although dividends are guaranteed not to go below a specified rate, they may rise with the profitability of the company.

Premium rates are usually the same for the entire period the insurance is in force. In fact, there are several characteristics of all nonterm policies: 1) uniform premium rate (except under special plans) for the force of the policy, 2) cash value, and 3) opportunity to earn dividends. Of course, the more benefits selected, the higher the premiums.

Limited payment life. This policy provides protection for the life of the insured, but the premiums are anticipated to be paid up before death. That is, the premiums are scheduled so that the insured pays for 20 years or some specified period and does not pay again. In the meantime, he is always insured for the face amount.

Endowment life. Endowment policies are forms of savings for the insured. They are often purchased for children and other dependents. The premiums and protection are only for a specified period of time. During that time, if the insured dies, the beneficiaries receive the face value of the policy. After the specified period, if still alive the policyholder receives the face amount.

Determining consumer costs. Premiums are determined by consideration of various factors by a mathematician specializing in insurance. That person is called an actuary. He or she is responsible for setting up the tables that spell out the premium rates for the various forms and amounts of protection. These tables are based upon *mortality* tables that show the number of expected deaths per thousand at each age level. The premium cost is based upon 1) the deaths expected at a certain age level, 2) the number of policyholders this company has at that age level, and 3) the face value of the policy.

Premiums alone, however, do not determine consumer value. Other factors discussed here must also be considered. Many states offer consumer guides for life insurance purchases that rate several companies according to premiums, face value, cash value, paid-up insurance, dividends, and a specified after-tax interest rate. These factors are used to establish an index that can be used as a guide to help compare insurance policies. (Term life policies are easiest to compare because protection and premium are the only dollar factors. Complicated policies that contain several forms of protection in various combinations are the most difficult to scrutinize.)

PROPERTY INSURANCE

Protection of the home and the objects in the home is provided under property insurance. There are two basic forms: one primarily for renters, the other for property owners. Within each of these, there are various options. Coverage includes real and personal property lost or damaged owing to certain specified causes (hail, wind, fire, theft, vandalism, etc.). Cost factors include amount of coverage, type of dwelling (frame homes cost 10%-15% more to insure), area in which dwelling exists, and more. Rates are reported on tables per $100 or $1000 of coverage.

There are certain coverage limits. The house structure itself is, of course, the biggest single item. Other property is covered on the basis of a certain percent of the house. For example, other structures on the premises like a garage, a toolshed, and the like (called *appurtenant structures*) are covered at 10% of the amount shown for the dwelling. Loss of furnishings, appliances, clothing, books (called *unscheduled personal property*) are covered at 50% of the dwelling coverage. Expenses for living away from home while repairs are made receive protection equal to 20% of the dwelling protection. And physical damage to other persons' property is covered at 1% of the dwelling amount. Thus, a property owner whose home is insured for $50,000 also has insurance of $5000 for his garage, $25,000 for his belongings, $10,000 for his living expenses, and $500 for damage to property of others.

One must be cautious, however, because there are certain other limits. First of all, the quoted amounts are maximums; that is, coverage is "up to" those amounts, and an insured person may receive far less depending on the extent of damage and other factors. The buyer ought to be well acquainted with the "other factors" before buying what might appear to be a cheaper insurance but is, in fact, more expensive. To illustrate, most insurance companies require that the dwelling be insured for at least 80% of its market value. Because of inflation, this percentage is often difficult to determine, so once an original amount is established, the insurance company will automatically adjust the coverage each year based upon the rate of inflation. Of course, the premium is also adjusted. If, however, the property is not insured at 80% of its market value, there is a clause in the policy requiring that reimbursement for any loss be reduced by a factor based upon the different amounts. That is to say, if a $100,000 home is insured for only $40,000, then a loss claim of $500 will be reduced as follows.

$$\frac{\text{Actual face value}}{\text{Expected face value}} \times \text{any claim amount}$$

80% of $100,000 or $80,000 is the expected face value

or
$$\frac{\$40,000}{\$80,000} \times \$500 = \frac{1}{2} \times \$500 = \$250$$

This is called a *coinsurance clause*, because to replace the $500 claim both the company and the insured must contribute a certain amount.

Furthermore, coverage is usually limited to the current value of the item (not the replacement value but the original value minus a certain amount for depreciation) and a certain *deductible* amount. A $100 deductible policy means that the insurance company will pay only for that loss which exceeds $100. A $125 loss pays only $25, while a $98 loss pays nothing. Deductibles can range from $50 to $500 or more. All of these limitations in coverage should be carefully examined before a policy is purchased. An index similar to that used for life insurance is also used by various states to rank property insurances.

HEALTH INSURANCES
Medical, hospital, dental, disability, and other forms of health insurance are offered to many workers as part of their employment. Some people do not receive this fringe benefit, and others must supplement their coverage at work with additional insurance. Generally, group plans are less expensive than individual plans. There are, just as with property insurance,

limits on various coverages. Doctor's fees that do not meet the insurance schedule must be paid out of the pocket. There are usually deductible amounts that must be paid before the insured is reimbursed, and most companies will pay only a certain percent (commonly 80%) of the remainder of the bill. Workmen's compensation coverage is usually paid by the employer, and it provides protection from loss due to accident or illness that results from the person's employment. State laws require certain basic coverage with specified limits (customarily $20,000 plus 2/3 of the weekly wages). These are often supplemented by additional insurance at the discretion of the employer or the option of the employee.

Other insurances. Almost anything can be insured, and there are even tailormade policies that are written for everything from airplanes to zoos. The wise consumer will realize that insurance is not the same as savings and that the value of insurance cannot be measured entirely by cost in and payment out.

RETIREMENT

The various sources of retirement income include savings, insurance, investments, social security, private pensions, and the like. *Social security* is the most prevalent monetary source for retirees. Payments into the social security fund have already been discussed. Benefits include retirement income, medicare, and survivors' income for households where death has left dependent children. The amount received from social security is dependent upon the worker's average yearly earnings and the number of years employed. These factors determine the insured's *primary insurance amount* (PIA). The monthly benefits to the worker and his dependents are calculated from the PIA.

Private retirement plans, like social security, are usually funded by employee and employer contributions. The money collected is normally used to invest in various mutual funds or bonds, so that it is not a part of the company's general operating revenue. However, some retirement plans do operate in the latter way, and there is always the danger that the pension fund will dry up if the company falls upon hard times. Pensions can also be in the form of profit sharing or stock bonuses.

Annuities, often sold by insurance companies, are contracts whereby a set amount of money is agreed to be paid in return for a guarantee that starting at a certain age a specified amount of money will be received for life. Annuities can be started at almost any age, in almost any amount. There are single premium annuities, annual premium annuities, and so on. There is a certain amount of protection in case of death both before and after the annuity is collected, and many annuities compare favorably with other savings programs.

IRA and *Keogh* plans have the advantage of providing a tax shelter. Both are similar concepts; IRA stands for "Individual Retirement Account" and is for individuals only. The Keogh plan is for the self-employed. Both are available only if the workers are not covered by another pension plan. They are tax sheltered because the amount paid into the plan each year may be subtracted from the wage earner's gross income so as to reduce his tax liability. In addition, earnings are not taxed either. There are limits, of course. IRA rules have allowed 15% or $1500, whichever is less, to be placed in an IRA (exceptions are allowed for certain married couples), and the funds cannot be withdrawn before age 59-1/2. If they are withdrawn earlier, a substantial tax penalty results (in addition to an interest penalty). Taxes must eventually be paid on the amount withdrawn at retirement, but at that time overall income is reduced, and the individual is probably in a lower tax bracket. Indeed, many people will find themselves without any tax liability at that time, because social security benefits are not taxable and there is an additional exemption for anyone 65 years old or older. When compared with a regular savings account or certificate, these advantages must be weighed against a higher interest rate.

Topic IV: Housing Costs

Buying, renting, paying utilities, property taxes, mortgage costs, and buying home furnishings all contribute to housing costs. When considering renting a place to live, consumers need to be aware of current local costs; average rents for various areas; whether or not a security deposit is required; and who is responsible for utilities, repairs, and general maintenance. Recently, consumer advocates have helped to establish laws guaranteeing tenants' rights, and current as well as prospective tenants should try to keep abreast of such developments. Public housing offers many the opportunity to live in rental property that is well maintained. Availability of public housing to families is dependent upon the income and size of the family. People in need of good low-cost housing should contact their local housing authority.

BUYING

Several cost factors need to be considered when determining whether to buy or to rent. One must consider the cost of tying up several thousand dollars that must be used as a down payment. That money could be invested where it will earn interest and perhaps gain substantially in value. On the other hand, the inflation rates of the past several years have surpassed the interest rates most likely to be earned in investments, and home buying has become a good investment itself. However, the following costs must be included when considering the total cost of home ownership.

1. *Closing costs.* These include attorney's fees, title insurance, recording of deed and mortgage, escrow fees (which are collected in anticipation of expenditures such as mortgage insurance premiums, taxes, property insurance premiums, and special assessments), property appraisal and inspection, survey of property, and points. Some of the escrow fees are repeated year after year, and they may be held by the bank in escrow so that the bank may pay them, or they may be paid by the homeowner himself. Points are paid only once, at the sale of the property. They are charged at the time of the sale by the lending institution to increase their yield on the money they are lending. The tighter the money market, the higher the points. One point equals 1%, so that if a home sells for $50,000, and 3 points are charged, 3% of $50,000 or $1500 must be paid to the lender. On Department of Housing and Urban Development (HUD) and Veterans Administration (VA) loans, the seller pays the points. On *conventional* (private lender) loans, the buyer pays.

2. *Down payment.* This, of course, reduces the amount of the mortgage. The greater the down payment, the lower the monthly payments. The mortgage company will usually require a certain percent of the selling price as a down payment. This is often a factor that limits many buyers.

3. *Insurances.* As mentioned in regard to the closing costs, there are two insurance costs. One is property (homeowner's) insurance. This is almost always required as part of the sale. The other is a form of term life insurance that pays the mortgage lender (who is the beneficiary) the amount of the mortgage in case of the death of

SPENDING AND SAVING

the home buyer. This assures the survivors the opportunity to keep their home without further cost. HUD and VA always require mortgage insurance. Conventional lenders may or may not require it.

4. *Repairs and decorating.* (See Chapter 8.) This expense is most difficult to determine and varies from house to house, depending upon the age, quality, and condition of the structure.

5. *Depreciation.* If a property loses value with age, it is said to depreciate. With recent inflation it does not appear that property loses value. However, the question is whether or not the market value of the property has kept pace with the rate of inflation. The property value may rise at a rate lower than the increase in property prices in another location. In terms of the purchasing power, the property may have actually depreciated even though the price is up.

6. *Mortgage costs.* Mortgages are amortized loans. The interest is charged on the unpaid balance. Interest rate and duration of mortgage loan determine the monthly payment. The monthly payment is always the same, but the portion that pays the interest grows smaller and the portion that pays the principal grows larger. Figure 10-3 shows a sample of an 8% amortized loan at various periods of the loan. Note the differences in the interest to principal ratio. These payments are determined by consulting tables such as those presented in Figures 10-2 and 10-3. The tables are based on the simple interest formula $i = pr$. To determine the interest for any one month, simply multiply the unpaid balance of the principal by the rate. Buyers often look with chagrin at the interest to principal ratio and shake their head in dismay at the small dent made on the principal in the early years of their mortgage. More shocking, however, is the total amount spent for the home after paying for 25 or 30 years. Depending upon the amount of the mortgage loan, the interest rate, and the duration of the loan, a home could cost 2, 3, or 4 times the original cost by the end of the mortgage term. (A sample printout of a simulated mortgage showing the interest-principal ratio is available at many banks for a nominal fee.)

Example 10: Find the amount of interest paid and the total cost of mortgage loans that have these terms and monthly payments (all are 12% loans).

A $60,000 home with a down payment of $10,000 gives
a $50,000 mortgage @ $1112.23 for 5 years.
@ $717.36 for 10 years.
@ $600.09 for 15 years.
@ $550.55 for 20 years.
@ $526.62 for 25 years.
@ $514.31 for 30 years.
@ $507.78 for 35 years.
@ $504.25 for 40 years.

Solution:

First, determine the number of months represented by the years shown. Second, multiply the number of months by the monthly payment to obtain the total spent on the mortgage. Add $10,000 (the down payment) to obtain the total cost of the home or subtract $50,000 to obtain the total interest paid.

The 5-year mortgage is:

5 X 12 = 60 months
60 X $1112.23 = $66,733.80 total spent on mortgage.
$66,733.80 + $10,000 = $76,733.80 total cost of $60,000 home.
$66,733.80 − $50,000 = $16,733.80 interest paid.

The home on a 10-year mortgage costs $96,083.23, of which $36,083.20 is interest.
The home on a 15-year mortgage costs $118,016.20, of which $58,016.20 is interest.
The home on a 20-year mortgage costs $142,132.00 of which $82,132.00 is interest.
The home on a 25-year mortgage costs $167,986.00, of which $107,986.00 is interest.
The home on a 30-year mortgage costs $195,151.60, of which $135,151.60 is interest.

What are the costs for 35 and 40 years?

Several variations of the standard mortgage have been suggested. Some are currently available; others are awaiting approval from state and federal authorities. These variations include *deferred loans,* which allow the mortgagee to pay only the interest for a short period at the beginning of the loan. *Open-end loans* allow certain other loans to be made periodically on the same property. *Graduated monthly payment loans (GMPs),* allow smaller payments to be made at the beginning of a loan when the mortgagee is earning less, and then increases the payments at later dates in the consideration that the mortgagee will be earning more. And there are also *renegotiable loans,* which can be renegotiated at higher or lower interest rates, depending upon the changes in the market. (As an example, mortgage rates during the 20-year period from 1960 to 1980 fluctuated from 6% and below to 16% and above.)

Still another consideration is the prepaid penalty imposed by some lenders if the mortgage is paid off early. The lender charges a certain percent (1/2% to 5%) of the mortgage balance in order to recoup some of the interest he may have lost owing to the early payment.

7. *Taxes.* There is both a tax advantage and a tax disadvantage to buying a home. Taking the good news first, the federal income tax can be reduced by itemized deductions that include deductions for state and local property taxes paid and interest paid on the mortgage. These two items alone are enough to give most people an advantage with homeownership. Even state income taxes may be reduced (in some states) by local property taxes paid. (Points paid by the buyer may also qualify as tax deductions.) These are tax advantages not available to renters, who may pay a monthly rent equal to, or even greater than, a homeowner's mortgage payment.

On the other hand, *real estate* (or *property*) *taxes* can be burdensome, even with the income tax writeoff. Property taxes are based upon the *assessed valuation* of the property and the *tax rate.* The tax rate is determined by the ratio of the taxes needed by the community to the total assessed valuation of all property in that community. Thus:

$$\text{Tax rate} = \frac{\text{Taxes needed}}{\text{Assessed valuation}}$$

Consequently, if a community requires $760,500 from property taxes and the total assessed valuation of property in the community is $23,400,000, the tax rate should be:

$$\text{Tax rate} = \frac{\$760,500}{\$23,400,000} = .0325$$

The tax rate is commonly expressed in mills (a mill equals $.001 or 1/10¢). The tax rate determined in this example, therefore, can be expressed as 32.5 mills. That is to say that the property in this community is taxed at a rate of $32.50 for each $1000 of assessed valuation. Other ways to express the same rate are:

$.0325 per dollar of assessed valuation

$3.25 per 100 dollars of assessed valuation

The property tax is determined by multiplying the assessed valuation of a property by the tax rate. For example, a property assessed at $27,000 and a tax rate of 32.5 mills has a property tax of:

$27,000 × $32.5 per $1000

$$\$27{,}000 \times \frac{\$32.5}{\$1000} = \frac{\$877500}{\$1000} = \$877.50$$

If the rate is expressed in other terms, it is important to show the ratio (per $1 or per $100, etc.) in the calculation.

UTILITIES AND FUELS

Gas, electricity, water, telephone, oil, or coal expenses make up a large percent of the total cost of running a home. Energy-conscious Americans are now seeking alternative means to energy sources for the home. Tax incentives make the potential savings even more attractive, and new technologies are being developed to make energy alternatives such as solar and wind more and more practical. The greatest impediment, the cost to install new equipment, is overcome somewhat by tax credits for installation of energy-saving devices and materials. Credits are given for such things as insulation, storm windows, thermostat controls, and furnace replacements. Contact the IRS and obtain Form 5695 and Publication 903, *Energy Credits for Individuals*, for more information.

Energy consumption is measured in various ways. Utilities that enter the home pass through simple meters that record the amount of electricity, gas, or water being used. These meters are either the direct-reading digital type or the multiple-dial type. Wise consumers will take the time to learn to read them so as to determine their energy consumption. The units commonly employed to measure the use of various utilities are:

electricity is measured in kilowatt hours (kwh.)

gas is measured in hundred cubic feet (cu. ft.)

water is measured in gallons or hundred gallons

Meters. Simply stated, a dial meter consists of several circular dials with numbers from 0 to 9 and a pointer that revolves with utility use. Each dial represents a single number according to the pointer position (the lower number is read if the pointer is between two numbers). The numbers are read from left to right. The amount of the utility (gas, electricity, water) used for a month is determined by subtracting the previous month's reading from the reading for the current month.

Direct-reading digital meters are read in the same way an automobile odometer is. The numbers are read from left to right. Some meters may be in tens or hundreds of units, and the reading must be multiplied by these factors to be accurate.

Electricity. The first step in calculation is to read the meter. Simplified meter-reading instructions are available from the utility or may be obtained by consulting a good practical mathematics text such as *Stein's Refresher Mathematics* (Allyn and Bacon, Inc., 1980). Utilities may also provide up-to-date information on how to calculate the bill. Generally, there is a flat service rate that is charged every month. This may be from $2 to $6 or more. Thereafter, a certain rate is charged for each unit (kwh.) used. That rate increases with increased usage. For example, the energy charges may be:

$.05 per kwh. for the first 400 kwh.

$.06 per kwh. for the next 400 kwh.

$.07 per kwh. for all use over 800 kwh.

In addition, there is a fuel adjustment allowance, which enables the utility to raise (or lower) the price according to the cost of fuel it purchases to produce the electricity. This "surcharge" varies from month to month, but a typical month's fuel adjustment surcharge might be about $2.50. Other charges might be an operating-and-maintenance surcharge, a system availability surcharge, and state and local taxes. Special rates are often given for hot-water heaters, air conditioning, residential space heating, senior citizens, and outdoor protective lighting.

Consumption of electricity can be controlled more effectively if people know how much energy each appliance uses. Federal law now requires that all appliances display their estimated energy costs for one year's operation. Air-cooling systems must carry an energy efficiency ratio (EER), which is a measure of the BTU rating (cooling power) compared with the wattage: (BTU rating)/Wattage. The higher the EER, the lower the energy cost. Watts are a measure of the power the appliance will use, and a simple calculation will determine the cost of running that appliance. Since the bills are calculated in kilowatt hours, the watts can be converted to kwh. as follows.

$$a\ kwh. = \frac{Watts}{1000} \times 1\ hour\ (kilo\text{-} = 1000).$$

a 100-watt light bulb burning for 1 hour uses:

$$\frac{100}{1000} \times 1\ hour = \frac{1}{10}\ or\ .1\ kwh.$$

a 1000-watt hairdryer operated for 20 minutes uses:

$$\frac{1000}{1000} \times \frac{1}{5}\ hour = \frac{1}{5}\ kwh.\ or\ .2\ kwh.$$

a 5000-watt clothesdryer run for 30 minutes uses:

$$\frac{5000}{1000} \times \frac{1}{2}\ hour = 2.5\ kwh.$$

The cost to operate each of these can be determined by multiplying by the average rate per kwh. (about 5.8¢ per kwh. using the figures previously reported and a 1000-kwh. usage). Based on a 5.8¢ ($.058) average cost per kwh., the appliances illustrated would use:

100-w. light bulb = .1 kwh. X $.058 = $.0058 (slightly more than 1/2¢)

1000-w. hairdryer = .2 kwh. X $.058 = $.0116

5000-w. clothesdryer = 2.5-kwh. X $.058 = $.145

Gas and water bills are calculated in a similar way. There is usually an initial flat-rate service charge and a charge based upon the quantity used. The greater the quantity, the greater the rate. The increased rate for high-volume users is instituted to encourage conservation. Previous rate schedules that charged less for high-volume users were encouraging greater waste. Fuel adjustment surcharges are also made for gas usage. Heating bills (gas) are dependent upon the climate, the amount of insulation, the efficiency of the furnace, the size of the home, and the height of the ceilings. Careful attention to those factors which can be controlled (insulation, closing off rooms, replacing inefficient units, etc.) can lead to substantial savings. Water leaks, dripping faucets, and running taps unnecessarily contribute to excessive water consumption. Expenses to repair these items should be balanced against the potential savings in the utility bills.

Telephone rates are based upon a monthly flat rate and a charge for calls over a certain allotment. Rates for long-distance calls are dependent upon the distance of the call, the zone to which they are made, the time of the day they are made, and the length of the call.

Other service charges for the home may include oil or coal for heat, soft-water treatments, cable television, and pest-control service. All of these services contribute to the overall cost of running a home.

HOME FURNISHINGS

Anyone who has purchased a brand-new home and furnished it from "scratch" knows the great expense involved here. Recall that the property insurance provision for home furnishings is 50% of the value of the dwelling. Mathematics plays a big role in determining needs and costs for items that require measurement such as carpeting, drapes, wallpaper, paint, and the like. Each room has some large ticket items, but the kitchen has the biggest expenses with a stove, refrigerator, dishwasher, and so on. Wise consumers will plan carefully so as not to be faced with too many replacements of these items at any one time.

Topic V: Buying and Maintaining a Car

An automobile is probably the second-largest expense for most consumers. It may, in fact, be the largest expense when one considers that home purchase is actually an investment that at least can bring a return on resale to enable purchase of a comparable home. When an auto is sold, the receipts do not come near the value of a new purchase, and people buy and sell several cars over a lifetime. In spite of the financial negatives, America has long had a fascination with the automobile. Cars have been more than just a way to get there; they have represented freedom and independence to millions. Three areas of cost will be considered in this section: the costs to buy, to finance, and to operate an automobile.

INITIAL COST

The first consideration in buying a car is to determine the general price range one is willing to spend within. Most buyers find that they must be flexible with their initial estimate because of the options and styles they may wish to purchase. Car buyers ought to look closely at what they want and determine what those choices cost. Considerations of car size, body style, power train, and special options are often so numerous that they are overwhelming. The following is a comprehensive list of choices that contribute to the initial and overall cost of an automobile purchase. Obviously, the selections determine the cost of the auto, but smart buyers will be cautious to select what is needed and affordable.

1. Choices in size include minicompact, subcompact, compact, intermediate, and full size.
2. Body-style considerations are sedans of two, three (hatchback), and four doors, two- and four-door hardtops, convertibles, station wagons, vans, specialty and sports cars, light trucks, and pickups.
3. Mechanical options are engine size, transmission, gear ratio, suspension systems, brakes, steering, power options (seats, windows, trunk lid, antenna), air conditioning, tinted glass, radio and stereo options, automatic speed control, rear window defroster-defogger, tilt-steering wheel, sun roof, remote mirrors, hood- and gas-tank locks, heavy-duty battery, intermediate windshield wipers, additional lights, trailer tow, theft alarm, wheels, tires, and so on.
4. Other options include color and trim, special paint and paint design, vinyl roof, wheel covers, type and upholstery of seats, floor covering, and special instrumentation.
5. Fuel economy is a major consideration.
6. Overall quality of the auto, reliability of the manufacturer, safety record and performance on safety tests, and previous owner evaluations are all important to car buying.
7. Test driving over a predetermined course, personally testing handling ease, control, and general "feel" of the auto, is essential.
8. Dealer reliability and convenience may be very important.
9. Warranty provisions and maintenance-and-service requirements are most certainly determining factors. However, these are difficult to pin down to specific costs because they are highly speculative, being based upon manufacturer's recommendations and the record of previous models.
10. Trade-in value of the car currently owned. Often it is financially to the owner's advantage to sell a car on his or her own rather than sell it to a dealer who intends eventually to profit from the purchase.
11. Fees, taxes, registrations, and licenses are required expenditures.
12. Finally, the cost of financing, as already demonstrated, adds considerably to the total cost of the auto.

There are several other suggestions of things to do when in the market for a car, such as: 1) shop around, investigate several makes and models; 2) use consumer's guides to determine average price and dealers' cost- to- selling-price ratio; 3) use the sticker price and cost- to- price ratio to bargain with the dealer.

COST TO OPERATE

Fuel economy, maintenance and service, insurance, fees, taxes, licenses, and depreciation all contribute to the overall cost of ownership of a car. Fuel economy was discussed in Chapter 5. Maintenance and service includes regularly occurring expenses such as insurance, registration (auto license), driver's license, parking fees, finance charges, and perhaps property ownership taxes. Other expenses that occur with regularity are gasoline and oil costs, tire purchases, repair and upkeep expenses, and the like.

Insurance expenses form a high percentage of the operating costs of an automobile. All states have a minimal requirement for all drivers, and most drivers carry additional coverage. Good drivers enjoy better rates than those whose driving records are poor. In fact, poor-risk drivers sometimes are required to pay insurance premiums that approach (or even exceed) the value of the car. Owing to their higher accident rate, younger drivers, particularly males, are penalized with high premiums even without having a bad record personally. Many states have a rating system for auto insurance that compares coverage with premiums. This rating can help consumers determine their best insurance buy. Generally speaking, insurance companies determine their rates based upon such considerations as driving record, kind of car, age, sex, marital status, who is allowed to drive the car, area in which driver lives, income, vehicle use, and health or accident coverage already in force on the driver.

Although state rating systems and general knowledge of insurance are helpful, it is difficult to compare premiums because of the various choices available for coverage. For example, many states have some form of "no-fault" insurance that needs to be understood completely so that additional coverage may be determined. These include bodily injury liability, property damage liability, comprehensive perils, collision, medical payments, uninsured motorists, accidental death, road service, and towing.

Depreciation is a major operating expense when one considers that such a large investment is likely to deteriorate to pure scrap in less than ten years. It is said that an average car loses as much as one-third of its original value as soon as it rolls off the showroom floor. The contribution of this and other expenses to overall cost of ownership is often expressed in terms of cost per year or per mile. The cost per mile figure is an average, and estimates vary according to several factors, particularly mileage (the greater the mileage on a newer car, the lower the average per mile cost). Projected into the future, these costs are expected to exceed $.30 per mile and soon soar to more than $.40 per mile.

LEARNING EXPERIENCES

THE DIAGNOSTIC SURVEY AND INDEPENDENCE SCALE

- Use the **Diagnostic Survey** on Reproduction Page 82 to estimate each student's degree of competence to instruction in the topics of this chapter.

- Upon completion of the Survey, use the **Independence Scale** on Reproduction Page 83 to identify those skills that need to be strengthened by further instruction and practice.

- Utilize the results of these two instruments to determine the appropriate assignments from the following list.

Topic I: Credit and Loan

- Use the **Diagnostic Survey,** problems 1-7, as examples of the types of problems one is likely to encounter with credit purchases and borrowing money. Use separate lessons for each skill as identified on the Independence Scale.

- Obtain literature on the truth in lending laws from the federal government or local consumers' groups. Ask students to investigate the value of the laws and their effectiveness. (e.g., How are complaints handled? Are there any prosecutions? Is money refunded to bilked consumers?)

- Have students obtain several credit contracts for various sources locally. Ask students to review each contract and determine if they comply with the truth in lending laws.

- Have students identify various lending institutions and obtain from them information regarding types of loans and methods of determining payments (e.g., lump sum or monthly payments).

- Have students write to several credit-card companies and obtain their credit-card applications and contracts. Ask students to scrutinize these contracts to determine how interest and payments are determined (e.g., adjusted, previous, and average balance methods).

- Obtain current amortization schedules from a lending institution. Include tables, formulas, and a sample printout of a typical home mortgage, showing the principal and interest for each payment during a period of 20 to 30 years.

- Have students scan newspapers and periodicals for ads of debt consolidation firms. Ask students to investigate several firms and determine how they operate and how much interest they actually charge. Compare these firms with banks, credit unions, and savings and loan associations.

- Ask students to write to various government and consumer related groups concerning the Equal Credit Opportunity Act, the Fair Credit Reporting Act, the Fair Credit Billing Act, the Fair Debt Collection Practices Act, as well as local and state legislation.

- Write to the Federal Reserve Bank of Philadelphia for a pamphlet on the Rule of 78s and review this interesting concept with students.

Topic II: Savings and Investing

- Use the **Diagnostic Survey**, problems 21-40 and 48-55, as examples of the types of problems one is likely to encounter with simple interest and compound interest savings.

- Have students scan newspapers and call local savings institutions to determine the annual interest rates and the true or effective interest rate (often called "effective yield"). Ask students to verify the effective rate by showing how it is determined.

- Have students prepare reports on commercial banks, mutual savings banks, credit unions, and savings and loan associations showing the services offered, funding, and ownership of each. Ask students to detail the distinguishing differences among the four.

- Have students investigate and report on the FDIC, FSLIC, and FSIA.

- Have students investigate and report the history, function, and everyday operation of stocks and bonds markets. Ask that they particularly bring out the procedures and policies with regard to quotations, buying and selling, fluctuations in market values, and making a profit. Have them include other markets in addition to the New York Stock Exchange.

- Create a simulated investment situation whereby each student invests $5000 in 10 different stocks and/or bonds. Have the students keep track of their investment each day, selling and buying new securities each day, paying brokers' fees and commissions, and finally determining their net worth (showing gain or loss from the $5000 start).

- Have students investigate and report on personal education as an investment, showing comparative costs at various schools, colleges, universities, training programs, and so on; income opportunities as a result of training; income loss while completing training; and the like.

Topic III: Insurance and Retirement

- Ask students to write to the state insurance bureau to obtain their ratings of cost to protection for various forms of insurance offered within the state. Have students report how the ratings are determined, the value of such ratings to consumers, and factors not included in the ratings but important to consumers.

- Have students investigate and report on life insurance programs showing methods of determining premiums; the work of actuaries; the relevance of probability; the use of mortality tables; the way to read premium, face-value, and nonforfeiture tables, and so on.

- Have students report the comparative costs, protection, and other benefits of term, whole life, limited-payment life, and endowment policies in various forms. Comparisons should be made within the same company as well as among various companies.

- Ask students to obtain samples of property insurances in various forms, comparing cost with protection and showing how to read rate schedules. Include both renter's and homeowner's policies.

- Review with the students the coverage necessary for a $97,000 home. Include recommended protection such as the 80% factor, the amount of coverage limits on garage, home furnishing, living expenses, and damage to other property. Also explain the pitfalls of not carrying complete coverage such as coinsurance, deductible, and depreciation clauses.

- Obtain and review information about various health-care programs. Have students investigate and report on health-care coverage available as a benefit to workers. Ask students to debate pros and cons of private versus government programs.

- Ask students to investigate and report on the costs and benefits of the social security system. Include methods of determining benefits using sample primary insurance amounts and other simulated data.

- Have students investigate and report on various pension plans offered by local private industries.

- Obtain information regarding annuities, how they are funded, the investment required, and the benefits returned; and have students report advantages and disadvantages of annuities over regular savings and investment programs.

- Ask students to investigate and report on IRA and Keogh tax-sheltered savings, demonstrating the tax advantages with specific examples as well as comparing these programs with other investment programs.

Topic IV: Housing Costs

- Use the **Diagnostic Survey**, problem 48, and the information on amortized loans and mortgages in Topics I and IV of the "Content Overview" of this chapter as well as Figures 10-2, 10-3, and 10-4 as guides for instruction about mortgage loans.

- Have students investigate and report on rental property available in the local area considering all options and features available. Include information on tenants' rights; tax benefits for renters; security deposits; utilities; repairs; general maintenance; and low-income, subsidized housing.

- Have students consider the interest lost owing to down payment when purchasing a home. Examine the pros and cons from various angles, including income potential from interest on savings and potential profit from investment of the money rather than using it for a down payment.

- Have students simulate the process of home purchase over the period of several years following step- by- step and including 1) down payment, 2) closing costs, 3) insurance, 4) repairs and decorating, 5) depreciation, 6) mortgage costs (interest), and 7) taxes.

- Obtain a sample printout of a mortgage loan (available from a loan institution for a nominal charge) and demonstrate how the ratio of the interest to the principal decreases over the term of the mortgage.

- Use Reproduction Page 84 and have students determine the amount of interest paid and the total cost of a home from the information given.

- Have students investigate and report on other than standard mortgages including deferred loans, open-end loans, graduated monthly payment loans (GMPs), and renegotiable loans.

- Have students inquire at local and state government agencies to determine property tax rates. Ask that they examine the total assessed valuation and the total tax collected from property taxes to determine the tax rate expressed in mills.

- Ask students to investigate and report on the income tax advantages of homeownership in terms of interest and property taxes paid.

- Use Reproduction Page 85, Part One, and have students determine the tax rate for the taxes needed and the assessed valuations shown.

- Use Reproduction Page 85, Part Two, and have students determine the property taxes to be paid based upon the information given.

SPENDING AND SAVING 211

- Have students investigate and report local utility billing procedures and rates.
- Refer to a text like *Stein's Refresher Mathematics* (Allyn and Bacon, Inc., 1980), pages 568-570, for practice assignments in reading meters and calculating utility bills.
- Have students read their own utility meters at home for a period of several days and then calculate the cost based upon actual rates in the area.
- Ask students to investigate and report on energy credits from federal and state governments.
- Ask students to investigate and report the energy used by various appliances and determine the cost to operate each at the current rates.
- Obtain information about the amount of gas required to heat a specific volume from a certain temperature to another temperature. Have students measure the volume of each room in their homes and determine the amount and cost of gas needed to heat their homes to that temperature.
- Have students make lists of all the pieces of furniture and appliances needed to furnish a seven-room home. Include floor covering and window decor. Ask students to find the cost of these items using catalogs, newspaper ads, and store surveys. As an additional assignment, students might be asked to write checks and keep a checking account record for all purchases.

Topic V: Buying and Maintaining a Car

- Using the list of choices presented in the text as a guide, have each student make a list of personal choices for automobile purchase. Have students visit new car showrooms to check sticker prices to obtain the cost for each selection. Ask that they consult consumer guides to find the cost- to- price ratio for various models and determine the likely bargaining range.
- Have students investigate and report on state regulations regarding legal limits and requirements for auto loans, fees, and licenses for owning and operating an automobile, as well as auto insurance requirements. Include state ratings and comparisons of auto insurance premiums in the state and local area.
- Have students investigate and report on finance charges for auto loans at various lending institutions. Ask that they compare initial and overall charges.
- Have students obtain current auto insurance premiums and ask that they investigate and report what those premiums were in the past. Have them trace the earliest insurance requirements and costs, compare them with current rates, and determine the reasons for such increases. Local auto clubs and associations are good sources for this information.
- Ask students to consult consumer guides for car buying such as *Car Buying Made Easier*, published by the Ford Motor Company, and other publications available through consumer and safety groups. Have students compare these with advertisements designed to attract buyers.

- Have students determine the depreciation of several models of cars over the past 1, 2, 3, 4, 5 years and more by consulting consumer guides and visiting used-car lots. Be certain that students factor inflation rates into the determination of actual depreciation in terms of current cost to replace an old model with a new one.
- Have students investigate and report on the latest data regarding the overall cost and cost per mile to own a car. Ask that they compare these data to those available from previous years, showing the increase in cost. Have them break down the cost to determine which contributing factors (original price of auto, gasoline prices, insurance premiums, repair costs, etc.) rose the most over the past several years.

ASSESSING ACHIEVEMENT OF OBJECTIVES

Ongoing Evaluation

The extent to which students have mastered the concepts covered under the five topics in this chapter can be measured by any of the activities assigned to class members individually.

Culminating Activity

An interesting and informative way to draw all experiences together at the end of this study is to create a simulation activity whereby each student is given a large sum of "money" (the amount may vary according to the student's performance on assignments) and instructions to *spend all the money in such a way as to improve his or her net worth.* For example, students may begin with $25,000 and be asked to spend all of it on 1) stock and bond investments, 2) insurance policies, 3) buying a home, 4) buying home furnishings, 5) buying a car, 6) purchasing a wardrobe of clothing for all seasons, 7) paying for food costs for one year, 8) going on vacation, 9) going to school, and 10) any other miscellaneous expenses. For credit purchases such as home and car have them figure down payment and monthly payments for one year. Teachers may wish to require "checks" to be written and recorded for these purchases. At the conclusion, each student's net worth can be determined based upon the dollar value of what was purchased or the exchange or bargaining power gained because of the purchase (e.g., education will probably yield greater income for the future).

Final Evaluation

For an overall evaluation of the students' mastery of the concepts in this chapter, if all topics in the chapter have been taught, a test constructed directly from the "Objectives" listed at the beginning of the chapter can be used. As an alternative, one might consider using the **Diagnostic Survey** as a final test.

RESOURCES FOR TEACHING ABOUT SPENDING AND SAVING

Below is a selected and annotated list of resources useful for teaching the topics in this chapter, divided into audiovisual materials, games, and print materials. Addresses of publishers or distributors can be found in the alphabetic list in Appendix B.

Audiovisual Materials

A. FOR LOW ABILITY AND SPECIAL EDUCATION STUDENTS

Apartment Hunting Series, 5 filmstrips and 5 audio-cassettes. Interpretive Education, 1977.

Applying for Credit, filmstrip and audio-cassette. Interpretive Education, 1977.

Automobile Insurance, filmstrip and audio-cassette. Interpretive Education, 1977.

Budgeting Series, 5 filmstrips, 5 audio-cassettes, and 20 student workbooks. Interpretive Education, 1977.

Consumer Education, 5 filmstrips, 5 audio-cassettes, and 20 student workbooks. Interpretive Education, 1977.

Credit Buying Series, 5 filmstrips and 5 audio-cassettes. Interpretive Education, 1977.

Dynamite Wheels, 4 audio-cassettes and 20 reading books. Interpretive Education, 1977.

How to Buy an Automobile, 5 filmstrips and 5 audio-cassettes. Interpretive Education, 1977.

Loans, filmstrip and audio-cassette. Interpretive Education, 1977.

B. FOR GENERAL MATHEMATICS STUDENTS

"How's Your Budget Math?"; "How's Your Savings Math?"; "How's Your 'Easy Credit' Math?"; "How's Your Borrowing Math?"; "How's Your Car Buying Math?"; "How's Your Insurance Math?"; "How's Your Housing Math?" *Consumer Math Cassettes* produced by F. Lee McFadden. The Math House, 1977.

Savings Accounts, film, 13 minutes. Consumers Union of the United States, 1978.

Budgeting, film, 11 minutes, Aetna Life and Casualty.

Managing Your Money, 4 filmstrips and 4 cassettes. Teaching Resources Films, 1974.

Consumer Education Series, 6 filmstrips and cassettes. Doubleday Multimedia, 1972.

"Money Talks," "Be Credit Wise," "Wheels, Deals and You," *Money Management Filmstrip Library*—kit, 4 filmstrips and 4 cassettes, workbooks, and transparencies. Household Finance Corporation, Money Management Institute, 1977.

Installment Buying, film, 13 minutes. BFA Education Media, 1971.

Money for Sale, film, 18 minutes. AIMS Instructional Media Services, 1974.

Consumer Studies: The Price of Credit, filmstrip. Guidance Associates, 1972.

Using a Charge Account, cassette, teacher's guide, student worksheets, and tests. Media Materials, Inc., 1978.

Protection Through Insurance, cassette, teacher's guide, student worksheets, and tests. Media Materials, Inc., 1978.

Economics: The Credit Card. 9-1/2 minutes. BFA Educational Media, 1971.

How to Buy a Used Car, 12 minutes. General Motors Corporation.

Insurance, 16 minutes. Aetna Life and Casualty, 1973.

Credit, 18 minutes. Aetna Life and Casualty, 1973.

Buying (Housing), 13 minutes. Aetna Life and Casualty, 1977.

Your Credit Is Good—Unfortunately, 10 minutes. Parthenon Pictures, 1976.

Savings Accounts, 13 minutes. Consumers Union of the United States, 1977.

Money Management and Financial Planning, 19 minutes. Aetna Life and Casualty, 1976.

Insurance: What's It All About, 25 minutes. Filmfair Communications, 1975.

I'll Only Charge You For The Parts, 10 minutes. Parthenon Pictures, 1976.

Games

Big Deal! Creative Teaching Associates, 1976.

Consumer. Western Publishing Company, 1971.

Cutting Corners developed by John Schatti. The Math Group, Inc., 1977.

"Managing Your Money." Credit Union National Association, 1969, 1970.

Budgeting Game. EMC Publishing, 1970.

Print

A. FOR LOW ABILITY AND SPECIAL EDUCATION STUDENTS

Useful Arithmetic, Volume I, by John D. Wool and Raymond J. Bohn. See especially pp. 18-24. Frank E. Richards Publishing Company, Inc., 1972.

Useful Arithmetic, Volume II, by John D. Wool. See especially pp. 45-72. Frank E. Richards Publishing Company, Inc., 1972.

Banking, Budgeting, and Employment by Art Lennox. See especially pp. 22-42. Frank E. Richards Publishing Company, Inc., 1979.

The Bank Book by John D. Wool. See especially pp. 8-27 and 65-89. Frank E. Richards Publishing Company, Inc. 1973.

About the Home, Buying Guides, Family Development Series by Stephen S. Udvari and Janet Laible. See entire book. Steck-Vaughn Company, 1978.

Skills for Everyday Living, Book 1, by David H. Wiltsie. See especially pp. 116-129 and 137-145. Motivational Development, Inc., 1976.

Skills for Everyday Living, Book 2, by David H. Wiltsie. See especially pp. 92-97, 102-126, 130-143. Motivational Development, Inc., 1978.

Using Money Series, Book IV Earning Spending and Saving, by John D. Wool. See especially pp. 32-33 and 46-62. Frank E. Richards Publishing Company, Inc., 1973.

It's Your Money, Book 1, by Lloyd L. Feinstein and Charles H. Maley. See especially pp. 75-98. Steck-Vaughn Company, 1973.

It's Your Money, Book 2, by Lloyd L. Feinstein and Charles H. Maley. See especially pp. 18-92. Steck-Vaughn Company, 1973.

Math for Today and Tomorrow by Kaye A. Mach and Allan Larson. See especially pp. 87-210. J. Weston Walch, Publisher, 1968.

Scoring High in Survival Math by Tom Denmark. See especially pp. 12-21. Random House, Inc., 1979.

Michigan Survival by Betty L. Hall and David Landers. Also available for all states. See especially pp. 100-118, 143-148, 152-155. Holt, Rinehart and Winston, 1979.

Mathematics for Today, Level Red, by Saul Katz, Ed.D.; Marvin Sherman; Patricia Klagholz; and Jack Richman. See especially pp. 85-94. Sadlier-Oxford, 1976.

Mathematics for Today, Level Blue, by Edward Williams; Saul Katz, Ed.D.; and Patricia Klagholz. See especially pp. 129-168. Sadlier-Oxford, 1976.

Mathematics for Today, Level Orange, by Wilmer L. Jones, Ph.D. See especially pp. 210-225. Sadlier-Oxford, 1979.

Mathematics for Today, Level Green, by Wilmer L. Jones, Ph.D. See especially pp. 168-193. Sadlier-Oxford, 1979.

B. FOR GENERAL MATHEMATICS STUDENTS

Stein's Refresher Mathematics, Seventh Edition, by Edwin I. Stein. See especially pp. 568-613. Allyn and Bacon, Inc., 1980.

Trouble-Shooting Mathematics Skills, Basic Competency Edition, by Allen L. Bernstein and David W. Wells. See especially pp. 346-384. Holt, Rinehart and Winston, 1979.

Consumer Mathematic with Calculator Applications by Alan Belstock and Gerald Smith. See especially pp. 174-317. McGraw-Hill Book Company, 1980.

Consumer Math, A Guide to Stretching Your Dollar, by Flora M. Locke. See especially pp. 53-271. John Wiley and Sons, Inc., 1975.

Consumer Mathematics, Third Edition, by William E. Goe. Activities book available. See especially pp. 1-48, 119-158, 241-300, 331-406. Harcourt Brace Jovanovich, 1979.

Mathematics for Today's Consumer by Jack Price, Olene Brown, Michael Charles, and Miriam Lien Clifford. Compiled from selections from *Mathematics for Everyday Life* and *Mathematics for the Real World*. See especially pp. 52-85, 100-139, and 176-194. Charles E. Merrill Publishing Company, 1979.

Mathematics for Everyday Life by Jack Price, Olene Brown, Michael Charles, and Miriam Lien Clifford. See especially pp. 100-139, 176-194, 312-331. Charles E. Merrill Publishing Company, 1978.

Mathematics for the Real World by Jack Price, Olene Brown, Michael Charles, and Miriam Lien Clifford. See especially pp. 52-85. Charles E. Merrill Publishing Company, 1978.

Mathematics for Daily Living by Harry Lewis. See especially pp. 1-98, 207-246, 264-476. McCormick-Mathers Publishing Company, 1975.

Mathematics Plus! Consumer, Business & Technical Applications by Bryce R. Shaw, Richard A. Denholm, and Gwendolyn H. Shelton. See especially pp. 133-200. Houghton Mifflin, 1979.

"Your Financial Plan," "Managing Your Credit," "Your Housing Dollar," "Your Home Furnishings Dollar," "Your Equipment Dollar," "Your Automobile Dollar," and "Your Savings and Investment Dollar," *Money Management Library*. Household Finance Corporation, Money Management Institute, 1978.

Consumer and Career Mathematics by L. Carey Bolster, H. Douglas Woodman, and Joella H. Gipson. See especially pp. 100-173, 196-259, 280-319, 360-381. Scott, Foresman and Company, 1978.

Business and Consumer Arithmetic by Milton C. Olson and A. E. McVelly. See especially pp. 117-148, 163-250, 265-278, 327-338, 353-372. Prentice-Hall, Inc., 1974.

Business Mathematics for the Consumer by Mearl R. Guthrie, William Selden, and Delbert Karnes. See especially pp. 155-222. Fearon-Pitman Learning, Inc., 1975.

Consumer Credit by Elsie Fetterman and Ruth Jordan. See entire book. Charles A. Bennett Company, Inc., 1976.

C. RESOURCE UNITS, PAMPHLETS, BROCHURES, ETC.

Applications of Algebra to Consumer Problems by Edward Segowski and Roger Strong, 1979. *Personal Banking Services* by James Cooper, 1979. *Commercial Banking Services and an Overview of Credit* by William R. McQuesten, 1978. *Caring for Your Car* by Clayton L. Evenden, 1978. *Consumer Credit* by Dan Kachnowski, 1979. *Business Mathematics* by Sally A. Loughrin, 1978. *Buying Furniture* by Patricia Arent, 1978. *Buying Home Furnishings* by Joyce Daughtery, 1978. *Consumer Rights and Responsibilities in the Marketplace* by Patricia Nazarian, 1979. *Budgeting* by James Cooper, 1978. Project Consumer, A Livonia Public Schools Project (with the Consumers' Education Office, Department of Health, Education, and Welfare).

Know Your Pension Plan, Department of Labor, 1973.

Saving for Retirement, Bankers Systems, Inc., 1977.

Insurance Consumer Alert (1-15), Michigan Department of Commerce, 1980.

Consumer's Guide To: Health Insurance in Michigan, 1980. *Consumer's Guide To: Life Insurance in Michigan*, 1980. *Consumer's Guide To: Auto Insurance in Michigan*, 1980. *Consumer's Guide To: Home Insurance in Michigan*, 1980. Michigan Department of Commerce.

A Family Guide to Auto and Home Insurance (one to a teacher), 1980. *Automobile Insurance Leaflet, Home Insurance Leaflet*, 1980. *Careers in Property and Liability Insurance*, 1980. *Chart on Home Insurance* (one to a classroom), 1980. *Chart on Auto Insurance* (one to a classroom), 1980. *Chances are . . .*, 1980. *Educator's Guide to Teaching Auto and Home Insurance* (one to a teacher), 1980. *Sample Insurance Policies/Intro-*

ductory Book, 1980. *Insurance Insights, List of Motion Picture Films,* 1980. Insurance Information Institute.

Insurance. The NOW Corporation, 1976.

Consumer Handbook to Credit Protection Laws, The Rule of 78's, and various other credit pamphlets from the Federal Reserve System, 1979.

All Kinds of People. Credit Union National Association, Inc., 1979.

Shell Answer Books. Shell Oil Company, 1976-1980.

Car Buying Made Easier. Ford Motor Company, 1980.

You and Your Community Bank. The NOW Corporation, 1976.

APPENDIX A

General Interest References

The following references are offered as resources of a general nature to the consumer mathematics teacher.

Audiovisual Materials

Lifeskills. Catalog of audiovisual materials. Michigan Products, Inc., 1980/82.

Consumer Education for Special Needs. Ten kits on various topics. EMC Publishing, 1979.

CTES Comprehensive Consumer Education Program. Six kits on various topics. EMC Publishing, subject to revision 1980, 1981.

Print

A. BOOKS, HARDBOUND AND SOFTBOUND

Buying with Sense by Carol King. A 1980 addition to the Pacemaker Practical Arithmetic Series for low ability students. Fearon-Pitman Learning, Inc., 1980.

Math for Everyday Living by Darlene Cornell, Judith Lawson, Timothy Lowe, and Thomas Paul. Ideal Publications, 1980.

Mathematics for the Consumer by Roswell E. Fairbank and Robert A. Schultheis. A textbook for the general mathematics student. South-Western Publishing Company, 1980.

Consumer Skills by Irene Oppenheim, Ph.D. Charles A. Bennett Company, Inc., 1977.

Consumer Credit by Elsie Fetterman, Ph.D., and Ruth Jordan. Charles A. Bennett Company, Inc., 1976.

Personal Finance for Consumers by Benjamin M. Trooboff and Fannie Lee Boyd. General Learning Press, 1976.

Smart Shopping and Consumerism by Rubie Saunders. Franklin Watts, Inc., 1973.

Overcoming Math Anxiety by Sheila Tobias. W. W. Norton and Company, 1978.

Family Money Management, Family Development Series About the Home, by Stephen S. Udvari and Janet Laible. Steck-Vaughn Company, 1978.

A Department Store in the Classroom by Sally R. Campbell. Sears, Roebuck and Company, 1979.

Age of Adaptation by Sally R. Campbell. Sears, Roebuck and Company, 1979.

Buying Guide Issue: Consumer Reports. Consumers Union of the United States, published for current year.

Brady on Bank Checks by Henry J. Baily. Warren, Gorham, and Lamont, 1978.

Classroom Ideas from Research on Computational Skills by Marilyn N. Suydam and Donald J. Dessart. National Council of Teachers of Mathematics, 1978.

B PAMPHLETS, BROCHURES, RESOURCE GUIDES, AND SOURCES

Consumer Economic Education Guidelines. Michigan Department of Education, 1979.

Consumer Education Guidelines and Activities. Kentucky Department of Education, 1979.

Consumer's Resource Handbook. The White House Office of the Special Assistant for Consumer Affairs, December, 1979. Consumer Information Center.

Teacher's Guide to Selected Consumer Films, Project Staff. Livonia Public Schools, 1978.

"Monthly Payment and Amortization Tables." Various booklets computed and published by Financial Publishing Company.

C. ARTICLES, PERIODICALS, NEWSLETTERS, AND SERVICES

"Math Skills for Survival in the Real World" by Ann Wilderman. See especially pp. 68-70. Enumerates, categorizes, describes, and correlates mathematics and consumer skills and discusses their implications for teaching strategies. Teacher Magazine, February 1977.

Consumer Education Resource Network (CERN). Provides periodic information regarding developments in consumer education. CERN has a computerized data bank of all publications and research up to date in consumer education and will provide a bibliography for related topics.

Michigan Consumer Education Skills. Michigan Consumers Council, 1977.

Consumer Reports, a periodical. Executive Director, Rhoda H. Karpatkin. Consumers Union of the United States, published monthly.

Consumer Information Catalog. Consumer Information Center, published quarterly.

Consumer Index to Product Evaluations and Information Sources. Provides an index and cross reference of periodicals for an almost unlimited selection of categories. Pierian Press, published quarterly.

Consumer News. Consumer Information Center, published monthly.

Current Consumer, a periodical. Curriculum Innovations, Inc., published monthly.

FDA Consumer, a periodical. Food and Drug Administration, published monthly.

Newsletter of the Michigan Consumer Education Center, published monthly.

APPENDIX B

Addresses of Producers of Resources

Aetna Life and Casualty
Public Relations
151 Farmington Avenue
Hartford, Connecticut 06156

AIMS Instructional Media Services
626 Justin Avenue
Glendale, California 91201

Allyn and Bacon, Inc.
470 Atlantic Avenue
Boston, Massachusetts 02210

BFA Educational Media
Division of Columbia Broadcasting System
2211 Michigan Avenue
Santa Monica, California 90404

Bankers Systems, Inc.
Box 1457
St. Cloud, Minnesota 56301

Barr Films
Post Office Box 5667
Pasadena, California 91107

Charles A. Bennett Company, Inc.
809 W. Detweiller Drive
Peoria, Illinois 61614

Bobbs-Merrill Company, Inc.
4300 West 62d Street
Indianapolis, Indiana 46206

Chamber of Commerce of the United States
1615 H Street Northwest
Washington, D.C. 20062

Churchill Films
662 North Robertson Boulevard
Los Angeles, California 90069

Clark Oil Corporation
8530 W. National Avenue
Milwaukee, Wisconsin 53227

Consumer Education Resource Network (CERN)
1555 Wilson Boulevard
Suite 600
Rosslyn, Virginia 22209

Consumer Information Center
Department 532 G
Pueblo, Colorado 81009

Consumers Union of the United States
256 Washington Street
Mount Vernon, New York 10550

Coronet Instructional Films
Coronet Building
65 East South Water Street
Chicago, Illinois 60601

Cost of Living Council
Bureau of Labor Statistics
Washington, D.C. 20262

Creative Teaching Associates
Post Office Box 7766
Fresno, California 93727

Credit Union National Association, Inc.
Post Office Box 431
Madison, Wisconsin 53701

APPENDIX B

Curriculum Innovations, Inc.
501 Lake Forest Avenue
Highwood, Illinois 60040

Delux Check Printers, Inc.
1080 West County Road F
Post Office Box 43399
St. Paul, Minnesota 55164

Doubleday Multimedia
Box 11607
1371 Reynolds Avenue
Santa Ana, California 92705

EMC Publishing
Changing Times Education Service
180 East Sixth Street
St. Paul, Minnesota 55101

Fearon-Pitman Learning, Inc.
6 Davis Drive
Belmont, California 04002

Federal Reserve Bank of Boston
600 Atlantic Avenue
Boston, Massachusetts 02106

Federal Reserve Bank of New York
33 Liberty Street
New York, New York 10045

Federal Reserve Bank of Philadelphia
100 N. 6th Street
Philadelphia, Pennsylvania 19105

Federal Reserve System
Office of Consumer Affairs
Washington, D.C. 20551

Filmfair Communications
10900 Ventura Blvd.
Studio City, California 91604

Financial Publishing Company
82 Brookline Avenue
Boston, Massachusetts 02215

Food and Drug Administration
5600 Fishers Lane
Rockville, Maryland 20857

Ford Motor Company
Box 1982
The American Road
Dearborn, Michigan 48121

General Learning Press
250 James Street
Morristown, New Jersey 07960

General Motors Corporation
Public Relations
G.M. Building
Detroit, Michigan 48202

Guidance Associates
757 Third Avenue
New York, New York 10017

Harcourt Brace Jovanovich
757 Third Avenue
New York, New York 10017

Holt, Rinehart and Winston
383 Madison Avenue
New York, New York 10017

Houghton Mifflin
2 Park Street
Boston, Massachusetts 02207

Household Finance Corporation
Money Management Institute
2700 Sanders Road
Prospect Heights, Illinois 60070

Ideal Publications
11000 South Lavergne Avenue
Oak Lawn, Illinois 60453

Insurance Information Institute
110 William Street
New York, New York 10038

Interpretive Education
2306 Winters Drive
Kalamazoo, Michigan 49002

Kentucky Department of Education
Consumer Education Consultant
Capital Plaza Tower
Frankfort, Kentucky 40601

Department of Labor
Office of Information and Reports
Washington, D.C. 20210

Livonia Public Schools
15162 Farmington Road
Livonia, Michigan 48150

The Math Group, Inc.
396 East 79th Street
Minneapolis, Minnesota 55420

The Math House
Division of Mosaic Media, Inc.
Glen Ellyn, Illinois 60137

McCormick-Mathers Publishing Company
135 W. 50th Street
New York, New York 10020

APPENDIX B

McGraw-Hill Book Company
1221 Avenue of the Americas
New York, New York 10020

Media Materials, Inc.
2936 Remington Avenue
Baltimore, Maryland 21211

Charles E. Merrill Publishing Company
1300 Alum Creek Drive
Columbus, Ohio 43216

Michigan Consumer Education Center
Eastern Michigan University
College of Education
217-A University Library
Ypsilanti, Michigan 48197

Michigan Consumers Council
414 Hollister Building
Lansing, Michigan 48933

Michigan Department of Commerce
Insurance Bureau
Post Office Box 30220
Lansing, Michigan 48909

Michigan Department of Education
Post Office Box 30008
Lansing, Michigan 48909

Michigan Products, Inc.
A Lakeshore Curriculum Center
1200 Keystone Avenue
Lansing, Michigan 48909

Motivational Development, Inc.
Post Office Box 427
Bishop, California 93514

National Council of Teachers of Mathematics
1906 Association Drive
Reston, Virginia 22091

W. W. Norton and Company
55 Fifth Avenue
New York, New York 10003

Oregon Department of Revenue
State Office Building
Salem, Oregon 97310

Parthenon Pictures
2625 Temple Street
Los Angeles, California 90026

Pierian Press
5000 Washtenaw Avenue
Ann Arbor, Michigan 48104

Prentice-Hall, Inc.
Englewood Cliffs, New Jersey 07632

Pyramid Films
Box 1048
Santa Monica, California 90406

Random House, Inc.
201 East 50th Street
New York, New York 10022

Frank E. Richards Publishing Company, Inc.
P.O. Box 66
Phoenix, New York 13135

Sadlier-Oxford
11 Park Place
New York, New York 10007

Scott, Foresman and Company
250 James Street
Morristown, New Jersey 07960

Sears, Roebuck and Company
Consumer Information Services
D/703
Sears Tower
Chicago, Illinois 60684

Securities and Exchange Commission
Office of Public Information
Washington, D.C. 20549

Shell Oil Company
Box 61609
Houston, Texas 77208

Social Security Administration
6401 Security Boulevard
Baltimore, Maryland 21235

South-Western Publishing Company
5101 Madison Road
Cincinnati, Ohio 45227

Steck-Vaughn Company
807 Box 2028
Austin, Texas 78768

Teacher Magazine
Post Office Box 8414
Philadelphia, Pennsylvania 19101

Teaching Resources Films
c/o Charles W. Clark Company, Inc.
564 Smith Street
Farmingdale, New York 11735

Department of the Treasury
Internal Revenue Service
Washington, D.C. 20224

United States Metric Board
1815 North Lynn Street
Arlington, Virginia 22209

APPENDIX B

Universal Education and Visual Arts
c/o Charles W. Clark Company, Inc.
564 Smith Street
Farmingdale, New York 11735

J. Weston Walch, Publisher
P.O. Box 658
Portland, Maine 04104

Warren, Gorham, and Lamont
210 South Street
Boston, Massachusetts 02111

Franklin Watts, Inc.
730 Fifth Avenue
New York, New York 10019

Western Publishing Company
850 Third Avenue
New York, New York 10022

John Wiley and Sons, Inc.
605 Third Avenue
New York, New York 10016

APPENDIX C

Reproduction Pages

The pages that follow have been provided to facilitate the reproduction of exercises and other materials needed for activities suggested in the preceding pages. Each page is perforated to make removal from this book easy. Once removed, the page can be used in any of three ways:

1. *For projection with an opaque projector.* No further preparation is necessary if the page is to be used with an opaque projector. The page may simply be inserted in the projector for viewing by the whole class.

2. *For projection with an overhead projector.* The Reproduction Page must be converted into a transparency for use with an overhead projector. To produce the transparency, overlay the Reproduction Page with a blank transparency and run both through a copying machine.

3. *For duplication with a spirit duplicator.* A master can be made from the Reproduction Page by overlaying it with a special heat-sensitive spirit master and running both through a copying machine. The spirit master can then be used to reproduce 50-100 copies on paper.

Please note that all material appearing on Reproduction Pages (as well as all other material in this book) is protected under the United States Copyright Law. Allyn and Bacon, Inc., grants to readers the right to make multiple copies of Reproduction Pages for nonprofit educational use only. All other rights are reserved.

REPRODUCTION PAGE 1

DIAGNOSTIC SURVEY ONE

I. A. Write with a $.   B. Write with a ¢.
   1) 26¢   6) 67.9¢   11) $.37   16) $.689
   2) 14¢   7) 72.93¢   12) $.13   17) $.7492
   3) 4¢   8) 4.5¢   13) $.08   18) $.075
   4) 128¢   9) 0.5¢   14) $1.34   19) $.003
   5) 40¢   10) 0.05¢   15) $.60   20) $.0003

II. Read each of these aloud and/or write them in words.
   21) 14¢   24) $.02   26) 1.4¢   29) $1.479
   22) $2.60   25) $3.06   27) 14.7¢   30) $.679
   23) 5¢   28) $1.47

III. A. Which is larger:
   31) 5¢ or .9¢?   33) $6.79 or $.689?   35) 65.9¢ or $.659?
   32) 13¢ or 1.7¢?   34) $7.29 or 73.9¢?

   B. Which is smaller:
   36) $1 or $.87?   38) $4.36 or $.439?   40) 72.9¢ or $.729?
   37) $13 or $6.52?   39) 23.9¢ or $2.39?

   C. Arrange in order of size (largest first).
   41) $.05, $18, $1.30, $.389   44) 36.9¢, $.38, 57.9¢, $3.79
   42) 7¢, 3.6¢, .46¢, 36¢   45) $.679, $6.79, 680¢, 68¢, 76.9¢
   43) $2, 8¢, $.09, 27¢

   D. Arrange in order of size (smallest first).
   46) 5¢, 18¢, 130¢, 38.9¢   49) $3.69, $.369, 36¢, $36.9
   47) $.07, $.036, $.46, $.36   50) $679, 679¢, $6.79, 67.9¢, $.67
   48) 2¢, $8, .9¢, $.27

Copyright © 1982 by Allyn and Bacon, Inc. Reproduction of this material is restricted to use with *A Guidebook for Teaching Consumer Mathematics*, by Peter A. Pascaris.

REPRODUCTION PAGE 1   DIAGNOSTIC SURVEY ONE

IV. Add. Express all answers with $.
    51) $.60 + $.38 + $1.50 + $3.25 + $9.70
    52) $.74 + $1.60 + $12.54 + $9.49 + $.99 + $.04
    53) 24¢ + $2.40 + 4¢ + $.02 + $16.21
    54) 19¢ + $19 + $1.90 + 90¢ + 9¢
    55) Find the sum of: 56¢, $3.21, 142¢, $.04, 7¢, $.56, $2.74

V. Subtract.
    56) $167.35 − $24.21   60) $10 − $4.60   63) $39 − 19¢
    57) $156.44 − $66.55   61) $13.41 − $9   64) $7 − 9¢
    58) $200 − $148.56   62) $29 − $19   65) $20 − 20¢
    59) $300.56 − $235.82

VI. A. Round to the nearest *cent*.
    66) 37.4¢   68) 58.5¢   70) $.278   72) $25.994   74) $10.001
    67) 67.7¢   69) 723¢   71) $.359   73) $17.488   75) $19.995

  B. Round to the nearest *ten* cents.
    76) 57¢   78) 81¢   80) 24¢   82) $.65   84) $3.45
    77) 79¢   79) 32¢   81) $.86   83) $1.53   85) $.97

  C. Round to the nearest *tenth* of a cent.
    86) $.3578   88) $21.528   90) $.7005   92) 72.94¢   94) 60.04¢
    87) $.6791   89) $.5687   91) 43.76¢   93) 68.88¢   95) 69.95¢

  D. Estimate these sums mentally.
    96) $3.91 + $2.52 + $2.48 + $.63
    97) $26.31 + $13.53 + $51.74 + $7.29 + $.52
    98) $39.78 + $43.52 + $66.49 + $120.90 + $270.38
    99) $372.46 + $.54 + $54.00 + $37.24 + $62.76
  100) $851.93 + $48.07 + $63.09 + $630.90 + $36.91

Copyright © 1982 by Allyn and Bacon, Inc. Reproduction of this material is restricted to use with *A Guidebook for Teaching Consumer Mathematics*, by Peter A. Pascaris.

REPRODUCTION PAGE 2

INDEPENDENCE SCALE ONE

Use the results of the Diagnostic Survey to rate your degree of independence for the skills listed below. Place an X under the heading that most closely matches your ability at this time.

SKILL	Problem on Survey	Always Depend on Others	Often Depend on Others	Sometimes Depend on Others	Often Work Alone	Always Work Alone
Changing from ¢ to $	I. A. (1-10)					
Changing from $ to ¢	I. B. (11-20)					
Writing $ and ¢ in words	II. (21-30)					
Comparing $ and ¢	III. (31-50)					
Adding $ and ¢	IV. (51-55)					
Subtracting $ and ¢	V. (56-65)					
Rounding to nearest cent	VI. A. (66-75)					
Round to nearest ten cents	VI. B. (76-85)					
Rounding to nearest tenth of a cent	VI. C. (86-95)					
Rounding to nearest dollar & estimating	VI. D. (96-100)					

Copyright © 1982 by Allyn and Bacon, Inc. Reproduction of this material is restricted to use with *A Guidebook for Teaching Consumer Mathematics*, by Peter A. Pascaris.

REPRODUCTION PAGE 3

A CENTSABLE ASSIGNMENT

Express these figures with a $.

1) 38¢ = $_____
2) 58¢ = $_____
3) 27¢ = $_____
4) 9¢ = $_____
5) 1¢ = $_____
6) 58.9¢ = $_____
7) 65.9¢ = $_____
8) 47.3¢ = $_____
9) 4.7¢ = $_____
10) 10¢ = $_____

11) 3.1¢ = $_____
12) 31.1¢ = $_____
13) .31¢ = $_____
14) .03¢ = $_____
15) 311.3¢ = $_____
16) 135¢ = $_____
17) 138.9¢ = $_____
18) 147.3¢ = $_____
19) 109.7¢ = $_____
20) 200.9¢ = $_____

Express these figures with a ¢.

21) $.42 = _____¢
22) $.74 = _____¢
23) $.33 = _____¢
24) $.07 = _____¢
25) $1.86 = _____¢
26) $.20 = _____¢
27) $.789 = _____¢
28) $.678 = _____¢

29) $.444 = _____¢
30) $.4444 = _____¢
31) $.044 = _____¢
32) $.004 = _____¢
33) $.0004 = _____¢
34) $.0094 = _____¢
35) $.0009 = _____¢

That seems to be very centsable!

Copyright © 1982 by Allyn and Bacon, Inc. Reproduction of this material is restricted to use with *A Guidebook for Teaching Consumer Mathematics,* by Peter A. Pascaris.

READING AND WRITING $ AND ¢

I. Read each of these aloud and/or write them in words.
   1. $4
   2. $6.00
   3. $26
   4. $32.00
   5. $172
   6. $1720.00
   7. $4396.
   8. $54,396.00
   9. $272,938.00
   10. $4,651,296,375,379
   11. 36¢
   12. $.88
   13. 79¢
   14. $.93
   15. $4.36
   16. $6.88
   17. $26.79
   18. $32.93
   19. $385.67
   20. $3856.74
   21. $38,567.49
   22. $385,674.99
   23. $3,856,749.91
   24. $38,567,499.12
   25. $385,674,991.20

II. Read each of these aloud and/or write them in words, on your own paper.

   A.
   1) 19¢
   2) $.18
   3) 87¢
   4) $.86
   5) 5¢
   6) $.14
   7) $8.09
   8) $14.27
   9) $128.32
   10) $301.02
   11) 19.3¢
   12) $.184
   13) $.589
   14) 57.9¢
   15) $5.673
   16) 37.9¢
   17) 15.0¢
   18) $15.
   19) $3.199
   20) 319.9¢

   B.
   1) 13.4¢
   2) 72.9¢
   3) $.859
   4) 57.4¢
   5) $.323
   6) 4.7¢
   7) $.065
   8) .4¢
   9) $.006
   10) .04¢
   11) $.0006
   12) .13¢
   13) $.0018
   14) 26.13¢
   15) $.2618
   16) $20.189
   17) $260.893
   18) $2,608.939
   19) $26,089.395
   20) $260,893.951

Copyright © 1982 by Allyn and Bacon, Inc. Reproduction of this material is restricted to use with *A Guidebook for Teaching Consumer Mathematics*, by Peter A. Pascaris.

REPRODUCTION PAGE 5

COMPARING MONEY FIGURES

Which is more?

1) $.039 or 39¢? _____
2) 59¢ or $.059? _____
3) $.089 or 89¢? _____
4) 24¢ or $.024? _____
5) $.65 or 64.9¢? _____
6) $.729 or 73¢? _____
7) 44.3¢ or $.44? _____
8) $.67 or 67.5¢ _____
9) 88.9¢ or $.89? _____
10) 1¢ or $.011? _____
11) $7.29 or 72.9¢? _____
12) 83.3¢ or $8.33? _____
13) $3.21 or 32.1¢? _____
14) 62.85¢ or $6.28? _____
15) 47.39¢ or $4.73? _____
16) $5.01 or 50.10¢? _____
17) 137.5¢ or $1.37? _____
18) $12.05 or 120.5¢? _____
19) 264.9¢ or $2.56? _____
20) $40.08 or 400.8¢ _____

Which is less?

21) 69¢ or $.069? _____
22) $.071 or 71¢? _____
23) 47¢ or $.047? _____
24) $.094 or 94¢? _____
25) 54.9¢ or $.55? _____
26) 83¢ or $.839? _____
27) $55 or 55.2¢? _____
28) 37.5¢ or $.37? _____
29) $.23 or 23.4¢? _____
30) $.022 or 2¢? _____
31) 52.9¢ or $5.29? _____
32) $3.88 or 38.8¢? _____
33) 43.2¢ or $4.32? _____
34) $8.91 or 8919¢? _____
35) $3.27 or 32.78¢? _____
36) 60.60¢ or $6.06? _____
37) $1.68 or 168.5¢ _____
38) 250.1¢ or $25.01? _____
39) $5.62 or 562.9¢? _____
40) 700.7¢ or $70.07? _____

Arrange in order of size, smallest first.

41) $.57, $.573, $5.70, and 5.7¢
42) $3.92, 3.9¢, $.392, and $.39
43) $2.87, $.287, 28¢, and $28.71

Arrange in order of size, largest first.

44) $.96, $.962, $9.60, and 9.6¢
45) $8.74, 8.7¢, $.874, and $.87

Copyright © 1982 by Allyn and Bacon, Inc. Reproduction of this material is restricted to use with *A Guidebook for Teaching Consumer Mathematics*, by Peter A. Pascaris.

ADDING $ AND ¢

Add these. Rewrite addends as necessary. Write all answers with a dollar sign.

1) $.22	2) 81¢	3) $.22	4) 42¢	5) $.78
.67	27¢	27¢	$.78	56¢
.37	3¢	$.37	$.85	38¢
.73	83¢	3¢	99¢	7¢
+.85	+62¢	+$.73	+ 5¢	+$.14

6) $2.82	7) $2.33	8) $47.82	9) $293.32	10) $247.49
.96	8.25	45.71	109.87	36¢
9.64	5.96	14.63	46.52	$575.37
4.72	9.43	18.65	$381.65	23¢
+ .51	+ 4.72	+ .68	+ 38¢	+ 19¢

(11-17) Show all work in columns as above.

11) $15 + $.67

12) $.23 + $.59 + $.73

13) $.25 + 36¢ + 14¢

14) 81¢ + $.17 + $.31 + 42¢

15) $1.72 + $3.81 + $4.38

16) $3.61 + 63¢ + 9¢ + $.04

17) 38¢ + $2.92 + $15.97 + 86¢ + 3¢ + $.06 + 72¢

18) $36.40 + 45¢ + $5.93 + 58¢ + $5.80 + 7¢ + $7

19) $22.85 + 67¢ + 13¢ + $34 + 34¢ + $16.20 + 70¢ + $.07

20) 85¢ + $85 + $125 + $12.50 + 12¢ + $1.25 + $.12

21) $186.58 + $57 + 99¢ + 84¢ + 39¢ + 57¢ + $84 + $1.86

22) $528.32 + 43¢ + $34 + $254 + $2.54 + 34¢ + 72¢

23) $781.01 + $7.81 + 78¢ + 780¢ + $78.10 + 7¢ + $.78

24) $5294.23 + 52¢ + 5¢ + $.05 + $52.94 + $529.42 + $5.29

Find the sum of:

25) $4.26, 76¢, $3, $6.04, $72, 18¢, $2.58, 4¢, $.03

Copyright © 1982 by Allyn and Bacon, Inc. Reproduction of this material is restricted to use with *A Guidebook for Teaching Consumer Mathematics*, by Peter A. Pascaris.

REPRODUCTION PAGE 7

SUBTRACTING $ AND ¢

Find the difference. Rewrite figures as necessary and write all answers with a dollar sign.

1) $38 − 14
2) $46 − 28
3) $83 − 23
4) $75 − 71
5) $53 − 46
6) $80 − 49

7) $.37 − .15
8) $.56 − .18
9) $.75 − .45
10) $.94 − .92
11) $.71 − .65
12) $.60 − .37

13) $4.36 − .22
14) $6.77 − 2.43
15) $8.24 − 3.02
16) $4.83 − 2.66
17) $5.42 − 3.47
18) $6.25 − 4.91

19) $3.26 − 2.84
20) $8.46 − 3.51
21) $7.49 − 1.56
22) $5.33 − 2.83
23) $8.85 − 3.36
24) $7.43 − 5.44

25) $9.06 − 7.49
26) $76.53 − 47.25
27) $39.62 − 14.87
28) $920.45 − 219.24
29) $682.47 − 443.95
30) $894.63 − 256.84

31) $604.81 − 395.92
32) $5,450.94 − 4,452.97
33) $4,460.00 − 3,759.87
34) $4,060.03 − 3,958.94

35) $42 − $.37 =
36) $53 − $.26 =
37) $83 − $.68 =
38) $92 − $.92 =
39) $71 − $.71 =
40) $4265 − $31.42 =
41) $3851 − $26.74 =
42) $2791 − $22.91 =

43) $8376 − $83.76 =
44) $5432 − $94.32 =
45) $59 − 29¢ =
46) $63 − 63¢ =
47) $98 − 89¢ =
48) $32 − 77¢ =
49) $5 − 7¢ =
50) $232 − 232¢ =

Copyright © 1982 by Allyn and Bacon, Inc. Reproduction of this material is restricted to use with *A Guidebook for Teaching Consumer Mathematics*, by Peter A. Pascaris.

MAKING SENSE OF MAKING CHANGE

Tell what your change would be for each of these purchases for the amount given to the clerk.

	Amount of Purchase	Amount Given to Clerk	Change Received
1.	$ .37	$ 1	
2.	$ .54	$ 1	
3.	$ .58	$ 1	
4.	$ 3.47	$ 5	
5.	$ 2.82	$ 5	
6.	$ 7.27	$ 10	
7.	$ 4.51	$ 10	
8.	$ 13.16	$ 20	
9.	$ 1.15	$ 20	
10.	$ 13.13	$100	
11.	$ 7.72	$ 10.02	
12.	$ 4.78	$ 20.03	
13.	$ 1.86	$ 5.01	
14.	$ 1.88	$ 2.03	
15.	$ .65	$ 20.15	
16.	$ 4.48	$ 5.48	
17.	$ .81	$ 1.06	
18.	$ 17.61	$ 20.11	
19.	$ 73.26	$100.01	
20.	$147.31	$160.06	

Copyright © 1982 by Allyn and Bacon, Inc. Reproduction of this material is restricted to use with *A Guidebook for Teaching Consumer Mathematics*, by Peter A. Pascaris.

REPRODUCTION PAGE 9

ROUNDING AND ESTIMATING WHILE SHOPPING

1. Make a survey at a store or use a catalog and obtain the actual price of each item listed.
2. Round the actual price as directed.
3. Estimate the total sum without adding.
4. Add the actual prices and compare the actual sum with your estimated sum.
5. Try the "Extra Practice" shown after Part Two.

PART ONE:

Items to Purchase	Actual Price	Round to Nearest Dollar
Tennis shoes		
Tennis racket		
Tennis balls (pkg. of 3)		
White socks		
Wristwatch		
Compact portable 8-track stereo player		
A cassette tape of your favorite recording artist		
Batteries for player		
Lightweight jacket		
Sunglasses		
Actual Sum and Estimated Sum		

Copyright © 1982 by Allyn and Bacon, Inc. Reproduction of this material is restricted to use with *A Guidebook for Teaching Consumer Mathematics*, by Peter A. Pascaris.

ROUNDING AND ESTIMATING WHILE SHOPPING   **REPRODUCTION PAGE 9**

PART TWO:

Items to Purchase	Actual Price	Round to Nearest Ten Cents
28-oz. can of baked beans		
10-3/4 oz. can of condensed soup		
32-oz. bottle of ketchup		
17-oz. can of peas		
1 dozen eggs		
1-lb. box of salt		
1 box macaroni & cheese dinner		
12-oz. jar of jelly		
18-oz. jar of peanut butter		
1 quart of milk		
Actual Sum and Estimated Sum		

EXTRA PRACTICE:

1. Compare the estimated sum with the actual sum for both Part One and Part Two.
2. Suppose this was a list of things to buy before a vacation trip and you found that you did not have enough money. How would you go about reducing the amount? Can you cut down expenses without eliminating items?

Copyright © 1982 by Allyn and Bacon, Inc. Reproduction of this material is restricted to use with *A Guidebook for Teaching Consumer Mathematics*, by Peter A. Pascaris.

REPRODUCTION PAGE 10

ROUNDING AND ESTIMATING

Express all answers with a dollar sign.

A. Round to the nearest cent.
   1) $.385 = _____
   2) 47.5¢ = _____
   3) $.563 = _____
   4) 68.3¢ = _____
   5) $.269 = _____
   6) 57.9¢ = _____
   7) $.017 = _____
   8) 3.7¢ = _____
   9) $.063 = _____
   10) 5.3¢ = _____
   11) $.096 = _____
   12) $17.562 = _____
   13) $37.094 = _____
   14) 586¢ =
   15) 66.49¢ = _____

B. Round to the nearest ten cents.
   16) 47¢ = _____
   17) 89¢ = _____
   18) 51¢ = _____
   19) 42¢ = _____
   20) 64¢ = _____
   21) $.86 = _____
   22) $.35 = _____
   23) $1.43 = _____
   24) $5.65 = _____
   25) $.98 = _____

C. Round to the nearest tenth of a cent.
   26) $.3782 = _____
   27) 57.85¢ = _____
   28) $.6293 = _____
   29) $.5635 = _____
   30) 56.35¢ = _____
   31) 37.87¢ = _____
   32) $15.5555 = _____
   33) 38.05¢ _____
   34) $.4004 = _____
   35) $30.09962 = _____

D. Round to the nearest dollar and then *estimate* their sum by adding mentally.
   36) $1.72 + $3.81 + $4.38
   37) $2.33 + $8.25 + $5.96 + $9.43 + $4.72
   38) $47.82 + $45.71 + $14.63 + $18.65
   39) $36.40 + 45¢ + $5.93 + 58¢ + $5.80
   40) $186.58 + $57 + 99¢ + 84¢ + 39¢ + 57¢ + $84 + $1.86 + $18.65 + 50¢

Copyright © 1982 by Allyn and Bacon, Inc. Reproduction of this material is restricted to use with *A Guidebook for Teaching Consumer Mathematics*, by Peter A. Pascaris.

DIAGNOSTIC SURVEY TWO

I. State three advantages of a checking account.

II. Match each term with the meaning that is most nearly correct for use with checking accounts.

   1. check
   2. balance
   3. register
   4. deposit
   5. endorsement
   6. entry
   7. outstanding
   8. overdraw
   9. statement
   10. reconcile

   a. way to get in
   b. procedure that verifies balance
   c. signature on face of check
   d. signature on back of check enabling it to be cashed
   e. left outside
   f. written order to pay money
   g. report of money collected by bank
   h. exact amount in account
   i. amount added to account
   j. place to record check
   k. report of transactions made in account
   l. to draw again
   m. item written on check register
   n. to write check for more than balance
   o. written checks not on statement

III. Use Reproduction Page 15 as a facsimile of a personal check. Identify the parts and tell how they are used.

IV. Write these dollar figures in words as they should appear on a check.
   a. $16.00   b. $25.31   c. $246.87

V. Use Reproduction Page 21, Part One, and identify each endorsement as blank, special, or restrictive. Which endorsement is least secure?

VI. Correctly record the following information, first using the check register, and second using the check stubs.
   a. Begin with a balance of $500.
   b. Pay Woody Doit $349.26 for car repairs.
   c. Deposit $296.84 from your paycheck.
   d. Pay J. C. Nickels $115.78 on your charge account.

Copyright © 1982 by Allyn and Bacon, Inc. Reproduction of this material is restricted to use with *A Guidebook for Teaching Consumer Mathematics*, by Peter A. Pascaris.

REPRODUCTION PAGE 11   DIAGNOSTIC SURVEY TWO

VII. Using the check register below and Reproduction Page 22 as a corresponding bank statement, reconcile the statement.

CHECKING ACCOUNT RECORD

Check No.	Date	Check Issued to And Purpose of Payment	Balance Forward:	5000	00
1	11/21	To: Chisler Corp. For: loan payment	Amount of Check (−) or Deposit (+)	−168	92
			Balance	4831	08
	11/30	To: DEPOSIT For:	Amount of Check (−) or Deposit (+)	+456	78
			Balance	5287	86
2	11/30	To: Homefold Finance For: down payment	Amount of Check (−) or Deposit (+)	−5250	00
			Balance	37	86
3	12/3	To: Otto Nobedder For: car repair	Amount of Check (−) or Deposit (+)	−24	58
			Balance	13	28
4	12/4	To: Jingle Bell Phone Co. For: phone bill	Amount of Check (−) or Deposit (+)	−7	57
			Balance	5	71
	12/10	To: DEPOSIT For:	Amount of Check (−) or Deposit (+)	+3810	07
			Balance	3815	78
5	12/10	To: Eddie Sun Co. For: electrical service	Amount of Check (−) or Deposit (+)	−31	89
			Balance	3783	89
6	12/10	To: Shipwreck Co. For: new boat	Amount of Check (−) or Deposit (+)	−3764	00
			Balance	19	89
	12/19	To: DEPOSIT For:	Amount of Check (−) or Deposit (+)	+67	92
			Balance	87	31
7	12/19	To: Bored Water Board For: water service	Amount of Check (−) or Deposit (+)	−45	99
			Balance	41	32
	12/26	To: DEPOSIT For:	Amount of Check (−) or Deposit (+)	+879	08
			Balance	920	40
8	12/28	To: Neal N. Hammer For: house repair	Amount of Check (−) or Deposit (+)	−690	00
			Balance	230	40

Balance to be Forwarded

Copyright © 1982 by Allyn and Bacon, Inc. Reproduction of this material is restricted to use with *A Guidebook for Teaching Consumer Mathematics*, by Peter A. Pascaris.

REPRODUCTION PAGE 12

INDEPENDENCE SCALE TWO

Use the results of the Diagnostic Survey to rate your degree of independence for the skills listed below. Place an X under the heading that most closely matches your ability at this time.

		To successfully complete this skill, I:				
SKILL	Problem on Survey	Always Depend on Others	Often Depend on Others	Sometimes Depend on Others	Often Work Alone	Always Work Alone
State three Advantages	I.					
Recognize Terms	II.					
Identify Parts of a Check	III.					
Numbers As Words	IV.					
Endorsements	V.					
Keeping Records A. Check Registers	VI. A.					
Keeping Records B. Check Stub	VI. B.					
Reconcile Bank Statement	VII.					

Copyright © 1982 by Allyn and Bacon, Inc. Reproduction of this material is restricted to use with *A Guidebook for Teaching Consumer Mathematics*, by Peter A. Pascaris.

REPRODUCTION PAGE 13

DEPOSIT TICKET FACSIMILE

DEPOSIT TICKET

CASH →
LIST CHECKS SINGLY

13-88
533

DATE_____19____

CHECKS AND OTHER ITEMS ARE RECEIVED FOR DEPOSIT SUBJECT TO THE PROVISIONS OF THE UNIFORM COMMERCIAL CODE OR ANY APPLICABLE COLLECTION AGREEMENT.

DEPOSITED IN

TOTAL FROM OTHER SIDE
TOTAL ITEMS
TOTAL

USE OTHER SIDE FOR ADDITIONAL LISTING.
↓ ENTER TOTAL HERE
BE SURE EACH ITEM IS PROPERLY ENDORSED

Non-Negotiable Bank
Knowhere, Michigan   007007
07734-07734

⑆5481⑈0138⑆   647⑊422⑋6⑋

CHECKS LIST SINGLY	DOLLARS	CENTS
1		
2		
3		
4		
5		
6		
7		
8		
9		
10		
11		
12		
13		
14		
15		
16		
17		
18		
19		
TOTAL		

ENTER TOTAL ON THE FRONT OF THIS TICKET

DEPOSIT TICKET

CASH →
LIST CHECKS SINGLY

13-88
533

DATE_____19____

CHECKS AND OTHER ITEMS ARE RECEIVED FOR DEPOSIT SUBJECT TO THE PROVISIONS OF THE UNIFORM COMMERCIAL CODE OR ANY APPLICABLE COLLECTION AGREEMENT.

DEPOSITED IN

TOTAL FROM OTHER SIDE
TOTAL ITEMS
TOTAL

USE OTHER SIDE FOR ADDITIONAL LISTING.
↓ ENTER TOTAL HERE
BE SURE EACH ITEM IS PROPERLY ENDORSED

Non-Negotiable Bank
Knowhere, Michigan   007007
07734-07734

⑆5481⑈0138⑆   647⑊422⑋6⑋

Copyright © 1982 by Allyn and Bacon, Inc. Reproduction of this material is restricted to use with *A Guidebook for Teaching Consumer Mathematics*, by Peter A. Pascaris.

REPRODUCTION PAGE 14

CHECKING ACCOUNT DEPOSIT TICKETS

The following are deposits made on the specified date to the Account of Dolly Dimples, 382638 Heavenly Lane, Knoweer, Michigan 007007. Her account number 0734-40. Complete a deposit ticket for each date shown (use the current year). Use Reproduction Page 13.

1. May 1: $79.85 cash.

2. May 7: A check for $193.47. Check number 7904, bank number $\frac{39-26}{31}$.

3. May 14: $35.15 cash and a check for $206.59. Check number 8091, bank number $\frac{39-26}{31}$.

4. May 22: Checks for.

   $8.00, check number 365, bank number $\frac{14-160}{31}$.

   $7.36, check number 293, bank number $\frac{14-160}{31}$.

   $4.00, check number 83, bank number $\frac{78-42}{31}$.

   $216.92, check number 8278, bank number $\frac{39-26}{31}$.

   Less $65.00 cash received for above checks.

5. May 28: Checks as follows.

   $37.81, check number 320, bank number $\frac{79-26}{31}$.

   $201.41, check number 8465, bank number $\frac{39-26}{31}$.

   $13.53, check number 171, bank number $\frac{79-42}{31}$.

   $8.00, check number 371, bank number $\frac{14-160}{31}$.

   $14.72, check number 299, bank number $\frac{14-160}{31}$.

   $8.00, check number 1741, bank number $\frac{15-160}{31}$.

   $8.00, check number 88, bank number $\frac{78-42}{31}$.

   Less $60.00 cash received from above checks.

Copyright © 1982 by Allyn and Bacon, Inc. Reproduction of this material is restricted to use with *A Guidebook for Teaching Consumer Mathematics,* by Peter A. Pascaris.

REPRODUCTION PAGE 15

CHECKING ACCOUNTS

Identify the various parts of the check as indicated below.

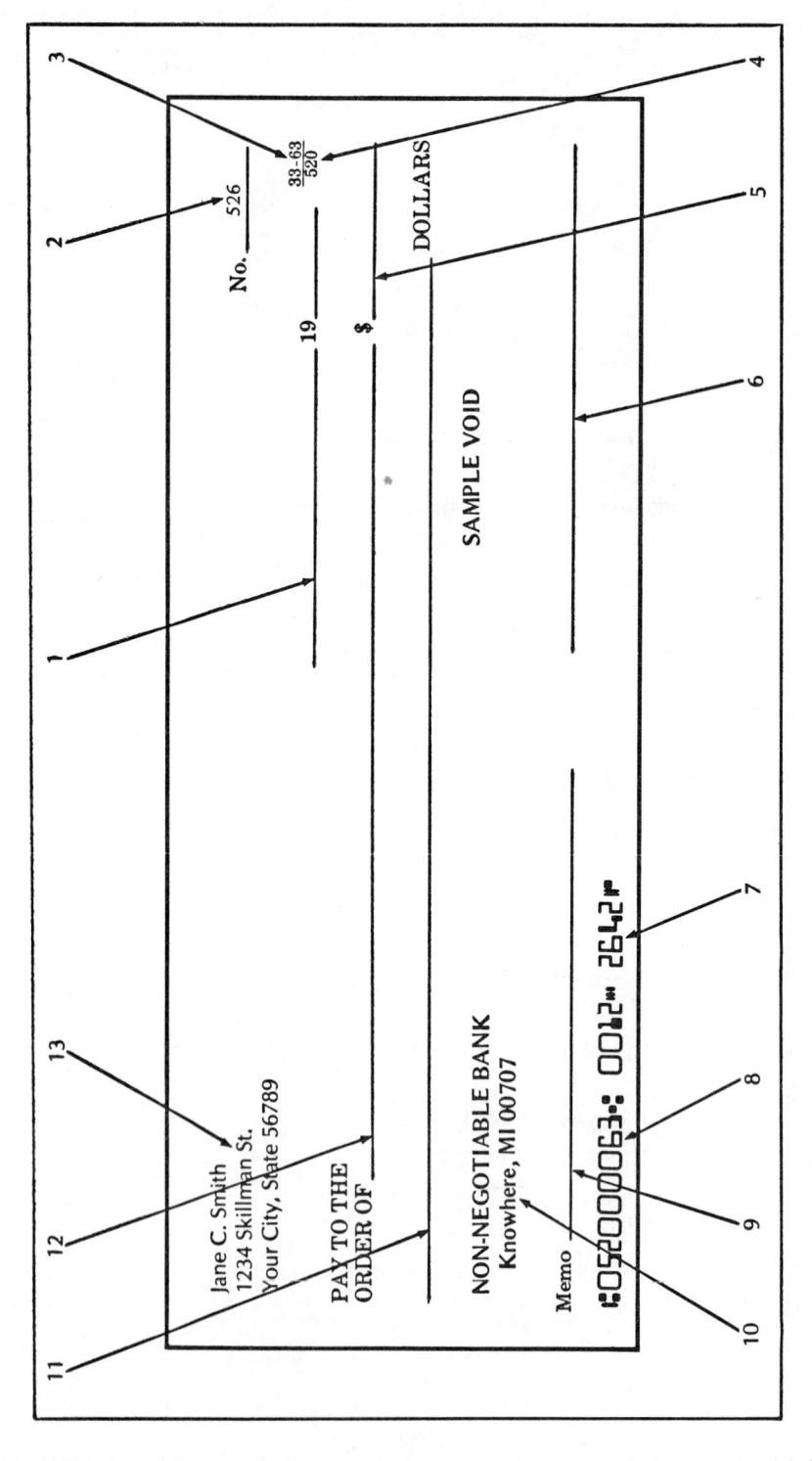

Copyright © 1982 by Allyn and Bacon, Inc. Reproduction of this material is restricted to use with *A Guidebook for Teaching Consumer Mathematics,* by Peter A. Pascaris.

REPRODUCTION PAGE 16

CHECKING ACCOUNTS: WORDS FOR NUMBERS

In the blank spaces below, write the words for the numbers given as they are to appear on a check. (Assume the word "dollars" will follow each number and it need not be written.)

$1.00 _____ $6.00 _____ $11.00 _____

$2.00 _____ $7.00 _____ $12.00 _____

$3.00 _____ $8.00 _____ $13.00 _____

$4.00 _____ $9.00 _____ $14.00 _____

$5.00 _____ $10.00 _____ $15.00 _____

$16.00 _____ $43.00 _____

$17.00 _____ $54.00 _____

$18.00 _____ $65.00 _____

$19.00 _____ $76.00 _____

$21.00 _____ $87.00 _____

$32.00 _____ $98.00 _____

$100.00 _____ $1000.00 _____

$276.00 _____

$3499.00 _____

$9.41 _____

$30.79 _____

$684.32 _____

$1900.83 _____

$2023.23 _____

$3004.04 _____

$10,652.67 _____

$.87 _____

Copyright © 1982 by Allyn and Bacon, Inc. Reproduction of this material is restricted to use with *A Guidebook for Teaching Consumer Mathematics*, by Peter A. Pascaris.

REPRODUCTION PAGE 17

SAMPLE CHECKS

| Check Number _____ | Date _____ 19___ |

Pay to the order of _____ $_____

_____Dollars

Non-Negotiable Bank
Knowhere, Michigan 007007

Memo _____   _____
                       07734-07734

Check Number _____   Date _____ 19___

Pay to the order of _____ $_____

_____Dollars

Non-Negotiable Bank
Knowhere, Michigan 007007

Memo _____   _____
                       07734-07734

Check Number _____   Date _____ 19___

Pay to the order of _____ $_____

_____Dollars

Non-Negotiable Bank
Knowhere, Michigan 007007

Memo _____   _____
                       07734-07734

Copyright © 1982 by Allyn and Bacon, Inc. Reproduction of this material is restricted to use with *A Guidebook for Teaching Consumer Mathematics,* by Peter A. Pascaris.

REPRODUCTION PAGE 18

CHECKS
(without deposits)

Write a check in full as directed. Use the current year and number the checks the same as the number of the problem.

1. Oct. 3. Pay T.V. Wonder $42.95 for fixing your television so you could watch the World Series.

2. Oct. 7. Pay N. D. Man $25.00 for repairing your storm windows while you watch the Series.

3. Oct. 8. Pay Pepper Roney $5.78 for delivering a pizza while you watched the ball game.

4. Oct. 9. Pay your Aunt Enna $22.50 for going on your roof and resetting your antenna after the wind blew it down. Aunt Enna's last name is Upp.

5. Oct. 9. Pay Pop Zoda $9.56 for a case of pop and a bag of peanuts.

6. Oct. 10. Pay Lon Mower $7.50 for cutting your grass while you drink, eat, and watch the ball game.

7. Oct. 11. Pay Cliff Hanger, who dropped over in the ninth inning of the last game and asked you to pay $47.89 for the speed drill you borrowed (and broke).

8. Oct. 12. Pay S. O. R. Loser $8.00 because you lost your bet on the World Series. (Don't take the world serious.)

DEFINE THESE TERMS.

check   register   stub   credit   debit   deposit   endorse

Copyright © 1982 by Allyn and Bacon, Inc. Reproduction of this material is restricted to use with *A Guidebook for Teaching Consumer Mathematics,* by Peter A. Pascaris.

REPRODUCTION PAGE 19

CHECKING ACCOUNTS
(with deposits)

Write a check for each payment below. Write a deposit ticket for each deposit. Use the dates shown and the current year. Number the checks consecutively, beginning with check number 1.

1. Nov. 9. Pay the Non-Negotiable Bank $360 for a loan.
2. Nov. 10. Pay Hugh's Cars, Inc., $476.50 for a car.
3. Nov. 10. Pay Sophie's Furniture $104.60 for a sofa.
4. Nov. 14. Deposit $570.72 from your paycheck.
5. Nov. 14. Make a rent payment of $194 to Ren-ten-tents, Inc.
6. Nov. 18. Pay Sew and Sew Tailors $79.82 for new clothes.
7. Nov. 21. Deposit $1432.50 from your paycheck.
8. Nov. 22. Pay Ray Deo's Music Store $275.98 for a stereo.
9. Nov. 24. Pay the Crop Shop $36.49 for groceries.
10. Nov. 24. Pay Eddie Son $25.78 for electric repairs.

DEFINE THESE TERMS

legal tender	money	outstanding	overdraw	payee
signature	statement	balance a statement		verify
void	withdraw	cancelled check		cashier's check
certified check		draft (or bank draft)		forged check
money order		traveler's check		checkbook balance
joint account	currency	deduction	drawer	payee

Copyright © 1982 by Allyn and Bacon, Inc. Reproduction of this material is restricted to use with *A Guide to Teaching Consumer Mathematics*, by Peter A. Pascaris.

REPRODUCTION PAGE 20

IMPROPER CHECKS

Examine the following checks and identify the errors and careless entries. If possible, show how the check can be altered.

Check Number __536__   Date __Aug. 14__ 19__82__
Pay to the order of __Sew and Sew Tailors__ $__24.98__
__Twenty-four .98__ _____ Dollars
Non-Negotiable Bank
Knowhere, Michigan   007007
Memo __repair trousers__
          07734-07734   _____

Check Number __537__   Date __April 15,__ 19__82__
Pay to the order of __Jenny Rader__ $__62.50__
__62 dollars and 50 cents__ _____ Dollars
Non-Negotiable Bank
Knowhere, Michigan   007007
Memo __fix generator__   __James__
          07734-07734

Check Number __538__   Date __Sept. 2__ 19__82__
Pay to the order of _____ $__50.00__
__Fifty and no/100__ _____ Dollars
Non-Negotiable Bank
Knowhere, Michigan   007007
Memo __cash__   __James C. Smith__
          07734-07734

Check Number __539__   Date __Oct. 18__ 19__82__
Pay to the order of __Harry Snip__ $__8.50__
__Eight fifty__ _____ Dollars
Non-Negotiable Bank
Knowhere, Michigan   007007
Memo __haircut__   __James C. Smith__
          07734-07734

Copyright © 1982 by Allyn and Bacon, Inc. Reproduction of this material is restricted to use with *A Guide to Teaching Consumer Mathematics*, by Peter A. Pascaris.

REPRODUCTION PAGE 21

ENDORSEMENTS

PART ONE. Identify the types of endorsement and explain their advantages or disadvantages. Assume the payee is Hiram Unaware.

1. | Hiram Unaware |

2. | Pay to order of Gaylen Carets / Hiram Unaware |

3. | For Deposit Only / First Local Bank / 1234-5421 / Hiram Unaware |

PART TWO. Properly endorse these checks with the given circumstances. Explain your action.

4. Your name is Bill Collector. You wish to deposit part of your paycheck in your account but still obtain some cash.

   Check Number 540   Date Jan. 17 1982
   Pay to the order of Bill Collector $761.43
   Seven hundred sixty one and 43/100 —— Dollars
   Non-Negotiable Bank
   Knowhere, Michigan 007007
   Memo _____
   07734-07734   James C. Smith

5. Your name is Jane Spelled. You wish to use all of this check to pay Mr. I. L. Cashette the money you owe him.

   Check Number 541   Date Feb. 3 1982
   Pay to the order of Ms. Jane Spelled $43.00
   Forty three and 00/100 —— Dollars
   Non-Negotiable Bank
   Knowhere, Michigan 007007
   Memo _____
   07734-07734   James C. Smith

6. Your name is Hugo Yourway. You received this check and wish to use only part of the money to pay a bill.

   Check Number 542   Date April 8 1982
   Pay to the order of Dan Jimdy $68.49
   Sixty eight and 49/100 —— Dollars
   Non-Negotiable Bank
   Knowhere, Michigan 007007
   Memo _____
   07734-07734   James C. Smith

   | Pay to order of Hugo Yourway / Dan Jimdy |

Copyright © 1982 by Allyn and Bacon, Inc. Reproduction of this material is restricted to use with *A Guide to Teaching Consumer Mathematics*, by Peter A. Pascaris.

REPRODUCTION PAGE 22

SAMPLE CHECK REGISTER

NAME _____

DATE _____ HOUR _____

CHECKING ACCOUNT RECORD

Check No.	Date	Check Issued to And Purpose of Payment	Balance Forward:		
		To:	Amount of Check (-) or Deposit (+)		
		For:	Balance		
		To:	Amount of Check (-) or Deposit (+)		
		For:	Balance		
		To:	Amount of Check (-) or Deposit (+)		
		For:	Balance		
		To:	Amount of Check (-) or Deposit (+)		
		For:	Balance		
		To:	Amount of Check (-) or Deposit (+)		
		For:	Balance		
		To:	Amount of Check (-) or Deposit (+)		
		For:	Balance		
		To:	Amount of Check (-) or Deposit (+)		
		For:	Balance		
		To:	Amount of Check (-) or Deposit (+)		
		For:	Balance		
		To:	Amount of Check (-) or Deposit (+)		
		For:	Balance		
		To:	Amount of Check (-) or Deposit (+)		
		For:	Balance		
		To:	Amount of Check (-) or Deposit (+)		
		For:	Balance		
		To:	Amount of Check (-) or Deposit (+)		
		For:	Balance		

Balance to be Forwarded

Copyright © 1982 by Allyn and Bacon, Inc. Reproduction of this material is restricted to use with *A Guide to Teaching Consumer Mathematics*, by Peter A. Pascaris.

REPRODUCTION PAGE 23

CHECK STUBS

CHECKING ACCOUNT Deposit Ticket		CASH	Currency		
			Coin		
Name _____		C H E C K S			
Address _____					
City _____	Be Sure Each Check Is Endorsed Properly				
State, Zip _____					
Date _____					
Non-Negotiable Bank		TOTAL			
Knowhere, Michigan 007007		LESS CASH RECEIVED			
07734-07734		NET DEPOSIT			

CHECKING ACCOUNT Deposit Ticket		CASH	Currency		
			Coin		
Name _____		C H E C K S			
Address _____					
City _____	Be Sure Each Check Is Endorsed Properly				
State, Zip _____					
Date _____					
Non-Negotiable Bank		TOTAL			
Knowhere, Michigan 007007		LESS CASH RECEIVED			
07734-07734		NET DEPOSIT			

CHECKING ACCOUNT Deposit Ticket		CASH	Currency		
			Coin		
Name _____		C H E C K S			
Address _____					
City _____	Be Sure Each Check Is Endorsed Properly				
State, Zip _____					
Date _____					
Non-Negotiable Bank		TOTAL			
Knowhere, Michigan 007007		LESS CASH RECEIVED			
07734-07734		NET DEPOSIT			

Copyright © 1982 by Allyn and Bacon, Inc. Reproduction of this material is restricted to use with *A Guide to Teaching Consumer Mathematics*, by Peter A. Pascaris.

REPRODUCTION PAGE 24

KEEP YOUR BALANCE

Start with a balance of $5000.00. Write a check for each payment indicated. Number each check consecutively, beginning with number 1. Write deposit tickets and keep a record.

March 1. Write a $42.00 check to Mr. Jack Hungry for groceries.

March 2. Pay $1250 to Hughs DiCar for a used car.

March 3. You got paid for two weeks' work. Deposit $200.

March 4. Pay Mrs. Ima Fur $538.25 for winter clothes.

March 5. Pay Ms. E. Lectric $2456.90 for a new stereo.

March 6. Write a check for a new air conditioner to Cold-Gold Department Store for $524.00.

March 7. You won the Broom Sweepstakes. Deposit $582.79.

March 8. You inherited money from Aunt Lotta Kash. Deposit $350.00.

March 29. You bought a new stove. Write a check to Heat 'n Eat Kitchen Store for $257.00.

March 30. The total cost to remodel your home is $6784.28. How much money must you borrow and deposit so that you end up with a balance of $100 after you pay Mr. Belvemoose $6784.28? Make the proper deposit and payment.

FRESHEN UP YOUR SKILLS

$431.79	$634.52	$493.04	$377.01	$800.20
− 243.65	− 587.67	− 283.51	− 177.11	− 593.32

Copyright © 1982 by Allyn and Bacon, Inc. Reproduction of this material is restricted to use with *A Guide to Teaching Consumer Mathematics*, by Peter A. Pascaris.

REPRODUCTION PAGE 25

CHECKS AND DEPOSITS

Begin with $350 in your checking account. Write the following checks in the order that they appear, numbering consecutively beginning with number 1. You must add deposits of $100 (cash) whenever your balance is too small for the next check. Write checks and deposit tickets and keep a record. Use the dates shown with the current year.

Jan. 1.   Pay Charro's Chairs $286.27 for furniture.

Jan. 10. Pay Knowhere Edison Co. $35.28 for electricity service.

Jan. 12. Pay Mark It's Food Store $35.21 for groceries.

Jan. 13. Pay the Dripoff Water Co. $6.36 for water service.

Jan. 15. Pay I. C. Daze Co. for a snowplow, $97.31.

Jan. 17. Pay the Non-Negotiable Bank $142 for a loan payment.

Jan. 17. Pay the Len D. Brook Library $5.85 for a lost book.

Jan. 20. Pay Snap Photo Service $4.67 for certain developments.

Jan. 30. Pay Slick Oil Co. $26.78 for gasoline credit.

Feb. 1.   Pay Cal Ender $13.47 for a five-year diary.

TEST YOUR WITS

How much money did the farmer have if he had 20 female pigs and 20 male deer?

Copyright © 1982 by Allyn and Bacon, Inc. Reproduction of this material is restricted to use with *A Guide to Teaching Consumer Mathematics,* by Peter A. Pascaris.

REPRODUCTION PAGE 26

MORE CHECKS; ETC.

Start with a balance of $5000. Write checks and deposit tickets and keep a record. Write checks in the order given and number consecutively. You must add a deposit whenever your balance is too small to cover the next check. The entire amount of the deposit must be used even if it is more than needed. Choose the appropriate time to add the deposit, but it must be deposited in the order given. Note that more than one amount may need to be deposited at one time. However, when there are enough funds to write the next check, no more may be deposited.

Nov. 21. Car payment . . . Chisler Corp. . . . $168.92.

Nov. 30. Home down payment . . . Homefold Finance . . . $5250.

Dec. 3. Auto repair . . . Otto Nobedder . . . $24.58.

Dec. 4. Phone bill . . . Jingle Bell Phone Co. . . . $7.57.

Dec. 10. Electricity service . . . Eddie Sun Co. . . . $31.89.

Dec. 10. New Boat . . . Shipwreck Co. . . . $3764.

Dec. 19. Water service . . . Bored Water Board . . . $45.99.

Dec. 28. House repair . . . Neal N. Hammer . . . $690.

Use these deposits in the order given, but more than one may be added at a time:
1) $456.78   2) $28.76   3) $136.51   4) $249.90   5) $3004.00   6) $390.90   7) $67.92
8) $879.08

Copyright © 1982 by Allyn and Bacon, Inc. Reproduction of this material is restricted to use with *A Guide to Teaching Consumer Mathematics*, by Peter A. Pascaris.

REPRODUCTION PAGE 27

CAN YOU MANAGE?

Use the same directions as "More Checks; Etc." on Reproduction Page 26. Remember to add the deposits only as you need the money for the next check but in the order given. Assume this is a new account and begin by depositing $500.

July 1.   Loan payment . . . Mrs. Lo Ann Shark . . . $150.

July 3.   Plumbers union . . . Mr. P. Lummer . . . $30.

July 5.   Home insurance . . . Miss Bea Kareful . . . $47.83.

July 5.   Credit examination . . . Mr. Cred Itt . . . $28.00.

July 5.   TV repair . . . Mr. T. V. Sea . . . $95.46.

July 6.   School fees . . . Miss Dee Tu Ition . . . $585.00.

July 6.   Late finance charges . . . Miss D. Intrest . . . $22.41.

July 7.   Car insurance . . . Mr. Otto Helpay . . . $176.40.

July 8.   Retirement plan . . . Mr. Pen Shun . . . $250.00.

July 10. Commission . . . Mr. A. Salzman . . . $55.00.

July 12. Hospital costs . . . Major M. E. Dical . . . $774.97.

July 15. Bookkeeper fee . . . Ms. A. C. Count . . . $29.79.

July 15. Car crash repairs . . . Mr. Ack Seedent . . . $100.00.

July 16. Taxes . . . Uncle Sam . . . $168.30.

July 28. Fire damage . . . Mr. Fi R. Burns . . . $50.00.

Use these deposits in the order given, but more than one may be added at a time:
1) $75.80   2) $153.72   3) $198.23   4) $500.39   5) $786.67   6) $30.00   7) $178.88
8) $70.00   9) $100.00

Copyright © 1982 by Allyn and Bacon, Inc. Reproduction of this material is restricted to use with *A Guide to Teaching Consumer Mathematics*, by Peter A. Pascaris.

BLANK BANK STATEMENT

NON-NEGOTIABLE BANK OF KNOWHERE

ACCOUNT NUMBER

STATEMENT PERIOD FROM   THRU

BEGINNING BALANCE
PLUS   DEPOSITS   ITEMS
LESS   CHECKS
LESS SERVICE CHARGE
ENDING BALANCE

			DEPOSITS	DATE	BALANCE

CHECKS OUTSTANDING

Check Number or Date of Issue	AMOUNT	Check Number or Date of Issue	AMOUNT
		TOTAL FORWARDED	
Total and forward to next column		TOTAL (Enter on Line 3 of Balance Form)	

BALANCE FORM

Enter balance shown on this statement here—Line 1	
Add any deposits made after the date of this statement here—Line 2	
(Use this space for any additional deposits made)	
Total of Line 1 and Line 2	
Enter total of checks outstanding and subtract here—Line 3	
Your checkbook should show this BALANCE!	

Copyright © 1982 by Allyn and Bacon, Inc. Reproduction of this material is restricted to use with *A Guide to Teaching Consumer Mathematics*, by Peter A. Pascaris.

REPRODUCTION PAGE 29

MODIFIED BANK STATEMENT
(Use with Reproduction Page 18)

NON-NEGOTIABLE BANK OF KNOWHERE

ACCOUNT NUMBER: R.P. 2-11

STATEMENT PERIOD FROM 8/4/82 THRU 8/16/82

BEGINNING BALANCE		$170.00
PLUS 0 DEPOSITS 0 ITEMS		0.00
LESS 8 CHECKS		169.18
LESS SERVICE CHARGE		0.00
ENDING BALANCE		.82

Checks			DEPOSITS	DATE	BALANCE
$42.95				8/4	$129.05
25.00	9.56			8/10	92.49
7.50	5.78	22.50		8/14	56.71
47.89				8/15	8.82
8.00				8/16	.82

CHECKS OUTSTANDING

Check Number or Date of Issue	AMOUNT	Check Number or Date of Issue	AMOUNT
NONE		TOTAL FORWARDED	
Total and forward to next column		TOTAL (Enter on Line 3 of Balance Form)	NONE

BALANCE FORM

Enter balance shown on this statement here—Line 1	$170.00
Add any deposits made after the date of this statement here—Line 2	NONE
(Use this space for any additional deposits made)	
Total of Line 1 and Line 2	
Enter total of checks outstanding and subtract here—Line 3	NONE
Your checkbook should show this BALANCE!	

Copyright © 1982 by Allyn and Bacon, Inc. Reproduction of this material is restricted to use with *A Guide to Teaching Consumer Mathematics*, by Peter A. Pascaris.

REPRODUCTION PAGE 30

MODIFIED BANK STATEMENT
(Use with Reproduction Page 19)

NON-NEGOTIABLE BANK OF KNOWHERE

STATEMENT PERIOD FROM 11-8-82 THRU 11-30-82

ACCOUNT NUMBER: R.P. 2-12

BEGINNING BALANCE	$1000.00
PLUS 2 DEPOSITS ITEMS	2003.22
LESS 8 CHECKS	1553.17
LESS SERVICE CHARGE	0.00
ENDING BALANCE	1450.05

Checks			Deposits	Date	Balance
$360.00				11/9	640.00
			570.72	11/14	1210.72
476.50	194.00			11/16	540.22
104.60				11/18	435.62
79.82			1432.50	11/21	1788.30
25.78	36.49			11/27	1726.03
275.98				11/30	1450.05

CHECKS OUTSTANDING

Check Number or Date of Issue	AMOUNT	Check Number or Date of Issue	AMOUNT
		TOTAL FORWARDED	

Total and forward to next column

TOTAL (Enter on Line 3 of Balance Form)

BALANCE FORM

Enter balance shown on this statement here—Line 1

Add any deposits made after the date of this statement here—Line 2

(Use this space for any additional deposits made)

Total of Line 1 and Line 2

Enter total of checks outstanding and subtract here—Line 3

Your checkbook should show this BALANCE!

Copyright © 1982 by Allyn and Bacon, Inc. Reproduction of this material is restricted to use with *A Guide to Teaching Consumer Mathematics*, by Peter A. Pascaris.

REPRODUCTION PAGE 31

MODIFIED BANK STATEMENT
(Use with Reproduction Page 24)

NON-NEGOTIABLE BANK OF KNOWHERE

ACCOUNT NUMBER: RP2-17

STATEMENT PERIOD FROM 5-1-82 THRU 5-21-82

BEGINNING BALANCE	$5000.00
PLUS 3 DEPOSITS ITEMS	1132.79
LESS 5 CHECKS	4811.15
LESS SERVICE CHARGE	0.00
ENDING BALANCE	1321.64

CHECKS			DEPOSITS	DATE	BALANCE
42.00				5/2	4958.00
			200.00	5/3	5158.00
538.25				5/6	4619.75
			582.79	5/7	5202.54
524.00	2,456.90		350.00	5/11	2571.64
1250.00				5/20	1321.64

CHECKS OUTSTANDING

Check Number or Date of Issue	AMOUNT	Check Number or Date of Issue	AMOUNT
3/29	257.00	TOTAL FORWARDED	
3/30	6784.28		
Total and forward to next column		TOTAL (Enter on Line 3 of Balance Form)	

BALANCE FORM

Enter balance shown on this statement here—Line 1	$5,000.00
Add any deposits made after the date of this statement here—Line 2	
(Use this space for any additional deposits made)	
Total of Line 1 and Line 2	
Enter total of checks outstanding and subtract here—Line 3	
Your checkbook should show this BALANCE!	

Copyright © 1982 by Allyn and Bacon, Inc. Reproduction of this material is restricted to use with *A Guide to Teaching Consumer Mathematics*, by Peter A. Pascaris.

REPRODUCTION PAGE 32

MODIFIED BANK STATEMENT
(Use with Reproduction Page 25)

NON-NEGOTIABLE BANK OF KNOWHERE

STATEMENT PERIOD FROM 1-10-82 THRU 1-27-82

ACCOUNT NUMBER: RP2-18

BEGINNING BALANCE	$350.00
PLUS 3 DEPOSITS ITEMS	300.00
LESS 7 CHECKS	577.67
LESS SERVICE CHARGE	0.00
ENDING BALANCE	72.33

Checks				DEPOSITS	DATE	BALANCE
$286.27					1/10	63.73
				100.00	1/13	163.73
35.21					1/14	128.52
6.36				100.00	1/15	222.16
97.31				100.00	1/17	224.85
5.85					1/20	219.00
4.67	142.00				1/27	72.33

CHECKS OUTSTANDING

Check Number or Date of Issue	AMOUNT	Check Number or Date of Issue	AMOUNT
		TOTAL FORWARDED	
Total and forward to next column		TOTAL (Enter on Line 3 of Balance Form)	

BALANCE FORM

Enter balance shown on this statement here—Line 1	$350.00	
Add any deposits made after the date of this statement here—Line 2		
(Use this space for any additional deposits made)		
Total of Line 1 and Line 2		
Enter total of checks outstanding and subtract here—Line 3		
Your checkbook should show this BALANCE!		

Copyright © 1982 by Allyn and Bacon, Inc. Reproduction of this material is restricted to use with *A Guide to Teaching Consumer Mathematics*, by Peter A. Pascaris.

REPRODUCTION PAGE 33

MODIFIED BANK STATEMENT
(Use with Reproduction Page 26)

NON-NEGOTIABLE BANK OF KNOWHERE

ACCOUNT NUMBER: RP2-19

STATEMENT PERIOD FROM 11-29-82 THRU 12-25-82

BEGINNING BALANCE	5000	00
PLUS 2 DEPOSITS ITEMS	4166	85
LESS 5 CHECKS	9122	38
LESS SERVICE CHARGE		70
ENDING BALANCE	43	77

			DEPOSITS	DATE	BALANCE
$5,250.00			456.78		206.78
31.89	7.57			12/9	167.32
168.92			3810.07	12/10	3808.47
3,764.00	70 s.c.			12/18	43.77

CHECKS OUTSTANDING

Check Number or Date of Issue	AMOUNT	Check Number or Date of Issue	AMOUNT
		TOTAL FORWARDED	
Total and forward to next column		TOTAL (Enter on Line 3 of Balance Form)	

BALANCE FORM

Enter balance shown on this statement here—Line 1	
Add any deposits made after the date of this statement here—Line 2	
(Use this space for any additional deposits made)	
Total of Line 1 and Line 2	
Enter total of checks outstanding and subtract here—Line 3	
Your checkbook should show this BALANCE!	

Copyright © 1982 by Allyn and Bacon, Inc. Reproduction of this material is restricted to use with *A Guide to Teaching Consumer Mathematics*, by Peter A. Pascaris.

REPRODUCTION PAGE 34

MODIFIED BANK STATEMENT
(Use with Reproduction Page 27)

NON-NEGOTIABLE BANK OF KNOWHERE

STATEMENT PERIOD FROM 7-1-82 THRU 7-15-82

ACCOUNT NUMBER: RP2-20

BEGINNING BALANCE		0	00
PLUS 4 DEPOSITS ITEMS		2244	81
LESS 9 CHECKS		2097	20
LESS SERVICE CHARGE		4	26
ENDING BALANCE		143	35

CHECKS			DEPOSITS	DATE	BALANCE
$150.00			500.00	7/1	350.00
28.00	30.00			7/4	292.00
47.83			928.14	7/5	1172.31
176.40	585.00		786.67	7/9	1197.58
			30.00	7/10	1227.58
55.00	774.97			7/14	397.61
250.00	4.26 S.C.			7/15	143.35

CHECKS OUTSTANDING

Check Number or Date of Issue	AMOUNT	Check Number or Date of Issue	AMOUNT
		TOTAL FORWARDED	
Total and forward to next column		TOTAL (Enter on Line 3 of Balance Form)	

BALANCE FORM

Enter balance shown on this statement here—Line 1		
Add any deposits made after the date of this statement here—Line 2		
(Use this space for any additional deposits made)		
Total of Line 1 and Line 2		
Enter total of checks outstanding and subtract here—Line 3		
Your checkbook should show this BALANCE!		

Copyright © 1982 by Allyn and Bacon, Inc. Reproduction of this material is restricted to use with *A Guide to Teaching Consumer Mathematics*, by Peter A. Pascaris.

DIAGNOSTIC SURVEY THREE

I. Multiply completely.

1) 7000
   × 4

2) a. 583 × 10 =
   c. 27 × 1000 =
   b. 621 × 100 =
   d. 349 × 10,000 =

3) a. 32 × 3
   b. 42 × 4
   c. 19 × 6
   d. 97 × 8

4) a. 323 × 3
   b. 814 × 2
   c. 237 × 3
   d. 918 × 5

   e. 187 × 9
   f. 895 × 7
   g. 702 × 4
   h. 608 × 8

5) a. 2341 × 2
   b. 8213 × 3
   c. 7956 × 8
   d. 3001 × 3

   e. 1006 × 7
   f. 9005 × 9

6) a. 43 × 21
   b. 386 × 43
   c. 863 × 439
   d. 683 × 903

   e. 803 × 702
   f. 8006 × 5009
   g. 6000 × 374

7) a. 21¢ × 4
   b. 87¢ × 9
   c. 73¢ × 17
   d. $3.42 × 2

   e. $8.32 × 3
   f. $7.15 × 7
   g. $5.94 × 8
   h. $3.21 × 93

   i. $9.57 × 76
   j. $8.09 × 53
   k. $6.04 × 509

II. Estimate these products. Whenever possible, do them mentally.

8) a. 317 × 6
   b. 218 × 5

9) a. 523 × 9
   b. 737 × 8

10) a. 314 × 28
    b. 537 × 46

11) a. $8.23 × 7
    b. $9.83 × 6

12) a. $16.47 × 18
    b. $53.54 × 31

13) a. $70.51 × 12
    b. $90.57 × 17

Copyright © 1982 by Allyn and Bacon, Inc. Reproduction of this material is restricted to use with *A Guidebook for Teaching Consumer Mathematics*, by Peter A. Pascaris.

DIAGNOSTIC SURVEY THREE   REPRODUCTION PAGE 35

III. Multiple purchases. Calculate the total cost.
   14) Translate each of these phrases into a shorter form using the @ symbol.
      a. The purchase of 7 buttons when each costs $.47.
      b. The bottle deposit for 3 cases of soda pop when each case contains 24 bottles and each bottle requires a ten-cent deposit.
      c. The same purchase as item b above when each case cost $7.56. (Do not include bottle deposit.)
   15) Calculate the total price for each purchase made in 14 a, b, and c above.

IV. Receiving change.
   16) Use the least number of standard U.S. bills and coins to make the following amounts.
      a. 72¢   b. Three dollars and ninety-one cents
      c. $9.33   d. $57.84

V. Sales tax.
   17) Change each of the following percent figures to its proper decimal fraction.
      a. 1% =   b. 1-1/2% =   c. 2% =
      d. 3-1/2% =   e. 5% =
   18) Use your "common sense" (intuition) to spot the most likely correct sales tax for the rate and purchase shown. Select the best answer from the choices given.
      A. 1% tax on a purchase of $8.91:   a) $.01   b) $8.91
                                          c) $.89   d) $.09
      B. 4% tax on a purchase of $99.89: a) $.04   b) $99.89
                                          c) $.99   d) $.40
                                          e) $4.00
   19) Figure the tax.
      a. $15.00 plus 1% tax.
      b. $17.88 plus 4% tax.
      c. $21.00 plus 3-1/2% tax.
      d. $37.53 plus 2-1/2% tax.
      e. $41.68 plus 1-1/2% tax.
      f. $27.44 plus 4-1/2% tax.
      g. $14.42 plus 3.5% tax.

VI. Using multiplication to compare costs.
   20) Assuming that the quality of the contents is the same, tell which package is the best buy.
      a. A package of 3 items for $.95 or a package of one item for $.33?
      b. A 38-ounce package for $1.17 or a 19-ounce package for $.59?
      c. An 840-g. box for $1.98 or a 210-g. box for $.49?
      d. A package of 2 18-ounce cans for $10.38 or a package of 2 3-ounce cans for $1.73?
      e. A package of 20 items for $2.59 or a package of 15 items for $1.95?

Copyright © 1982 by Allyn and Bacon, Inc. Reproduction of this material is restricted to use with *A Guidebook for Teaching Consumer Mathematics*, by Peter A. Pascaris.

REPRODUCTION PAGE 36

INDEPENDENCE SCALE THREE

Use the results of the Diagnostic Survey to rate your degree of independence for the skills listed below. Place an X under the heading that most closely matches your ability at this time.

		To successfully complete this skill, I:				
SKILL	Problem on Survey	Always Depend on Others	Often Depend on Others	Sometimes Depend on Others	Often Work Alone	Always Work Alone
Multiplication: Multiples of ten	1, 2					
One-digit multiplier	3, 4, 5					
Two-digit or more multiplier	6					
$ and ¢	7					
Estimate Products: Whole numbers	8, 9, 10					
$ and ¢	11, 12, 13					
Multiple Purchases: Using @	14					
Calculating	15					
Receiving Change:	16					
Sales Tax: Change % to decimal	17					
Common sense figure tax	18					
No rounding	19 (a)					
With rounding	19 (b-g)					
Compare Costs:	20					

Copyright © 1982 by Allyn and Bacon, Inc. Reproduction of this material is restricted to use with *A Guidebook for Teaching Consumer Mathematics*, by Peter A. Pascaris.

ESTIMATING PRODUCTS

PART ONE: Rounding Off Prior to Estimating.
1. Round off to the nearest ten.
   a) 8   b) 16   c) 93   d) 205   e) 1254
2. Round off to the nearest hundred.
   a) 361   b) 849   c) 774   d) 1032   e) 2050
3. Round off to the nearest dollar.
   a) 72¢   b) 51¢   c) $.84   d) $7.39
   e) $14.36   f) $43.56   g) $70.51
4. Round each to the nearest dollar, ten dollars, and hundred dollars.
   a) $762.47   b) $7845.50   c) $10,499.49

PART TWO: Estimate. Whenever Possible Estimate Mentally.

1) 317	2) 317	3) 93	4) 205	5) 254
X 6	X 8	X 16	X 47	X 14

6) 361	7) 849	8) 774	9) 1032	10) 2050
X 24	X 128	X 39	X 96	X 87

Write answers with a $ and decimal point.

11) 72¢	12) 51¢	13) $.84	14) $7.39	15) $14.36
X 14	X 94	X 136	X 45	X 54

16) $43.56	17) $70.51	18) $762.47	19) $7845.50	20) $10,499.49
X 27	X 104	X 16	X 24	X 12

Copyright © 1982 by Allyn and Bacon, Inc. Reproduction of this material is restricted to use with *A Guidebook for Teaching Consumer Mathematics*, by Peter A. Pascaris.

REPRODUCTION PAGE 38

MULTIPLE PURCHASES

Number Purchased	Article of Clothing	Price of One	Total Cost	4% Sales Tax	Cost Plus Tax	If You Paid This Amount in Cash	Figure the Change You Will Receive
2	Jackets	@ $23.00	$46.00	$1.84	$47.84	$50.00	$2.16
2	Neckties	@ 8.00	16.00	_____	_____	20.00	3.76
2	Shirts	@ 8.50	_____	_____	_____	20.00	_____
8	Pairs of socks	@ .80	_____	_____	_____	7.00	_____
2	Pair of shoes	@ 18.50	_____	_____	_____	40.00	_____
3	Handkerchiefs	@ 1.75	_____	_____	_____	10.51	_____
2	Dresses	@ 18.78	_____	_____	_____	50.06	_____
2	Coats	@ 35.98	_____	_____	_____	75.09	_____
4	Pairs of gloves	@ 5.75	_____	_____	_____	25.02	_____
2	Hats	@ 6.50	_____	_____	_____	20.02	_____
2	Sweaters	@ 15.29	_____	_____	_____	35.00	_____
4	Slacks	@ 4.89	_____	_____	_____	50.00	_____
3	Bracelets	@ 13.79	_____	_____	_____	45.02	_____

Verify the sum in the tax box by figuring 4% tax on the sum of the Total Cost column.

Copyright © 1982 by Allyn and Bacon, Inc. Reproduction of this material is restricted to use with *A Guidebook for Teaching Consumer Mathematics*, by Peter A. Pascaris.

COUNTING CHANGE

I. Use the *least number* of the coins and/or bills named below to make the designated total. However, you must use at *least one* of each coin or bill named, and you may not use any denomination that is not named.

1) Use only dimes and nickels to make 40¢.
2) Use only pennies, nickels, and quarters to make 73¢.
3) Use only nickels, dimes, and quarters to make $1.45.
4) Use only quarters and nickels to make $3.00.
5) Use only quarters and dimes to make $2.25.
6) Use only half dollars, quarters, dimes, and pennies to make $2.75.
7) Use only 1$ coins, quarters, and pennies to make $2.00.
8) Use only $1 and $2 bills, half dollars, and nickels to make $5.50.
9) Use only $2 bills, 1$ coins, quarters, dimes, and pennies to make $9.05.
10) Use only $2 and $10 bills, 1$ coins, nickels, and pennies to make $18.00.

II. Use the *least* number of coins and bills named to obtain the designated total. Select only those coins and bills that are needed.

Bills of $20, $10, and $2.
11) 40¢
13) $1.45
15) $2.65
17) $27.42
19) $103.29

Coins of $1, q, d, and p.
12) 73¢
14) $3.33
16) $5.55
18) $95.19
20) $139.94

III. You may use any of the standard U.S. coins and bills in the problems below. Use exactly the number of coins and bills stated to obtain the monetary value shown.

1) 4 coins to make 30¢.
2) 5 coins to make 65¢.
3) 7 coins to make 45¢.
4) 12 coins to make 80¢.
5) 19 coins to make $3.15.
6) 4 coins to make $1.35.
7) 5 coins to make 95¢.
8) 7 coins to make $1.05.
9) 12 coins to make $2.00.
10) 19 coins to make 85¢.
11) 5 bills and 17 coins to make $13.00.
12) 8 bills and 13 coins to make $66.75. Hint — bills = $65.
13) 10 bills and 20 coins to make $17.20. Hint — bills = $15.
14) 6 bills and 9 coins to make $44.85. Hint — bills = $40.
15) 9 bills and 76 coins to make $128.00. Hint — bills = $115.

Copyright © 1982 by Allyn and Bacon, Inc. Reproduction of this material is restricted to use with *A Guidebook for Teaching Consumer Mathematics*, by Peter A. Pascaris.

REPRODUCTION PAGE 39 COUNTING CHANGE

BONUS PROBLEM: There are many ways to give change equal to a dollar: four quarters; or three quarters, two dimes, and a nickel; one half dollar and five dimes; and so on. But can you have more than a dollar in coins and still not have change for a dollar? If the answer is yes, what is the *most* amount of money you can have in coins and still not be able to change a dollar? Name the total and the specific amounts of each coin as well as the total number of coins.

RECEIVING CHANGE

Tell what your change would be for each of these purchases for the amount given to the clerk. Name the change specifically (how many standard U.S. bills and coins) so that the smallest number of bills and coins is used.

	Amount of Purchase	Amount Given to Clerk	
Example A	$ .48	$ 1.	= *2* p and *1* h-d = $.52
Example B	$ 33.56	$ 50.	= Bills of *1* @ $10, *1* @ $5, *1* @ $1; coins of *1* q, *1* d, *1* n, *4* p = $16.44
1.	$ .37	$ 1.	
2.	$ .54	$ 1.	
3.	$ .58	$ 1.	
4.	$ 3.47	$ 5.	
5.	$ 2.82	$ 5.	
6.	$ 7.27	$ 10.	
7.	$ 4.51	$ 10.	
8.	$ 13.16	$ 20.	
9.	$ 1.15	$ 20.	
10.	$ 13.13	$100.	
11.	$ 7.72	$ 10.02	
12.	$ 4.78	$ 20.03	
13.	$ .86	$ 5.01	
14.	$ 1.88	$ 2.03	
15.	$ .65	$ 20.15	
16.	$ 4.48	$ 5.48	
17.	$ .81	$ 1.06	
18.	$ 17.61	$ 20.11	
19.	$ 73.26	$100.01	
20.	$147.31	$160.06	

Copyright © 1982 by Allyn and Bacon, Inc. Reproduction of this material is restricted to use with *A Guidebook for Teaching Consumer Mathematics*, by Peter A. Pascaris.

REPRODUCTION PAGE 41

PREPARING TO USE SALES TAX

I. A. Change the following % figures to a decimal fraction.
   1) 1%   2) 2%   3) 3%   4) 4%   5) 5%
   6) $\frac{1}{2}$%   7) $1\frac{1}{2}$%   8) $2\frac{1}{2}$%   9) $3\frac{1}{2}$%   10) $4\frac{1}{2}$%

   B. (Optional) Change the following decimal figures to percents.
   1) .03   2) .01   3) .045   4) .025   5) .005

II. Round off these figures to the nearest cent.
   1) $.17905   2) $.08245   3) $8.245   4) $17.905   5) $34.7219
   6) $42.2791   7) $48.4545   8) $55.5445   9) $86.095   10) $99.996

III. Use common sense (intuition) to select the best answer.
1) 1% tax on a purchase of $6.93.
a) $.69   b) $.01   c) $.07   d) $6.93

2) 2% tax on a purchase of $14.31.
a) $.29   b) $2.86   c) $.03   d) $28.62

3) 3% tax on a purchase of $21.17.
a) $63.51   b) $.06   c) $.64   d) $6.35

4) 4% tax on a purchase of $16.12.
a) $6.45   b) $.06   c) $64.48   d) $.64

5) 5% tax on a purchase of $40.10
a) $.20   b) $2.01   c) $.05   d) $20.05

6) 4% tax on a purchase of $87.88.
a) $3.52   b) $.04   c) $35.51   d) $351.52

7) 1-1/2% tax on a purchase of $6.93.
a) $10.39   b) $1.04   c) $.015   d) $.10

8) 2-1/2% tax on a purchase of $14.31.
a) $.36   b) $3.52   c) $.04   d) $.025

9) 3-1/2% tax on a purchase of $21.17.
a) $.035   b) $.74   c) $7.40   d) $.07

10) 4-1/2% tax on a purchase of $16.12.
a) $7.25   b) $.045   c) $.73   d) $.07

IV. Calculate the tax and determine the total price.
   1) 1% tax on $7.49.
   2) 1% tax on $24.50.
   3) 2% tax on $4.24.
   4) 2% tax on $62.25.
   5) 3% tax on $2.17.
   6) 3% tax on $20.84.
   7) 4% tax on $6.26.
   8) 4% tax on $24.62.
   9) 4% tax on $9.87.
   10) 4% tax on $37.63
   11) 5% tax on $8.59
   12) 5% tax on $140.09.
   13) 1-1/2% tax on $7.49.
   14) 1-1/2% tax on $24.50.
   15) 2-1/2% tax on $4.24.
   16) 2-1/2% tax on $62.25
   17) 3-1/2% tax on $2.17.
   18) 3-1/2% tax on $20.84.
   19) 4-1/2% tax on $6.26.
   20) 4-1/2% tax on $24.62.

Copyright © 1982 by Allyn and Bacon, Inc. Reproduction of this material is restricted to use with *A Guidebook for Teaching Consumer Mathematics*, by Peter A. Pascaris.

COMPARING PRICES

For each of the examples below, use multiplication to determine which package is the best buy.

1) A package of 3 items for $1.00 or 1 item for $.34?

2) A package of 5 items for $4.39 or 1 item for $.89?

3) A package of 4 items for $2.79 or 1 item for $.69?

4) A package of 7 items for $6.89 or 1 item for $.99?

5) A package of 4 items for $1.79 or 1 item for $.45?

6) A package of 6 items for $5.18 or 1 item for $.83?

7) A 48-oz. package for $.89 or a 16-oz. package for $.29?

8) A 48-oz. package for $1.15 or a 12-oz. package for $.29?

9) A 900-g. package for $3.71 or a 150-g. package for $.61?

10) A 13-oz. package for $1.49 or a 6-1/2-oz. package for $.76?

11) A 48-oz. can for 85¢ or a 6-oz. can for 10¢?

12) An 840-g. package for $1.59 or a 120-g. package for $.23?

13) An 8-oz. can for $.96 or a 3-oz. can for $.36?

14) A package of 12 items for $.95 or a package of 10 items for $.79?

15) A 24-oz. package for $.98 or a 16-oz. package for $.65?

16) A package of 42 items for $3.29 or a package of 56 items for $4.29?

17) An 800-g. package for $2.75 or a 300-g. package for $1.04?

18) A 100-g. package for $.67 or a 75-g. package for $.51?

19) A package of 4 24-oz. items for $2.89 or a package of 2 8-oz. items for $.49?

20) A package of 4 10-oz. items for $1.46 or a package of 3 5-oz. items for $.55?

Copyright © 1982 by Allyn and Bacon, Inc. Reproduction of this material is restricted to use with *A Guidebook for Teaching Consumer Mathematics,* by Peter A. Pascaris.

REPRODUCTION PAGE 43

DIAGNOSTIC SURVEY FOUR

I. A. Divide.
   1) 22)84   2) 3)906   3) 7)882   4) 13)286
   5) 15)1520   6) 4)$8.44   7) 8)$7.52

   B. Find the quotient and round off to the nearest cent.
   8) 4)$953   9) 3)$89   10) 13)$435.57

   C. Find the range and approximation.
   11) 27)162   12) 69)276,207

   D. Find the quotient.
   13) 8).736   14) 278)17.514   15) 214).1284

II. 16. Which of the following is the same as the arithmetic "average"?
    a) normal   b) mean   c) median   d) mode
    17. Find the mean of these numbers.
    a) 14, 21, 7, 8, 21, 19, 15, 26, 4
    b) $3.27, $1.16, $4.03, $.79, $1.16, $1.16, $3.27
    18. Find the median of the numbers shown in items 17a and 17b above.
    19. Find the mode of the numbers shown in items 17a and 17b above.

III. Use restaurant menus supplied by your instructor and use the form on Reproduction Page 46. Plan a complete balanced breakfast, lunch, and dinner for one person for one day. You must spend between $12 and $14 altogether.
    20) List the items ordered and their price.
    21) Determine 4% sales tax on the total.
    22) Determine 15% service tip on the total.
    23) Add the food cost, sales tax, and service tip and record.
    24) Determine the average cost per meal (include the tax and tip).

Copyright © 1982 by Allyn and Bacon, Inc. Reproduction of this material is restricted to use with *A Guidebook for Teaching Consumer Mathematics*, by Peter A. Pascaris.

DIAGNOSTIC SURVEY FOUR

IV. The following table indicates the cost of selected food items that make up a balanced meal and the number of servings available from each item. Use the table to answer the questions below. Round off to the nearest tenth of a cent.
   25) a) Determine the cost per serving of each item.
       b) Determine the cost per meal for one person.
   26) If you are to serve four people the same items, what would be the total cost for three meals for all four people?

Breakfast:	Cost per Item	Servings per Item
Juice	$ .96	12
Eggs	1.08	6
Bread	.77	16
Milk	.68	4
Cereal	1.49	15
Coffee	3.95	40
Lunch:		
Soft Drink	2.29	8
Macaroni and Cheese	.67	2
Broccoli	.88	5
Ice Cream	2.35	16
Dinner:		
Tea	1.99	48
Lettuce	.69	4
Rolls	.92	12
Pork Chops	2.45	4
Corn	.92	6
Applesauce	.95	5
Pudding	.49	4

Copyright © 1982 by Allyn and Bacon, Inc. Reproduction of this material is restricted to use with *A Guidebook for Teaching Consumer Mathematics*, by Peter A. Pascaris.

REPRODUCTION PAGE 44

INDEPENDENCE SCALE FOUR

Use the results of the Diagnostic Survey to rate your degree of independence for the skills listed below. Place an X under the heading that most closely matches your ability at this time.

		To successfully complete this skill, I:				
SKILL	Problem on Survey	Always Depend on Others	Often Depend on Others	Sometimes Depend on Others	Often Work Alone	Always Work Alone
I. Division: No Remainder	1-7					
Round Off	8-10					
Range	11					
Approximation	12					
Using Zeros	13-15					
II. Mean	16-17					
Median	18					
Mode	19					
III. Restaurant: Select Items	20					
Sales Tax	21					
Service Tip	22					
Add $ Figures	23					
Average Cost	24					
IV. Budget Cost Per Serving	25					
Multiple Costs	26					

Copyright © 1982 by Allyn and Bacon, Inc. Reproduction of this material is restricted to use with *A Guidebook for Teaching Consumer Mathematics*, by Peter A. Pascaris.

REPRODUCTION PAGE 45

MEAN, MEDIAN, MODE

1. Define each of these terms and identify the term that is synonymous with "average."
   a) Mean b) Median c) Mode

2. Find the mean, median, and mode for each of these lists.
   a) 5, 7, 1, 13, 5, 9, 7, 21, 7
   b) 62, 14, 96, 99, 62, 21, 17, 19, 62, 14
   c) $2.91, $15.26, $1.17, $26.14, $38.00, $1.17, $15.25
   d) $317.19, $876.90, $131.14, $876.90, $317.91, $319.17, $113.41
   e) $80.76, $807.63, $76.80, $80.76, $807.63, $76.81

3. When calculating the "average," why is it false to say that all figures are "close to the average"? Explain your answer and cite examples.

4. Just for Fun: Try these and see if you can recognize a pattern.
   a) 22)$11.22 b) 34)$17.34 c) 46)$23.46 d) 58)$29.58
   e) 62)$31.62 f) 74)$37.74 g) 86)$43.86 h) 98)$49.98
   i) $16.32 ÷ what number? = $.51   j) what number? ÷ 26 = $.51

Copyright © 1982 by Allyn and Bacon, Inc. Reproduction of this material is restricted to use with *A Guidebook for Teaching Consumer Mathematics*, by Peter A. Pascaris.

REPRODUCTION PAGE 46

I. AVERAGES

A. Find the average expense per day for each problem. Show all your work neatly on loose-leaf paper. Round off all answers to the nearest cent.

1. $93.00, the sum for 4 days.
2. $167.70, the sum for 6 days.
3. $185.62, the sum for 8 days.
4. $274.21, the sum for 12 days.
5. $359.14, the sum for 15 days.
6. $418.16, the sum for 18 days.
7. $465.65, the sum for 21 days.
8. $500.12, the sum for 23 days.
9. $606.58, the sum for 26 days.
10. $639.73, the sum for 28 days.

B. Find the cost per person.

11. $235.13 for 10 people.
12. $2,351.34 for 100 people.
13. $23,513.42 for 1000 people.
14. $235,134.23 for 10,000 people.
15. $2,351,342.31 for 100,000 people.

C. See if you can discover a shortcut way to find the averages for numbers 11-15 above. Write an explanation of your shortcut and show it to your teacher.

II. AVERAGE AND MEDIAN

A. Below are examples of typical restaurant expenses for a single person traveling on vacation. Find the average expense per day for each problem and find the median expense per day.

1. $ 6.37 1st day
   5.91 2d day
   13.11 3d day

2. $11.33 1st day
   9.27 2d day
   15.69 3d day
   21.47 4th day
   10.25 5th day
   16.94 6th day
   20.76 7th day

3. $13.95 1st day
   17.90 2d day
   16.32 3d day
   10.20 4th day
   8.99 5th day
   7.52 6th day
   16.09 7th day
   11.03 8th day
   13.55 9th day
   27.65 10th day
   12.08 11th day

4. Consider these expenses for consecutive days:
$16.79, $18.31, $12.27, $9.99, $13.42, $19.07,
$15.27, $18.81, $12.62, $15.03, $16.09, $13.21,
$21.52, $6.39, $13.91, $19.34, $21.13, $9.57, $23.42.

B. Compare the median and the average. Which is a more accurate figure to use for planning the next day's expense if the traveler is on a tight budget? Explain.

Copyright © 1982 by Allyn and Bacon, Inc. Reproduction of this material is restricted to use with *A Guidebook for Teaching Consumer Mathematics*, by Peter A. Pascaris.

REPRODUCTION PAGE 47

RESTAURANT MATH

Meal	Items Ordered	Price	Determine ____ % Tax	Determine ____ % Tip	Total Cost with Tax and Tip
	Food Total				
	Food Total				
	Food Total				

Copyright © 1982 by Allyn and Bacon, Inc. Reproduction of this material is restricted to use with *A Guidebook for Teaching Consumer Mathematics*, by Peter A. Pascaris.

REPRODUCTION PAGE 48

BUDGETING FOR A PARTY

Below is a list of items to be served at a dinner party along with the number of servings available per item and the cost of each item. Assuming each person takes one serving of each item, determine the cost per serving of each item and the total cost to serve a party of:

a) 8   b) 12   c) 15   d) 20   e) 64

Item	Servings per Item	Cost per Item
8 lb. or 128 oz. Ham (3.63 kg.)	64	8 lb. for $23.84
1 lb. or 16 oz. Cheese (454 g.)	8	1 lb. for $2.94
5 lbs. or 80 oz. Turkey Roll (2.27 kg.)	40	5 lb. for $10.25
2 lbs. or 32 oz. Cabbage Salad (.91 kg.)	8	2 lb. for $3.12
5 lbs. or 80 oz. Potato Salad (2.27 kg.)	16	5 lb. for $4.45
1-1/2 lb. or 24 oz. Pickle Variety (681 g.)	24	1-1/2 lb. for $1.49
1 doz. or 12 Dinner Rolls	6	1 doz. for $1.57
1 lb. or 16 oz. Coffee (454 g.)	40	1 lb. for $4.29
1 gal. or 128 oz. Ice Cream (3.79 l.)	28	1 gal. for $4.93
2 liters of Soft Drink	8	2 liters for $1.59
14 oz. Mixed Nuts (397 g.)	14	14 oz. for $4.19

For Extra Credit: Determine the amount of each item needed to serve the same numbers of people as stated above. If the items have to be purchased only in the quantities stated, how much left over would there be for each item?

Copyright © 1982 by Allyn and Bacon, Inc. Reproduction of this material is restricted to use with *A Guidebook for Teaching Consumer Mathematics*, by Peter A. Pascaris.

DIAGNOSTIC SURVEY FIVE

Show all work

I. Miles Traveled (odometer readings).
Determine the number of miles traveled based on these odometer readings. For numbers 3 and 4, round to the nearest mile.

	(1)	(2)	(3)	(4)
Mileage at the *start*	2320	26712	32706.3	34429.7
Mileage at the *conclusion*	7546	32501	34429.7	36218.2

II. Fuel Purchases
Find the correct answer to replace the question marks for the problems below.

A. Type I Fuel Problems. Determine the total cost. Round to the nearest cent.
5) 10 gal. @ 132¢ = ?
6) 15.6 gal. @ 130¢ = ?
7) 12.3 gal. @ 133.8¢ = ?
8) 8.7 gal. @ 147.8¢ = ?

B. Type II Fuel Problems. Determine the cost per gallon. Round to the nearest tenth of a cent.
9) 10 gal. @ ___?___ = $13.60.
10) 12 gal. @ ___?___ = $17.50.
11) 13.8 gal. @ ___?___ = $19.84.
12) 11.3 gal. @ ___?___ = $16.70.

C. Type III Fuel Problems. Determine the number of gallons purchased. Round to the nearest tenth of a gallon.
13) ___?___ gal. @ 136¢ = $14.96.
14) ___?___ gal. @ 155.8¢ = $18.70.
15) ___?___ gal. @ 149.8¢ = $28.92.
16) ___?___ gal. @ 147.8¢ = $22.76.

III. Miles Per Gallon (mpg)
A. In each of the following problems, the "miles" figure tells how many miles a car has driven and the "fuel" figure indicates the number of gallons of fuel that the car used for that journey. *Find the miles per gallon (mpg) for each car.* Round answers to the nearest mile.

	(17)	(18)	(19)	(20)
miles	160	306	174	226
fuel (gal.)	16	18	12	8.6

B. Find the amount of fuel used. Round to the nearest tenth of a gallon.

	(21)	(22)	(23)	(24)	(25)
miles	180	210	288	490	603
mpg	10	21	18	28	44

IV. Mentally estimate these *without* paper and pencil or calculator.
26) Estimate the total cost. 10.1 gal. @ $1.439.
27) Estimate the total cost. 9.5 gal. @ $1.379.
28) Estimate the price per gallon. 12.1 gal. @ ___?___ = $16.90.
29) Estimate the number of gallons. ___?___ gal. @ $1.428 = $28.76.
30) Estimate the miles per gallon. 309 miles on 9.7 gal.

Copyright © 1982 by Allyn and Bacon, Inc. Reproduction of this material is restricted to use with *A Guidebook for Teaching Consumer Mathematics*, by Peter A. Pascaris.

REPRODUCTION PAGE 50

INDEPENDENCE SCALE FIVE

Use the results of the Diagnostic Survey to rate your degree of independence for the skills listed below. Place an X under the heading that most closely matches your ability at this time.

SKILL	Problem on Survey	Always Depend on Others	Often Depend on Others	Sometimes Depend on Others	Often Work Alone	Always Work Alone
I. Determining Mileage	1-4					
II. A. Compute Total Fuel Cost Type I	5-8					
II. B. Compute Cost Per Gal. Type II	9-12					
II. C. Compute Gal. Purchased Type III	13-16					
III. A. Determine mpg	17-20					
III. B. Determine Fuel Consumption	21-25					
IV. Mentally Estimate	26-30					
V. Round Off To: A. Nearest Mile	3-4 17-20					
B. Tenth of a Cent	9-12					
C. Tenth of a Gallon	13-16 21-25					

Copyright © 1982 by Allyn and Bacon, Inc. Reproduction of this material is restricted to use with *A Guidebook for Teaching Consumer Mathematics*, by Peter A. Pascaris.

TRANSPORTATION COMPUTATION

I. Below are odometer readings for a car used on a nine-day vacation.
   A. Find the miles traveled each day when the starting mileage was 34,113.
      Mileage at end of:

Day 1	Day 2	Day 3	Day 4	Day 5
34,235	34,347	34,568	34,889.7	35,183.9

Day 6	Day 7	Day 8	Day 9
35,312.7	35,798.8	36,111.2	36,400.0

   B. Find the total miles traveled for all nine days.

   C. How many miles were traveled the first four days?

   D. How many miles were traveled from the fifth through the seventh days?

   E. Find the average miles traveled per day.

II. Rounding off and estimating.

   A. Round to nearest mile.
      1) 257.4   2) 346.7   3) 760.5   4) 389.4   5) 259.6

   B. Round to the nearest *tenth* of a cent. Write each figure with a $.
      6) 146.86¢   7) 142.94¢   8) $1.4385   9) $1.39909   10) $1.5888

   C. Round to the nearest tenth of a gallon.
      11) 11.64   12) 12.85   13) 13.56   14) 14.04   15) 15.76

   D. 16) Mentally estimate the sum of the rounded mile figures in item A above.
      17) Mentally estimate the sum of the rounded gallon figures in item C above.

Copyright © 1982 by Allyn and Bacon, Inc. Reproduction of this material is restricted to use with *A Guidebook for Teaching Consumer Mathematics*, by Peter A. Pascaris.

REPRODUCTION PAGE 52

THE COST OF FUEL

I. **Type I:** Find the total cost for the following amounts of gasoline if each gallon costs 147.9¢. Round to the nearest cent.
1) 13 gal.   2) 17 gal.   3) 15.3 gal.   4) 19.1 gal.   5) 16.6 gal.

II. **Type II:** Find the cost per gallon. Round to the nearest tenth of a cent.
1) 11 gal. @ _____ = $15.73.   2) 13.8 gal. @ _____ = $16.52.
3) 13.8 gal. @ _____ = $18.8.

III. **Type III:** Find the number of gallons purchased.
A. Find the number of gallons purchased. Round to the nearest tenth of a gallon.
1) _____ gal. @ 137.9¢ = $13.79.   6) _____ gal. @ 145.9¢ = $23.76.
2) _____ gal. @ 137.8¢ = $20.68.   7) _____ gal. @ 145.9¢ = $10.78.
3) _____ gal. @ 137.8¢ = $16.54.   8) _____ gal. @ 135.8¢ = $24.58.
4) _____ gal. @ 151.9¢ = $29.14.   9) _____ gal. @ 137.8¢ = $15.30.
5) _____ gal. @ 151.9¢ = $13.52.   10) _____ gal. @ 149.8¢ = $16.32.

B. Mentally estimate the number of gallons purchased to the nearest gallon.
11) _____ gal. @ 145.9¢ = $14.59   14) _____ gal. @ 145.8¢ = $29.28.
12) _____ gal. @ 139.8¢ = $27.90   15) _____ gal. @ 149.9¢ = $30.08
13) _____ gal. @ 139.8¢ = $20.98.

Copyright © 1982 by Allyn and Bacon, Inc. Reproduction of this material is restricted to use with *A Guidebook for Teaching Consumer Mathematics*, by Peter A. Pascaris.

THE COST OF FUEL

I. Type I: Find the total cost for fuel.

A. Find the total expense. Round to the nearest cent.

	Gallons of Gas	@	Find Total Expense		Gallons of Gas	@	Find Total Expense
1.	11	136¢		6.	9.1	134.9¢	
2.	14	137.8¢		7.	15.5	133.9¢	
3.	11.4	137.9¢		8.	20.8	139.9¢	
4.	13.6	145.9¢		9.	22.2	131.9¢	
5.	17.3	149.9¢		10.	12.7	138.9¢	

B. Mentally estimate the total expense.

11.	10.4	136¢	14.	10.7	136¢
12.	10.2	137.9¢	15.	17.7	143.8¢
13.	9.5	149.9¢			

II. Type II: Find the cost per gallon.

A. Find the cost for each gallon. Round to the nearest tenth of a cent.

1) 10 gal. @ _____ = $13.99.
2) 15 gal. @ _____ = $20.86.
3) 12 gal. @ _____ = $17.26.
4) 19.2 gal. @ _____ = $28.08.
5) 16.8 gal. @ _____ = $23.16.
6) 16.3 gal. @ _____ = $23.76.
7) 8.9 gal. @ _____ = $11.74.
8) 11.1 gal. @ _____ = $16.40.
9) 14.8 gal. @ _____ = $19.22.
10) 9.6 gal. @ _____ = $13.24.

B. Mentally estimate the cost for each gallon to the nearest cent.

11) 10.4 gal. @ _____ = $13.99.
12) 9.6 gal. @ _____ = $14.58.
13) 13.3 gal. @ _____ = $18.28.
14) 14.8 gal. @ _____ = $21.04.
15) 11.5 gal. @ _____ = $17.92.

Copyright © 1982 by Allyn and Bacon, Inc. Reproduction of this material is restricted to use with *A Guidebook for Teaching Consumer Mathematics*, by Peter A. Pascaris.

FUEL EFFICIENCY

I. Miles per gallon (mpg).

   A. Find the mpg. Round to the nearest whole number.

   1) 200 miles, 10 gal.
   2) 260 miles, 13 gal.
   3) 143 miles, 11 gal.
   4) 186 miles, 12.4 gal.
   5) 375 miles, 16.8 gal.
   6) 158 miles, 13.6 gal.
   7) 353 miles, 20.3 gal.
   8) 277 miles, 9.3 gal.
   9) 459 miles, 11.9 gal.

   B. Mentally estimate the mpg to the nearest whole number.

   10) 280 miles, 10 gal.
   11) 395 miles, 19.6 gal.
   12) 297 miles, 10.4 gal.
   13) 304 miles, 14.5 gal.
   14) 357 miles, 18.4 gal.
   15) 757 miles, 15.3 gal.

II. Below is a record of miles traveled for an eight-day vacation trip.

   Day 1: *122* miles   2: *112* miles   3: *221* miles   4: *321* miles
   5: *294* miles   6: *129* miles   7: *486* miles   8: *313* miles

   A. If the car averaged 20 mpg, find the gallons used each day.

   B. If the fuel costs 138¢ per gallon, find the daily fuel cost.

   C. Find the total expense for fuel.

   D. Determine the average spent on fuel per day.

III. Find the fuel used (nearest tenth of a gallon) and the total fuel expense for each problem below.

	Miles Traveled	mpg	Cost per Gal.		Miles Traveled	mpg	Cost per Gal.
1)	122	12	143.9¢	6)	129	18	147.8¢
2)	112	14	145.9¢	7)	486	44	139.9¢
3)	221	19	151.9¢	8)	313	9	149.5¢
4)	321	31	141.9¢	9)	289	21	145.8¢
5)	294	27	148.3¢	10)	608	37	153.9¢

Copyright © 1982 by Allyn and Bacon, Inc. Reproduction of this material is restricted to use with *A Guidebook for Teaching Consumer Mathematics*, by Peter A. Pascaris.

REPRODUCTION PAGE 55

RENTING A CAR

I. Below is a list of car rental fees. Using the figures shown, calculate how much it will cost to rent each car for seven days, traveling a total of 825 miles, and figuring 4% sales tax.

       1) Bunny: $12 per day, 12¢ per mile.
       2) Palomino: $13 per day, 13¢ per mile.
       3) Malarké: $15 per day, 15¢ per mile.
       4) Fenderbird: $18 per day, 18¢ per mile.
       5) Oceanental: $19 per day, 19¢ per mile.
       6) Rolls Nice: $75 per day, 75¢ per mile.

II. When you rent a car, you must also buy your own gasoline. Calculate the number of gallons of gas you must buy for each of the following cars using the listed miles per gallon and 825 miles traveled for each car.

1) Bunny, 44 mpg.   2) Palomino, 31 mpg.
3) Malarké, 26 mpg.   4) Fenderbird, 20 mpg.
5) Oceanental, 18 mpg.   6) Rolls Nice, 12 mpg.

III. If each gallon of fuel cost $1.479, calculate the total fuel cost for each car.

IV. Now determine the complete cost for using a rental car including cost for 7 days, 825 miles, sales tax, and gasoline.

V. The following list includes some of the features you may want in a car. Rank-order each item listing the most important first, the least important last.

Rental Cost	Model Name	Size (outside)
Roominess (inside)	Body Style	General Appearance
Color	Speed and Power	Gas Mileage

Copyright © 1982 by Allyn and Bacon, Inc. Reproduction of this material is restricted *to use with A Guidebook for Teaching Consumer Mathematics,* by Peter A. Pascaris.

DIAGNOSTIC SURVEY SIX

I. A. Rewrite these from minutes to common fractions of an hour.
   1) 30 min.   2) 15 min.   3) 45 min.
   4) 10 min.   5) 20 min.   6) 6 min.

B. Express these common fractions in lowest terms.

7) $\frac{12}{60}$ 8) $\frac{24}{60}$ 9) $\frac{36}{60}$ 10) $\frac{42}{60}$ 11) $\frac{65}{60}$ 12) $\frac{25}{60}$

13) $\frac{35}{60}$ 14) $\frac{50}{60}$ 15) $\frac{55}{60}$ 16) $\frac{4}{60}$ 17) $\frac{16}{60}$ 18) $\frac{52}{60}$

C. Express these fractions in 60ths.

19) $\frac{1}{2}$ 20) $\frac{1}{4}$ 21) $\frac{3}{4}$ 22) $\frac{1}{3}$ 23) $\frac{2}{3}$ 24) $\frac{1}{6}$

25) $\frac{1}{10}$ 26) $\frac{5}{10}$ 27) $\frac{4}{5}$ 28) $\frac{11}{15}$ 29) $\frac{1}{12}$ 30) $\frac{3}{20}$

D. Change the common fractions of B above to decimal fractions. (31-42)

E. Multiply these common fractions.

43) $\frac{1}{5} \times \frac{1}{3}$ 44) $\frac{1}{2} \times \frac{2}{3}$ 45) $\frac{3}{4} \times \frac{8}{15}$ 46) $\frac{5}{6} \times 10$

47) $48 \times \frac{7}{12}$ 48) $5\frac{5}{6} \times 4\frac{4}{5}$ 49) $3\frac{9}{10} \times 2\frac{2}{3}$ 50) $3\frac{9}{12} \times 2\frac{14}{15}$

F. Divide these common fractions.

51) $\frac{1}{3} \div \frac{3}{4}$ 52) $\frac{3}{5} \div \frac{9}{10}$ 53) $\frac{5}{6} \div \frac{7}{12}$ 54) $\frac{3}{4} \div 6$

55) $2 \div \frac{3}{5}$ 56) $1\frac{4}{5} \div \frac{9}{10}$ 57) $2\frac{5}{6} \div 4\frac{3}{12}$ 58) $7 \div 4\frac{9}{10}$

II. A. Change these feet measurements to decimal fractions of a mile (to three decimal places).
   59) 2640 ft.   60) 1320 ft.   61) 660 ft.   62) 1760 ft.

B. Change these meters to kilometers.
   63) 278 m.   64) 521 m.   65) 893 m.   66) 333 m.

C. Find the distance traveled (nearest whole unit).
   67) 50 mph for 4 hours   68) 38 mph for 3-1/2 hours
   69) 40 mph, 4 hr. 6 min.   70) 80 kph for 4 hours
   71) 61 kph for 3-1/2 hours   72) 64 kph, 4 hr. 6 min.

D. Find the rate of speed (nearest whole unit).
   73) 255 miles in 5 hours   74) 75 miles in 2-1/2 hours
   75) 410 km. in 5 hours   76) 120 km. in 2-1/2 hours

E. Find the time traveled (nearest minute).
   77) 156 miles, 52 mph   78) 559 miles, 43 mph
   79) 252 kilometers, 84 kph   80) 897 kilometers, 69 kph

Copyright © 1982 by Allyn and Bacon, Inc. Reproduction of this material is restricted to use with *A Guidebook for Teaching Consumer Mathematics*, by Peter A. Pascaris.

REPRODUCTION PAGE 57

INDEPENDENCE SCALE SIX

Use the results of the Diagnostic Survey to rate your degree of independence for the skills listed below. Place an X under the heading that most clearly matches your ability at this time.

		To successfully complete this skill, I:				
SKILL	Problem on Survey	Always Depend on Others	Often Depend on Others	Sometimes Depend on Others	Often Work Alone	Always Work Alone
Change minutes to fractions of an hour	1-6					
Reduce fractions	7-18					
Express in higher terms	19-30					
Change to decimals	31-42					
Multiply	43-50					
Divide	51-58					
Feet to miles	59-62					
Meters to kilometers	63-66					
Distance	67-72					
Rate	73-76					
Time	77-80					

Copyright © 1982 by Allyn and Bacon, Inc. Reproduction of this material is restricted to use with *A Guidebook for Teaching Consumer Mathematics*, by Peter A. Pascaris.

CONVERTING UNITS

I. Change these feet measurements to decimal fractions of a mile (to three decimal places).
1) 2640 ft.   2) 1320 ft.   3) 660 ft.   4) 330 ft.   5) 528 ft.
6) 264 ft.   7) 132 ft.   8) 1056 ft.   9) 1584 ft.   10) 2112 ft.
11) 3168 ft.   12) 3696 ft.   13) 4752 ft.   14) 1980 ft.   15) 3300 ft.

II. Change the decimal answers from above to their simplest common fraction form.

III. Change these feet measurements first to common fractions in simplest form, and second to decimal fractions to three decimal places.
1) 1760 ft.   2) 3520 ft.   3) 880 ft.   4) 4400 ft.   5) 440 ft.
6) 2740 ft.   7) 2000 ft.   8) 1000 ft.   9) 2200 ft.   10) 1687 ft.

IV. Change these meters to kilometers.
1) 1000 m.   2) 667 m.   3) 333 m.   4) 871 m.   5) 901 m.
6) 500 m.   7) 50 m.   8) 5 m.   9) 5280 m.   10) 10,000 m.

V. Change these kilometers to meters.
1) .271 km.   2) .953 km.   3) .006 km.   4) .291 km.   5) .047 km.
6) 1 km.   7) 8 km.   8) 6.5 km.   9) 65 km.   10) 3.75 km.

VI. Change the kilometer measures in Part V to miles.

VII. Change these mile measures to kilometers.
1) 1 mile   2) 5 miles   3) 27 miles   4) 9.5 miles   5) 0.8 mile

Copyright © 1982 by Allyn and Bacon, Inc. Reproduction of this material is restricted to use with *A Guidebook for Teaching Consumer Mathematics*, by Peter A. Pascaris.

D = RT

A. Find the distance traveled (units may be metric or customary).

	Rate of Speed	Time		Rate of Speed	Time		Rate of Speed	Time
1)	50	4 hours	6)	50	4-1/2 hours	11)	40	6-1/2 hours
2)	45	4 hours	7)	48	3-1/4 hours	12)	52	2-1/4 hours
3)	47	4 hours	8)	52	6 hr. 30 min.	13)	54	6 hr. 20 min.
4)	36	6 hours	9)	52	4 hr. 45 min.	14)	44	5 hr. 45 min.
5)	51	8 hours	10)	45	5 hr. 10 min.	15)	51	4 hr. 40 min.

B. Find the rate of speed (units may be metric or customary).

	Distance	Time		Distance	Time		Distance	Time
1)	200	5 hours	6)	168	3-1/2 hours	11)	125	2-1/2 hours
2)	90	2 hours	7)	225	6-1/4 hours	12)	130	3-1/4 hours
3)	405	9 hours	8)	286	5 hr. 30 min.	13)	129	2-3/4 hours
4)	320	7 hours	9)	143	3 hr. 15 min.	14)	406	5 hr. 30 min.
5)	318	6 hours	10)	189	2 hr. 6 min.	15)	285	4 hr. 45 min.

C. Find the time (units may be metric or customary).

	Distance	Rate of Speed		Distance	Rate of Speed		Distance	Rate of Speed
1)	450	45	6)	260	40	11)	144	36
2)	400	50	7)	117	52	12)	242	44
3)	559	43	8)	132	48	13)	279.5	86
4)	360	45	9)	342	54	14)	320	48
5)	405	45	10)	192	60	15)	324	45

Copyright © 1982 by Allyn and Bacon, Inc. Reprodcution of this material is restricted to use with *A Guidebook for Teaching Consumer Mathematics*, by Peter A. Pascaris.

THREE-DAY VACATION TRIP

I. Auto
   A. Find the miles traveled each day, total mileage, and average miles per day.
Start	1st Day End	2d Day End	3d Day End
2170	2381	2512	2981

   B. Determine the fuel costs.
   1st Day 11.1 gal. @ 141.9¢. Find the total spent.
   2d Day 6.5 gal. @ ___?___ = $9.07. Find the cost per gallon.
   3d Day ___?___ gal. @ 145.7¢ = $29.43. Find the gallons purchased.

   C. Determine the mpg for each day and the average mpg.

II. Food
   A. Assuming a 4-1/2% tax and 15% tip, determine the cost for each of these meals and for the total restaurant cost.
1st Day	2d Day	3d Day
$8.85	$13.55	$7.20 and $23.45

   B. If $37.84 was also spent for groceries, determine the total food cost and the average cost per day.

III. Lodging: Assuming a 5-1/4% tax (5.25%), determine the cost for lodging each day, the total lodging cost, and average cost per day.
   1st Day: Motel, $29.85   2d Day: Camp, $4.50   3d Day: Camp, $5.75

IV. Entertainment and Recreation: Determine the total expense.
   Movie: $9.50; Amusement Park: $25.50; Souvenirs: $13.75 plus 3-1/2% tax;
   Film: $3.58 plus 4% tax; Tanning Lotion: $3.15 plus 4-1/4% tax; Books: $4.50.

V. Summary: Determine the total spent for all three days, the average spent per day, and average spent per mile.

Copyright © 1982 by Allyn and Bacon, Inc. Reproduction of this material is restricted to use with *A Guidebook for Teaching Consumer Mathematics*, by Peter A. Pascaris.

EXPENSES PREPARING FOR VACATION

I. Auto:
   A. "New" purchases:

   B. Maintenance and repair:

II. Necessities (food, clothing, equipment, and miscellaneous):

III. Summary of all expenses:

REPRODUCTION PAGE 62

VACATION TRIP LOG

I. Auto:

    A. Record the miles traveled this day, the total mileage so far, and the average miles per day.

    B. Determine the fuel costs.

    C. Determine the mpg for this day and the average mpg up to date.

II. Food:

III. Lodging:

IV. Entertainment and Recreation:

V. Summary: Determine the total spent today, the total spent up to date, the average spent per day, and the average spent per mile.

Copyright © 1982 by Allyn and Bacon, Inc. Reproduction of this material is restricted to use with *A Guidebook for Teaching Consumer Mathematics,* by Peter A. Pascaris.

DIAGNOSTIC SURVEY SEVEN

I. **Measurement**
   A. Write the standard units of length, area, volume, capacity, and weight (or mass) for both the customary system and the metric system.
   B. Give a concrete example of the approximate size of these units.
   1) inch   2) meter   3) liter   4) gallon   5) millimeter
   6) yard   7) ounce   8) pound   9) kilogram   10) centimeter
   11) mile   12) quart   13) gram   14) kilometer   15) milliliter
   C. What do these prefixes mean?
   16) kilo-   17) centi-   18) milli-
   D. Fill in the blanks.
   19) _____ inches =   20) _____ foot = 1 yard
   21) _____ millimeter =   22) _____ centimeter = 1 meter
   23) _____ ounces =   24) _____ quart = 1 gallon
   25) _____ milliliters = 1 liter
   26) _____ grams = 1 kilogram   27) _____ ounces = 1 pound

II. **Fractions**
   A. Express these in feet using fractions (reduced).
   28) 6 ft. 6 in.   29) 8 ft. 9 in.   30) 5 ft. 3 in.   31) 9 ft. 7 in.
   B. Express these in quarts using fractions (reduced).
   32) 4 qt. 16 oz.   33) 7 qt. 24 oz.   34) 2 qt. 8 oz.   35) 8 qt. 23 oz.
   C. Express these in pounds using fractions (reduced).
   36) 9 lb. 4 oz.   37) 3 lb. 6 oz.   38) 5 lb. 12 oz.   39) 7 lb. 15 oz.
   D. Add these fractions

   40) $7\frac{5}{16}$ $+3\frac{7}{16}$ 41) $16\frac{5}{8}$ $+23\frac{7}{8}$ 42) $6\frac{7}{8}$ $+3\frac{29}{32}$ 43) $4\frac{11}{12} + 3\frac{3}{8} + 1\frac{9}{16}$

   E. Subtract these fractions.

   44) $8\frac{7}{10}$ $-3\frac{1}{2}$ 45) 9 $-2\frac{17}{32}$ 46) $7\frac{5}{16} - 6\frac{3}{4}$ 47) $11\frac{3}{4}$ from $12\frac{9}{16}$

Copyright © 1982 by Allyn and Bacon, Inc. Reproduction of this material is restricted to use with *A Guidebook for Teaching Consumer Mathematics*, by Peter A. Pascaris.

REPRODUCTION PAGE 63   **DIAGNOSTIC SURVEY SEVEN**

III. A. Use common multiples to determine the best buy.
    48) 6 l. for $1.59 or 10 l. for $2.59?
    49) 16 oz. for $.59 or 18 oz. for $.65?
  B. Use proportions to find the best buy.
    50) 24 oz. for $.98 or 16 oz. for $.65?
    51) 100 g. for $.67 or 75 g. for $.51?
  C. Find the unit price and determine the best buy.
    52) 1 lb. 4 oz. at 67¢ or 5 lb. 4 oz. at $2.59?
    53) 567 g. at 67¢ or 2.4 kg. at $2.59?
  D. Complete the following table by filling in the blanks (54-61).

Food	Cost per kg.	Allowable Serving	Estimated Waste per kg.	Edible Servings per kg.	Cost per Serving
ground beef	$4.36	100 g.	none	10	$.436
whole chicken	$2.10	125 g.	375 g.	5	
chicken breast	$4.15	125 g.	75 g.	7.4	
steak (w/bone)	$6.59	100 g.	100 g.		
fresh whole fish	$5.43	175 g.	250 g.		
frozen fish fillet	$6.90	175 g.	none		

IV. A. Find the cost of the following.
    62) Cementing a driveway 3.5 m. by 4.9 m. at $19.75 per square meter.
    63) Carpeting a floor 16 ft. 3 in. by 14 ft. 4 in. at $9.95 per square foot.
  B. Determine the capacity of the freight cars.
    64) 12.7 m. by 3.2 m. by 3.1 m.
    65) 36 ft. 3 in. by 9 ft. 8 in. by 9 ft. 2 in.

Copyright © 1982 by Allyn and Bacon, Inc. Reproduction of this material is restricted to use with *A Guidebook for Teaching Consumer Mathematics*, by Peter A. Pascaris.

REPRODUCTION PAGE 64

INDEPENDENCE SCALE SEVEN

Use the results of the Diagnostic Survey to rate your degree of independence for the skills listed below. Place an X under the heading that most closely matches your ability at this time.

	\multicolumn{6}{c}{To successfully complete this skill, I:}
SKILL	Problem on Survey	Always Depend on Others	Often Depend on Others	Sometimes Depend on Others	Often Work Alone	Always Work Alone
Standard Units	Part I A & 1-27					
—Size	1-15					
—Prefixes	16-18					
—Converting	19-27					
Fractions —Parts of Units	28-39					
—Adding	40-43					
—Subtracting	44-47					
Best Buy —Multiples	48-49					
—Proportions	50-51					
—Unit Price	52-53					
—Per Serving	54-61					
Measurement —Area	62-63					
—Volume	64-65					

Copyright © 1982 by Allyn and Bacon, Inc. Reproduction of this material is restricted to use with *A Guidebook for Teaching Consumer Mathematics*, by Peter A. Pascaris.

REPRODUCTION PAGE 65

THE RELATIONSHIP OF UNITS

PART ONE: Change these customary units to the units indicated.
1) 27 in. to ft.
2) 3-1/3 ft. to in.
3) 10 ft. 6 in. to yd.
4) 7-2/3 yards to ft.
5) 9-7/12 ft. to yd.
6) 12-1/6 ft. to in.
7) 5-1/4 lb. to ounces
8) 56 oz. to lb.
9) 3-3/8 lb. to oz.
10) 98 ounces to lb.
11) 4 lb. 7 oz. to oz.
12) 301 oz. to lb.
13) 2-3/4 qt. to oz.
14) 2-3/4 gal. to qt.
15) 2-3/4 gal. to oz.
16) 338 oz. to qt.
17) 338 oz. to gal.
18) 3-1/16 qt. to oz.
19) 8.2 ft. to in.
20) 5.7 yd. to ft.
21) 9.6 lb. to oz.
22) 4.6 qt. to oz.
23) 16.9 qt. to gal.
24) 2.8 gal. to qt.

PART TWO: Change these metric units to the units indicated.
25) 36.8 m. to cm.
26) 591 mm. to cm.
27) 43 km. to m.
28) 3864 cm. to m.
29) .863 m. to mm.
30) 587.4 mm. to m.
31) 17.3 kg. to g.
32) 26.57 g. to mg.
33) 95.4 g. to kg.
34) 3176 mg. to g.
35) 35 mg. to g.
36) .017 kg. to g.
37) 4.6 l. to ml.
38) 1352 ml. to l.
39) .832 l. to ml.
40) 791 ml. to l.
41) 3.7 ml. to l.
42) .001 l. to ml.

PART THREE: Arrange in order of size, from greatest to least.
43) a) 3 inches   b) the width of a man's thumb   c) 1/3 ft.   d) 7/16 in.
44) a) a man's long pace   b) 31 in.   c) 3 yd.   d) 4-1/3 ft.
45) a) 479 mm.   b) 4.79 m.   c) the height of a bicycle   d) 479 cm.
46) a) the thickness of sharpened pencil lead   b) 3 mm.   c) .017 m.   d) .4 mm.
47) a) 5 city blocks   b) 5 km.   c) 4000 m.   d) .4219 km.
48) a) 2.2 cm.   b) width of one line loose-leaf paper   c) .023 m.   d) 2000 mm.
49) a) 8 raisins   b) 5 small paper clips   c) 7.3 g.   d) .049 kg.
50) a) an eyedropper full   b) 14 ml.   c) 1.4 l.   d) .14 ml.

Copyright © 1982 by Allyn and Bacon, Inc. Reproduction of this material is restricted to use with *A Guidebook for Teaching Consumer Mathematics,* by Peter A. Pascaris.

USING MULTIPLES, PROPORTIONS, AND UNIT PRICING

PART ONE: Use multiples to determine the best buy. Be sure to check units.

1) 3 oz. for 67¢ 8 oz. for $1.75	2) 6 l. for $5.73 12 l. for $11.29	3) 3 kg. for $1.27 5 kg. for $2.27
4) 8 lb. for $3.19 4 lb. for $1.69	5) 10 yards for $8.50 12 yards for $9.50	6) 1 lb. for $3.27 18 oz. for $3.59
7) 25 ft. for $1.88 20 ft. for $1.58	8) 15 oz. for 39¢ 20 oz. for 49¢	9) 24 oz. for $1.19 1 qt. for $1.56
10) 25 ml. for $3.95 0.1 l. for $14.95	11) 56 g. for $2.91 42 g. for $2.29	12) 84 oz. for $5.25 6 lb. for $5.76
13) .75 l. for $2.97 100 ml. for 37¢	14) 105 g. for 29¢ .150 kg. for 42¢	15) 72 ml. for 39¢ 108 ml. for $.65 .216 l. for $1.08

PART TWO: (16-30) Use proportions to solve the same problems as above. Be sure to check units.

PART THREE: A. (31-45) Verify the best buy for problems 1-15 by finding the unit prices. Be sure to check units.
B. Find the unit price and determine the best buy.

46) 10 items for 73¢ 85¢ per dozen	47) 6.5 oz. for 75¢ 12 oz. for $1.39	48) 7-3/4 oz. for $1.55 13 oz. for $2.55
49) 1 gal. for $5.49 24 oz. for $1.09	50) 3 yd. 2 ft. for $13.09 5-1/2 ft. for $6.60	51) 2.26 kg. for 96¢ 907 g. for 37¢
52) 227 g. for 51¢ 1.362 kg. for $295	53) .94 l. for 63¢ 2.2 l. for $1.49	

Copyright © 1982 by Allyn and Bacon, Inc. Reproduction of this material is restricted to use with *A Guidebook for Teaching Consumer Mathematics*, by Peter A. Pascaris.

REPRODUCTION PAGE 67

CATALOG SHIPPING AND HANDLING CHARGES

Use Figure 7-2 to determine the shipping and handling (s/h) charges for the weights and zones shown below. Find the charges for each item and also on the basis of total weight. Compare the two calculations. (Obtain Figure 7-2 from your teacher.)

I. Ship all these items: 2 lb. 15 oz., 15 lb. 1 oz., 9 oz., 24 lb. 13oz., 3 oz., 28 oz., 54 oz.
   To these zones (consider each zone a separate problem):
   1) Local 2) Zone 2 3) Zone 4 4) Zone 6 5) Zone 8

II. Ship all these items: 2-1/4 lb., 16-1/8 lb., 23-5/8 lb., 1/4 lb., 7/8 lb., 37 oz., 61 oz., 12 lb. 15 oz.
    To these zones:
    6) Local 7) Zone 1 8) Zone 3 9) Zone 5 10) Zone 7

III. Use a map of the United States and determine the s/h charges for the items shown to each of the cities listed. Assume that the distribution center is in St. Louis, Missouri.

   Ship these items: 1-3/4 lb., 13-3/8 lb., 19-1/16 lb., 1/8 lb., 5/8 lb., 15/16 lb., 57 oz., 41 oz., 21 lb. 13 oz.

   To these cities:
   1) Utica, New York
   2) Hot Springs, Arkansas
   3) Dayton, Ohio
   4) Lansing, Michigan
   5) International Falls, Minnesota
   6) Portland, Oregon
   7) Oklahoma City, Oklahoma
   8) Amarillo, Texas
   9) Anchorage, Alaska
   10) Edwardsville, Illinois

Copyright © 1982 by Allyn and Bacon, Inc. Reproduction of this material is restricted to use with *A Guidebook for Teaching Consumer Mathematics,* by Peter A. Pascaris.

REPRODUCTION PAGE 68

DIAGNOSTIC SURVEY EIGHT

I. 1. Define *ratio*.

2. Express the ratio of 4 to 8 in words, using the colon, as a fraction, a decimal, and a percent.

3. There are 80 g. of protein and 20 g. of fat in 100 g. of hamburger. What is the ratio of:
   a) protein to fat?
   b) fat to protein?
   c) protein to entire portion?
   d) fat to entire portion?

4. Which of the following pairs are equivalent ratios?
   a) $\frac{6}{9}$ and $\frac{6}{8}$
   b) $\frac{6}{10}$ and $\frac{27}{45}$
   c) $\frac{4}{6}$ and $\frac{6}{9}$
   d) $\frac{10}{15}$ and $\frac{16}{24}$
   e) $\frac{12}{80}$ and $\frac{15}{100}$
   f) $\frac{12}{15}$ and $\frac{35}{40}$
   g) $\frac{30}{60}$ and $\frac{18}{35}$
   h) $\frac{35}{42}$ and $\frac{15}{18}$

5. Define *proportion*.

6. The fractions 9/12 and 6/8 are equivalent ratios. a) Write a proportion in two different ways, using 9/12 and 6/8. b) Write the expression in words so that the same proportion is read in two different ways. c) Name the 1st, 2d, 3d, and 4th terms. d) What two products are always equal in any proportion? (Demonstrate.)

7. Write these percents as ratios.
   a) 17%   b) 23%   c) 39%   d) 57%   e) 36%

8. Express these as percents.
   a) 19 hundredths   b) 33 out of 100   c) 8/100   d) .37   e) .059

9. Write each percent in item 7 as a fraction and a decimal.

10. Determine the missing number for these proportions.
    a) $\frac{16}{25} = \frac{?}{100}$
    b) $\frac{36}{40} = \frac{?}{100}$
    c) $\frac{4}{9} = \frac{?}{100}$
    d) $\frac{11}{8} = \frac{?}{100}$
    e) $\frac{1}{?} = \frac{25}{100}$
    f) $\frac{3}{?} = \frac{15}{100}$
    g) $\frac{68}{?} = \frac{16}{100}$
    h) $\frac{?}{220} = \frac{5}{100}$

11. Express these as percents.
    a) $\frac{7}{100}$
    b) $\frac{3}{50}$
    c) $\frac{19}{20}$
    d) $\frac{27}{32}$
    e) $\frac{12}{23}$
    f) $\frac{59}{80}$
    g) $\frac{37}{39}$

Copyright © 1982 by Allyn and Bacon, Inc. Reproduction of this material is restricted to use with *A Guidebook for Teaching Consumer Mathematics*, by Peter A. Pascaris.

REPRODUCTION PAGE 68   DIAGNOSTIC SURVEY EIGHT

II. Find the following.
   12) 25% of 96   13) 18% of 54   14) 15% of $243.60
   15) 82% of $64.27   16) 3.5% of $8.80   17) 3.05% of $72.19
   18) .33-1/3 of $36.69   19) .093% of $47,000
   20. If a store has a sale of 25% off on coats, how much would an $84 coat cost?

III. Determine the following.
   21) 44 is what % of 50?   22) 19 is what % of 25?   23) 16 is what % of 24?
   24) 57 is what % of 57?   25) $70 is what % of $75?   26) $.75 is what % of $18.75?
   27) $3.75 is what % of $15?   28) $13.50 is what % of $17.10?
   29. Gail bought a dress for $60.42. The salesclerk said that the price represented a savings of $19.08.
       Find the percent discount from the original price.

IV. Find the number that goes in the blank.
   30) 15% of __?__ is 18.   31) 85% of __?__ is 17.   32) 28% of __?__ is 28.
   33) 60% of __?__ is 46.8.   34) 8% of __?__ is $180.   35) 70% of __?__ is $667.80.
   36) 26.3% of __?__ is $18.41. 37) 37-1/2% of __?__ is $8.13.
   38. What was the original price of an item if it was marked down 20% to a sale price of $13.80?

V. A. Estimate the answer mentally to the nearest ten dollars.
   39) 5% of $289.42   40) 78% of $107.93   41) 6.535 % of $477.29
   42) 11% of __?__ is $7.43   43) 19% of __?__ is 70.15   44) 65% of __?__ is 48.56
   B. For each problem, select the best estimate.

   45) $\frac{37}{80}$ = ?%   a) 37   b) 80   c) 45

   46) $\frac{99}{150}$ = ?%   a) 99   b) 150   c) 65

   47) __?__ of 78 is 4.   a) 5%   b) 10%   c) 50%
   48) __?__ of 74 is 23.   a) 33%   b) 15%   c) 23%
   49) 24% of __?__ is $12.32.   a) $48   b) $480   c) $3
   50) 41% of __?__ is $28.13.   a) $56   b) $12   c) $70

Copyright © 1982 by Allyn and Bacon, Inc. Reproduction of this material is restricted to use with *A Guidebook for Teaching Consumer Mathematics*, by Peter A. Pascaris.

INDEPENDENCE SCALE EIGHT

Use the results of the Diagnostic Survey to rate your degree of independence for the skills listed below. Place an X under the heading that most closely matches your ability at this time.

	To successfully complete this skill, I:					
SKILL	Problem on Survey	Always Depend on Others	Often Depend on Others	Sometimes Depend on Others	Often Work Alone	Always Work Alone
Ratios:						
Meaning	1-3					
Equivalents	4					
Proportions:						
Meaning	5, 6					
Solving	10					
Percents:						
Meaning	8					
as Ratios	7, 11					
as Decimals	9					
as Fractions	9, 11					
Type One	12-20					
Type Two	21-29					
Type Three	30-38					
Word Problems	20, 29, 38					
Estimating	39-50					

Copyright © 1982 by Allyn and Bacon, Inc. Reproduction of this material is restricted to use with *A Guidebook for Teaching Consumer Mathematics*, by Peter A. Pascaris.

REPRODUCTION PAGE 70

RATIOS AND PROPORTIONS

PART ONE: Find the ratio of the following.
1) 5 to 9   2) 6 to 25   3) 1 to 8   4) 5 to 11   5) 7 to 16
6) 2 to 8   7) 4 to 6   8) 25 to 30   9) 48 to 80   10) 72 to 96
11) 6 to 2   12) 24 to 8   13) 64 to 16   14) 75 to 25   15) 96 to 24
16) 11 to 6   17) 8 to 3   18) 17 to 2   19) 20 to 7   20) 14 to 3
21) 10 to 4   22) 18 to 8   23) 25 to 15   24) 80 to 24   25) 90 to 27
26) n to 14   27) a to 100   28) 7 to b 29) c to 12   30) 100 to n
31) 4 inches to 1 ft.   32) 45 min. to 1 hr.   33) one nickel to one dollar
34) 250 g. to 1 kg.   35) one dollar to one dime

PART TWO: Which of the following pairs are equivalent ratios?
1) $\frac{4}{6}$ and $\frac{3}{4}$ 2) $\frac{6}{10}$ and $\frac{27}{45}$ 3) $\frac{6}{8}$ and $\frac{4}{6}$ 4) $\frac{10}{16}$ and $\frac{15}{24}$ 5) $\frac{6}{15}$ and $\frac{40}{100}$
6) $\frac{28}{12}$ and $\frac{35}{15}$ 7) $\frac{9}{20}$ and $\frac{24}{40}$ 8) $\frac{15}{35}$ and $\frac{18}{42}$ 9) $\frac{27}{25}$ and $\frac{45}{72}$ 10) $\frac{16}{36}$ and $\frac{28}{63}$
11) $\frac{12}{15}$ and $\frac{80}{100}$ 12) $\frac{12}{35}$ and $\frac{15}{40}$ 13) $\frac{18}{30}$ and $\frac{35}{60}$ 14) $\frac{15}{35}$ and $\frac{18}{42}$ 15) $\frac{18}{108}$ and $\frac{4}{24}$

PART THREE: Solve these proportions.
1) $\frac{n}{10} = \frac{9}{15}$ 2) $\frac{n}{15} = \frac{3}{5}$ 3) $\frac{n}{24} = \frac{3}{4}$ 4) $\frac{a}{100} = \frac{12}{25}$ 5) $\frac{a}{100} = \frac{27}{40}$
6) $\frac{2}{n} = \frac{15}{90}$ 7) $\frac{16}{n} = \frac{2}{7}$ 8) $\frac{4}{n} = \frac{10}{25}$ 9) $\frac{60}{n} = \frac{3}{5}$ 10) $\frac{17}{n} = \frac{9}{30}$
11) $\frac{18}{14} = \frac{n}{21}$ 12) $\frac{12}{21} = \frac{n}{14}$ 13) $\frac{1}{6} = \frac{n}{5}$ 14) $\frac{35}{100} = \frac{c}{20}$ 15) $\frac{75}{100} = \frac{c}{32}$
16) $\frac{5}{15} = \frac{8}{n}$ 17) $\frac{6}{54} = \frac{7}{n}$ 18) $\frac{63}{84} = \frac{18}{n}$ 19) $\frac{84}{100} = \frac{21}{b}$ 20) $\frac{95}{100} = \frac{38}{b}$
21) $\frac{a}{100} = \frac{12}{23}$ 22) $\frac{a}{100} = \frac{\$19.08}{\$79.50}$ 23) $\frac{37.5}{n} = \frac{\$888.75}{\$23.70}$ 24) $\frac{24}{100} = \frac{c}{\$18.50}$
25) $\frac{85}{100} = \frac{\$166.60}{b}$

Copyright © 1982 by Allyn and Bacon, Inc. Reproduction of this material is restricted to use with *A Guidebook for Teaching Consumer Mathematics*, by Peter A. Pascaris.

REPRODUCTION PAGE 71

TYPE ONE PERCENTS

Directions: Translate each of the following into a number sentence or a proportion of the type.

$$a\% \text{ of } b = c \quad \text{or} \quad \frac{a}{100} = \frac{c}{b}$$

where a and b are known. Then, find c.

1. Mr. Flaming had a fireplace constructed that cost him $2451.49. If the tax rate is 3-1/2% find the total spent.

2. Ms. Erable purchased a dress which was marked down 24% from an original price of $67.99. What did she pay?

3. Mrs. DaBoat placed $500 in a savings certificate that paid at a rate of 6.875% per year. How much did she have after one year?

4. Ernie Morebucks was paid $378 per week but had several taxes withheld from each check. How much did he actually take home in one year if the following taxes were withheld: federal, 16%; state 4.65%; city, 1-1/2%; and social security, 6.05%?

5. Eve N. Moremoney was paid $614 per week and paid the same taxes except that after she reached $20,000 her federal tax rate went up to 19% and after she made $29,000 she did not pay any social security tax. How much did she take home in one year?

6. Mary Bynow paid for everything in cash to avoid a finance charge. Al Paylater bought many things on credit. Both purchased a stereo for $457.32. Each was charged a sales tax of 5.5%. Mary was given a 2% discount on the original price because she paid cash. Al put 25% down and was charged an 18% interest fee on the balance. How much did Mary save compared with Al?

Copyright © 1982 by Allyn and Bacon, Inc. Reproduction of this material is restricted to use with *A Guidebook for Teaching Consumer Mathematics*, by Peter A. Pascaris.

REPRODUCTION PAGE 72

TYPE TWO PERCENTS

Directions: Translate each of the following into a number sentence or a proportion of the type:

$$a\% \text{ of } b = c \quad \text{or} \quad \frac{a}{100} = \frac{c}{b}$$

where b and c are known. Then, find a.

1. If Mr. Flaming put $490.30 down on his $2451.49 fireplace, what percent of the original price was the down payment?

2. Ms. Erable found another dress that originally cost $67.99 and was now $48.95. What was the percent reduction?

3. Mrs. DaBoat cashed in a $500 savings certificate for $541.73 after one year. What was the rate of interest earned?

4. Ernie Morebucks noticed a change in his weekly take-home pay even though his actual (gross) pay remained the same. He now paid $52.92 for federal, $20.79 for state, $8.51 for city, and $24.00 for social security taxes. What is the rate of each tax?

5. Eve N. Moremoney shopped at the same store as Ms. Erable and found a dress that was reduced to $65.62. If this represented a savings of $24.27, who received the better buy?

6. Al Paylater made a purchase of a large cement mixer as a concrete example. If he had a balance of $719.99 after putting $192.00 down, what percent of the original price does the balance represent?

7. Mary Bynow paid only one half the normal sales tax when she bought a car because of a government incentive to spur sales. She wrote a check for $8750.59 and saved $205.09. What is the normal sales tax rate?

Copyright © 1982 by Allyn and Bacon, Inc. Reproduction of this material is restricted to use with *A Guidebook for Teaching Consumer Mathematics*, by Peter A. Pascaris.

REPRODUCTION PAGE 73

TYPE THREE PERCENTS

Directions: Translate each of the following into a number sentence or a proportion of the type.

$$a\% \text{ of } b = c \quad \text{or} \quad \frac{a}{100} = \frac{c}{b}$$

where a *and* c *are known. Then, find* b.

1. Mr. Flaming is burned up. He paid a tax of $16.47 on a new brass fireplace set and he can't recall the original price. If the tax rate is 5%, what was the original price?

2. Ms. Erable saved $59.46 on a set of cracked pots that were marked down 43% because of damage. What was the original price?

3. Mrs. DaBoat wanted to purchase a savings certificate that was paying 7.75% effective interest annually. How much money must she invest if she wishes to have at least $1000 by the end of one year?

4. Ernie Morebucks noticed his paycheck was less than anticipated, and after checking with the other workers he found that the tax rates had remained the same as before. Therefore, his gross pay must have been less. If the tax rates were 16%, 4.65%, 1-1/2%, and 6.05%, respectively, for federal, state, city, and social security taxes, and his deductions totaled $99.27, what was his new gross pay?

5. Mary Bynow and Al Paylater each purchased a dining room set for the same price. Al put 5% down and Mary saved 6% because she paid cash. Mary's savings was $32.58 more than Al's down payment. What was the original full price?

6. Eve N. Moremoney bought a chair that was marked down to $222.68 after being reduced by 12-1/2%. What was the original price?

Copyright © 1982 by Allyn and Bacon, Inc. Reproduction of this material is restricted to use with *A Guidebook for Teaching Consumer Mathematics*, by Peter A. Pascaris.

REPRODUCTION PAGE 74

PERCENTS

Directions: Translate each of the following into a number sentence or a proportion of the type.

$$a\% \text{ of } b = c \quad \text{or} \quad \frac{a}{100} = \frac{c}{b}$$

and, after determining the known variables, solve for the unknown.

1. Mr. Flaming paid 4-3/4% sales tax on a purchase of firewood. He found later that his state exempts firewood from sales tax. He returned to the woodcutter and demanded a $7.79 refund. If he bought 2 cords of wood, what did he pay for each cord?

2. Mrs. DaBoat slipped on the floor of a department store and knocked over some china. The manager said she had to pay for the damage but he would accept 40% of the price that had already been marked down 25%. If the china originally sold for $160, what did Mrs. DaBoat have to pay?

3. Ernie Morebucks compared his payroll deductions with Eve N. Moremoney's and was alarmed to find she made $14,820 more than he did. If his salary is $13,780, that represents what percent of Eve's salary?

4. Mary Bynow finally purchased an item without paying cash. When she decided to buy a house, she put $13,492.50 down and mortgaged the remainder. If her mortgage was $75,607.50, what percent of the original cost did she put down?

5. Al Paylater earns $19,800 a year. He plans a budget to spend 25% on food, 20% on shelter, 15% on clothing, 15% on an automobile, and 10% on savings. The remainder he calls miscellaneous. However, he forgot to include 25% for taxes. So he plans to take the taxes off the earned pay and plan his budget with the same percents on his earnings after taxes. Find the amount budgeted for each category.

Copyright © 1982 by Allyn and Bacon, Inc. Reproduction of this material is restricted to use with *A Guidebook for Teaching Consumer Mathematics*, by Peter A. Pascaris.

REPRODUCTION PAGE 75

DIAGNOSTIC SURVEY NINE

I. A. Define these terms.

   1. payroll deductions
   2. gross pay
   3. dependents
   4. withholding taxes
   5. net pay
   6. exemptions
   7. FICA
   8. W-4 form
   9. allowances
   10. progressive tax
   11. regressive tax

   B. List as many examples of payroll deductions as you can.

   C. Determine the net pay.

Gross Pay	FICA	Federal Tax	State Tax	City Tax	Misc.	Net Pay
$596.21	$36.55	$119.50	$24.85	$8.25	$24.25	12) ?
$223.57	$13.70	$ 22.80	$ 7.55	$2.66	$17.50	13) ?

   D. Determine the effective rate of payroll deduction (comparing gross pay with the amount of deduction) for each category in item C above (problems 14-23).

   E. Determine the amount of payroll deduction for each of the deductions named below. Use the effective rates that are shown. Upon completion, determine the net pay.

Gross Pay	6.70% FICA	13% Federal Tax	3-1/4% State Tax	1-1/2% City Tax	6% Misc.	Net Pay
$372.26	24)	25)	26)	27)	28)	29)
$814.43	30)	31)	32)	33)	34)	35)

   F. Use the tax tables provided by your teacher and determine the federal tax (**Figure 9-2**) and the state tax (**Figure 9-3**) when given the following data.

   36 and 37) Gross pay is $381.93; and 5 allowances or 5 exemptions claimed.
   38 and 39) Gross pay is $526.41; and 7 allowances or 5 exemptions claimed.

   G. Use the schedule in **Figure 9-1** (from your teacher) to calculate the federal withholding tax for these incomes.

   40) $596.21   41) $223.57   42) $814.43   43) $1520

Copyright © 1982 by Allyn and Bacon, Inc. Reproduction of this material is restricted to use with *A Guidebook for Teaching Consumer Mathematics*, by Peter A. Pascaris.

REPRODUCTION PAGE 75 **DIAGNOSTIC SURVEY NINE**

II. A. Determine the weekly pay based upon the hourly pay scales, given the following.

44) 40 hours @ $3.95
45) 40 hours @ $6.45, plus 4-1/2 hours @ time and a half overtime.
46) 40 hours @ $6.45 + (5% of $6.45), 4-1/2 hours @ time and a half of the regular pay.

B. Determine the weekly pay based upon the *piecework* scales, given the following.

47) 372 items @ 57¢
48) 39 items @ $5.21
49) 372 items @ 47¢ for the first 150 items
　　　　57¢ for the next 100 items
　　　　69¢ for the next 50 items
　　　　84¢ for the next 50 items
　　　　$1.04 for all items thereafter.

C. Determine the weekly pay based upon the *commission* rate specified.

50) A real estate salesperson earns 3-1/2% of the sale of a home that is listed by her company. She earns 1-3/4% of the sale on homes listed by another company. What is her total commission if she sells two homes, one listed by her company and sold for $79,500 and another listed by another company and sold for $85,400?

51) A salesman is paid 17% of all sales above $800. How much does he earn if his sales amount to $690, $975, $1450, and $1810?

52) A saleslady is paid on a graduated scale of 7% on the first $900 she sells, 9% of the next $700, 12.6% of the next $500, and 16-1/2% on sales over $2100. How much does she earn if her total sales are $3298?

III. A. Define these terms as they pertain to annual income tax returns.

53. gross income
54. adjusted gross income
55. deductions
56. W-2 form
57. IRS
58. IRS Form 1040
59. IRS Form 1040A
60. schedules
61. tables
62. zero bracket amount
63. filing status
64. tax liability

B.

65. Obtain a Federal Income Tax Form 1040A and Tax Table A from your teacher and complete the tax return given the following data: Tex Weary is single and earned $5214 in wages and $146 from a savings account. He had $298.40 withheld from his wages.

66. Obtain a Federal Income Tax Form 1040 and Tax Table B from your teacher and complete the tax return given the following data: Steven and Susan Long are married and have three children. Steven earned $31,305 as an auto mechanic and night watchman. Susan's only income was a $2500 prize she won in a raffle. They had a savings account that earned $160, and their stock paid a dividend of $293. Steven's two jobs resulted in an overpayment of $159.68 to FICA. His employers withheld a combined amount of $5037 for federal tax.

Copyright © 1982 by Allyn and Bacon, Inc. Reproduction of this material is restricted to use with *A Guidebook for Teaching Consumer Mathematics*, by Peter A. Pascaris.

INDEPENDENCE SCALE NINE

Use the results of the Diagnostic Survey to rate your degree of independence for the skills listed below. Place an X under the heading that most closely matches your ability at this time.

		To successfully complete this skill, I:				
SKILL	Problem on Survey	Always Depend on Others	Often Depend on Others	Sometimes Depend on Others	Often Work Alone	Always Work Alone
Definitions	1-11 and 53-64					
Net Pay	12, 13					
Effective Rate	14-23					
Payroll Deductions By Computation	24-35					
By Tax Table	36-39					
By Schedule	40-43					
Hourly Pay	44-46					
Piecework	47-49					
Commissions	50-52					
Annual Tax Return 1040A	65					
1040	66					

Copyright © 1982 by Allyn and Bacon, Inc. Reproduction of this material is restricted to use with *A Guidebook for Teaching Consumer Mathematics*, by Peter A. Pascaris.

REPRODUCTION PAGE 77

NET PAY

I. Determine the net pay.

	Gross Pay	FICA	Federal Tax	State Tax	City Tax	Misc.	Net Pay
A.	$212.00	$14.20	$40.28	$5.72	$4.24	$10.60	1)_____
B.	$476.18	$31.90	$71.00	$14.58	$4.76	$47.62	2)_____
C.	$391.19	$26.21	$39.12	$16.15	$6.45	$28.17	3)_____
D.	$207.49	$13.90	$30.57	$6.92	-0-	$31.50	4)_____
E.	$856.53	-0-	$191.14	$34.27	$16.51	$96.00	5)_____
F.	$554.07	$37.12	$65.41	$28.40	$5.04	$36.14	6)_____
G.	$328.61	$22.02	$48.86	$13.26	$2.88	$29.55	7)_____
H.	$726.06	$17.21	$79.66	$31.62	$7.68	$153.00	8)_____
I.	$592.51	$39.70	$146.93	$31.25	$6.92	$57.75	9)_____
J.	$179.14	$12.00	$19.60	$3.07	$1.94	$12.50	10)_____
K.	$89.76	$6.01	$9.10	$4.37	$.65	$7.50	11)_____
L.	$408.66	$27.38	$37.94	$17.86	$38.48	$36.22	12)_____
M.	$793.53	-0-	$151.46	$42.16	$74.32	$176.67	13)_____
N.	$526.97	$35.31	$78.67	$20.19	$57.39	$89.49	14)_____
O.	$78.13	$5.23	$8.14	$3.27	$.70	$68.54	15)_____
P.	$200.98	$13.47	$30.08	$10.05	$1.53	$14.25	16)_____
Q.	$626.32	$41.96	$68.63	$31.31	$5.32	$108.61	17)_____
R.	$711.11	$47.64	$114.41	$30.15	$7.11	$96.58	18)_____
S.	$657.97	$44.08	$121.18	$28.56	$7.11	$143.21	19)_____
T.	$503.21	$33.72	$53.01	$25.01	$3.01	$83.57	20)_____

II. A. Compare the gross pay with the amount of each deduction shown above and determine the effective rate of each deduction. (Your teacher may wish to assign all of the above or perhaps select several different payroll items.)

B. Compare the net pay with the gross pay and determine the percent of gross pay represented by net pay.

Copyright © 1982 by Allyn and Bacon, Inc. Reproduction of this material is restricted to use with *A Guidebook for Teaching Consumer Mathematics*, by Peter A. Pascaris.

PAYROLL DEDUCTIONS

PART ONE

Determine the amount of payroll deduction for each of the gross pays listed below. Use the effective rates that are shown. Upon completion, determine net pay.

FICA = 6.70%, Federal Tax = 13%, State Tax = 3-1/4%, City Tax = 1-1/2%, Misc. = 6%

Gross Pays are as follows.

1) $97.50	2) $231.81	3) $676.57	4) $153.56	5) $796.83
6) $394.04	7) $532.97	8) $402.02	9) $829.49	10) $67.49

PART TWO

Use the tax tables provided by your teacher and determine the federal tax (Figure 9-2) and the state tax (Figure 9-3) for the gross pays, allowances, and exemptions listed.

A.	(1)	(2)	(3)	(4)	(5)
Gross Pay	383	537	480.15	392.67	528.03
Allowances	4	3	9	-0-	2
Exemptions	4	3	9	-0-	2
B.	(6)	(7)	(8)	(9)	(10)
Gross Pay	467	501	496.31	508.01	539.86
Allowances	8	7	10	12	12
Exemptions	5	4	3	5	7

PART THREE

Use the schedule in Figure 9-1 (from your teacher) to calculate the federal withholding tax for these incomes and allowances. Incomes (1 & 2) have 3 allowances each, (3-5) have 5 allowances each, (6 & 7) have 7 allowances, and (8 & 9) have 11 and 12 allowances, respectively.

1) $83.91	2) $172.15	3) $291.93	4) $308.65	5) $382.77
6) $463.21	7) $542.98	8) $859.49	9) $1057.83	10) $1643.19

Copyright © 1982 by Allyn and Bacon, Inc. Reproduction of this material is restricted to use with *A Guidebook for Teaching Consumer Mathematics*, by Peter A. Pascaris.

REPRODUCTION PAGE 79

HOURLY PAY

I. Determine the weekly pay based upon 40 hours per week at the hourly scales shown.

1) $3.89	2) $4.15	3) $5.65	4) $7.93	5) $6.95
6) $4.09	7) $6.52	8) $9.19	9) $10.65	10) $14.85

II. Find the weekly wage for the following people.

	\multicolumn{5}{c}{Hours Worked}		
	M	T	W	Th	F	Hourly Rate	
Mr. Juan	7	8	7	8	8	8.35	
Ms. Tew	7	7	$7\frac{1}{2}$	8	$7\frac{1}{2}$	12.25	
Mrs. Thray	7	7	$6\frac{1}{2}$	6	$6\frac{1}{2}$	10.15	
Ms. Fore	$4\frac{1}{2}$	4	$4\frac{1}{4}$	4	$4\frac{1}{2}$	3.95	
Mr. Fife	8	8	7	9	10	12.25	plus time and a half after 8 hours per day.
Mr. Sicks	8	9	$9\frac{1}{2}$	$8\frac{1}{2}$	$9\frac{1}{4}$	8.47	
Ms. Savin	7	$5\frac{1}{2}$	-0-	$8\frac{1}{2}$	$9\frac{3}{4}$	6.39	
Mr. Aite	7	$4\frac{1}{2}$	8	5	9	7.77	plus time and a half after 40 hours per week.
Mr. Knyne	8	$9\frac{1}{2}$	$7\frac{3}{4}$	$9\frac{1}{4}$	$8\frac{1}{4}$	5.17	
Ms. Tenn	8	9	$7\frac{1}{2}$	$8\frac{3}{4}$	$9\frac{1}{4}$	7.05	

III. (11-20) All of the above people worked on Friday during a bad storm and were given a 12-1/2% incentive for the hours they worked on Friday. Determine their weekly wage with the new rate.

IV. (21-30) The company has a policy that penalizes tardiness beyond 10 minutes. The worker is penalized 1/4 of his hourly pay for each 10-minute segment (or fraction thereof) beyond the first 10 minutes. The minutes tardy are accumulated for the week. Determine the dock-pay penalty for each worker. (Use regular hourly pay.)

1) Mr. Juan: 7 min. late each day.
2) Ms. Tew: 8 min. M, 2 min. Th.
3) Mrs. Thray: 4 min. T, 7 min. W.
4) Ms. Fore: 5 min. Th, 8 min. F.
5) Mr. Fife: 3-1/2 min. M, T, Th, F.
6) Mr. Sicks: 5 min. M, 4-1/2 min. T, 3 min. W.
7) Ms. Savin: 12 min. late W.
8) Mr. Aite: 12 min. M, 14 min. W, 5 min. Th.
9) Mr. Knyne: 9 min. late T, W, F.
10) Ms. Tenn: 1 min. M, 7 min. T, 13 min. Th.

Copyright © 1982 by Allyn and Bacon, Inc. Reproduction of this material is restricted to use with *A Guidebook for Teaching Consumer Mathematics*, by Peter A. Pascaris.

PIECEWORK AND COMMISSIONS

I. Determine the weekly pay based on the piecework scales, given the following.

1) 278 items @ 93¢.
2) 843 items @ 58¢.
3) 521 items @ 46¢.
4) 197 items @ $1.89 for the first 150 and $2.15 thereafter.

5) 843 items @ 52¢ for first 100
   @ 55¢ for next 100
   @ 58¢ for next 100
   @ 70¢ for next 100
   @ 94¢ for next 100
   @ $1.20 for next 100
   @ $1.92 thereafter.

6) 521 items @ 41¢ for first 100
   @ 43¢ for next 100
   @ 47¢ for next 100
   @ 55¢ for next 50
   @ 71¢ for next 50
   @ $1.03 for next 50
   @ $1.67 thereafter.

II. Determine the weekly pay based upon the commission rate specified.

Salesperson	Gross Sales	Commission Rate and Schedule of Commission
Mr. Aae	$1121	4-1/2% on total sales
Ms. Bee	$2946	8-3/4% on sales over $500
Ms. Sea	$83,472	1-1/2% on first $50,000; 2-1/4% thereafter
Mr. Deeh	$3215	5% of first $1000 6-1/4% of next $750 8-1/2% of next $750 10-3/4% of next $750
Mrs. Eehee	$978	23% of first $150 28-1/2% of next $150 35% of next $125 47% of next $100 52% thereafter plus an additional 5% of Gross Sales if over $950

Copyright © 1982 by Allyn and Bacon, Inc. Reproduction of this material is restricted to use with *A Guidebook for Teaching Consumer Mathematics*, by Peter A. Pascaris.

REPRODUCTION PAGE 81

COMBINATION PAY SCALES

I. Determine the weekly pay for these combination pay scales. (assume a 40-hour week).

1) $4.65 per hour plus 278 items produced @ 57¢ for each item over 150.

2) $5.18 per hour plus 521 items produced @ 12¢ for each item.

3) $4.91 per hour plus 197 items @ 32¢ for the first 150 and 91¢ thereafter.

4) $3.73 per hour plus

    843 items @ 4¢ for first 100
              @ 7¢ for next 100
              @ 13¢ for next 100
              @ 25¢ for next 100
              @ 49¢ for next 100
              @ 85¢ for next 100
              @ $1.57 thereafter.

5) $4.41 per hour plus

    521 items @ 2¢ for first 100
              @ 4¢ for next 100
              @ 8¢ for next 100
              @ 16¢ for next 50
              @ 32¢ for next 50
              @ 64¢ for next 50
              @ $1.28 thereafter.

6) $4.97 per hour plus 1/4 of the commission rates determined on Reproduction Page 80, Part II.

7) A waitress averages 12-1/2% tips at a restaurant that has 4 other waitresses. If she is paid $3.47 per hour and the restaurant has a gross of $376 per hour, how much can she make in one week? (Assume a 40-hour week.)

II. Determine which job pays the most. (Assume a 40-hour week.)

8) A whatsit maker who earns $5.83 plus 38¢ per whatsit, if he makes 413 whatsits a week, or

A salesperson who earns $4.09 per hour plus 9-3/4% on sales over $1000, if she has $5291 in sales?

9) A secretary with an annual salary of $14,832 or a laborer who makes $7.14 per hour?

10) Job A: $17,226 annually.   Job B: $1436.10 per month.
     Job C: $717.71 bimonthly.   Job D: $663.44 biweekly.
     Job E: $331.25 weekly.   Job F: $8.29 per hour.

Copyright © 1982 by Allyn and Bacon, Inc. Reproduction of this material is restricted to use with *A Guidebook for Teaching Consumer Mathematics*, by Peter A. Pascaris.

DIAGNOSTIC SURVEY TEN

I. A. Find the *interest paid*, given the following information.

1) $1000 loan paid back in 12 months @ $89.26.
2) $1000 loan paid back in 18 months @ $61.39.
3) $1000 loan paid back in 24 months @ $47.49.
4) $1000 loan paid back in 36 months @ $33.64.
5) $1000 loan paid back in 48 months @ $26.77.

B. Find the *amount of the loan*, given the following information.

6) 10 months @ $106.06 with a total interest charge of $60.60.
7) 18 months @ $60.35 with a total interest charge of $86.30.
8) 24 months @ $23.14 with a total interest charge of $55.36.
9) 30 months @ $265.50 with a total interest charge of $965.
10) 4 years paying $360.16 per month with a total interest charge of $4287.68.

C. Find the *amount of the monthly payments..*

11) $1000 loan plus $70.10 interest paid back in 10 months.
12) $1000 loan plus $270.92 interest paid back in 42 months.
13) $400 loan plus $30.92 interest paid back in 18 months.
14) $17,000 loan plus $2132.80 interest paid back in 20 years.
15) $44,000 loan plus $128,134 interest paid back in 30 years.

D. Find the *number of monthly payments*.

16) $1000 loan plus $70.10 interest at $107.01 per month.
17) $1000 loan plus $311.84 interest at $27.33 per month.
18) $300 loan plus $24.96 interest at $27.08 per month.
19) $9000 loan plus $1848.30 interest at $361.61 per month.
20) $38,000 loan plus $162,529 interest at $477.59 per month.

II. A. 21) State the simple interest formula and identify each variable.

B. Use the simple interest formula to calculate the *interest*, given the following information.

22) principal is $1000, rate is 6% for 1 year.
23) principal is $2800, rate is 9.25% for 1 year.
24) principal is $5400, rate is 17-1/2% for 3 years.
25) principal is $450, rate is 10% for 7 months.
26) principal is $2120, rate is 12.7% for 3 years, 8 months.

C. Use the simple interest formula to determine the *principal*.

27) interest is $64.00, rate is 8% for 2 years.
28) interest is $160.00, rate is 16% for 1 year.
29) interest is $315.00, rate is 12% for 3-1/2 years.
30) interest is $168.75, rate is 12-1/2% for 9 months.
31) interest is $88.50, rate is 14-3/4% for 18 months.

Copyright © 1982 by Allyn and Bacon, Inc. Reproduction of this material is restricted to use with *A Guidebook for Teaching Consumer Mathematics*, by Peter A. Pascaris.

REPRODUCTION PAGE 82 **DIAGNOSTIC SURVEY TEN**

D. Use the simple interest formula to determine the *rate*.

32) interest is $18.00 on a principal of $300 for 1 year.
33) interest is $26.00 on a principal of $400 for 1 year.
34) interest is $84.00 on a principal of $600 for 21 months.
35) interest is $180.50 on a principal of $1900 for 8 months.

E. Use the simple interest formula to determine the *time*.

36) interest is $10 on a principal of $200 at 5%.
37) interest is $40 on a principal of $800 at 6-1/4%.
38) interest is $652.50 on a principal of $1500 at 14.5%.
39) interest is $157.50 on a principal of $1750 at 12%.
40) interest is $71.25 on a principal of $2000 at 14.25%.

III. Find the *effective rate* of interest (or *true rate*), given the following information.
   41) principal of $900, interest paid is $81, paid back in 12 months at $81.75 per month. Each monthly payment includes $75 for the principal and $6.75 for the interest.
   42) principal of $900, interest paid is $260.40, paid back in two years at $48.35 per month. Each monthly payment includes $37.50 for the principal and $10.85 for the interest.

IV. (Optional) To find the true rate of interest, given the information below, use the following constant ratio formula.

$$r = \frac{2mi}{p(1 + n)}.$$

   43) the principal is $1700, the interest is $204, and there are 12 monthly payments made over a period of 1 year.

V. 44) Use a sample loan of $900 at 9% for one year to distinguish between an add-on-interest loan and a discounted loan.

VI. A credit customer charged three $600 purchases on three different charge cards. Each came due at the same time, and each advertised a 1.7% interest rate per month. On the fifteenth day of the payment period the customer paid $500 on each account. Assuming no more purchases were made, how much interest is charged on each account at the next billing if the 1.7% monthly interest is determined as follows?
   45) Charge account A used the *adjusted balance method*.
   46) Charge account B used the *previous balance method*.
   47) Charge account C used the *average balance method*.

VII. 48) Use a loan of $1000 at a rate of 12% per year paid back in 12 months at $88.85 per month to describe an amortized loan.

Copyright © 1982 by Allyn and Bacon, Inc. Reproduction of this material is restricted to use with *A Guidebook for Teaching Consumer Mathematics*, by Peter A. Pascaris.

DIAGNOSTIC SURVEY TEN

49) Find the amount of interest paid and the total cost of these mortgage loans based upon a $60,000 home with a down payment of $10,000 leaving a $50,000 mortgage and paying as follows.
   a) $1112.23 per month for 5 years.
   b) $717.36 per month for 10 years.
   c) $550.55 per month for 20 years.
   d) $514.31 per month for 30 years.
   e) $504.25 per month for 40 years.

VIII. 50) Determine *the interest earned and the new account balance* on June 30 for a savings account that pays interest semiannually at 6%, assuming the following.
   the account is opened January 2 with $1000,
   on March 1, there is a deposit of $375,
   on April 1, there is a withdrawal of $117.

51) Find the *interest* and the *new balance* on the following account.
   interest rate is 6% paid by the daily balance method,
   account is opened on January 2 with $850,
   on February 14, there is a withdrawal of $798,
   on March 1, there is a deposit of $64,
   interest paid on March 31 is ___?___

IX. 52) Explain *compound interest*.

53) Use the compound interest table provided by your teacher (Figure 10-4) and determine how much $700, deposited for 4 years, will be worth for interest compounded as stated.
   a) $700 deposited for 4 years and paying 6% compounded annually.
   b) $700 deposited for 4 years and paying 6% compounded quarterly.

54) What is the *effective annual rate* for item 52b?

55) Use the formula $b = p(r + 1)^n$ to determine the new balance and the amount of interest earned.
   a) of a $1000 savings certificate that earns 6% interest compounded quarterly and held for two years.
   b) of $500 held for seven years and earning 10% interest compounded quarterly.

Copyright © 1982 by Allyn and Bacon, Inc. Reproduction of this material is restricted to use with *A Guidebook for Teaching Consumer Mathematics*, by Peter A. Pascaris.

REPRODUCTION PAGE 83

INDEPENDENCE SCALE TEN

Use the results of the Diagnostic Survey to rate your degree of independence for the skills listed below. Place an X under the heading that most closely matches your ability at this time.

		To successfully complete this skill, I:				
SKILL	Problem on Survey	Always Depend on Others	Often Depend on Others	Sometimes Depend on Others	Often Work Alone	Always Work Alone
Loan Agreements						
interest	1-5					
loan	6-10					
payments	11-15					
months	16-20					
$i = prt$	21					
i	22-26					
p	27-31					
r	32-25					
t	36-40					
True Rate	41-42					
$r = \dfrac{2mi}{p(1+n)}$	43 (Optional)					
Add-On Interest And Discount	44					
Adjusted Balance	45					
Previous Balance	46					
Average Balance	47					
Amortization	48-49					
Savings Paid						
semiannually	50					
daily	51					
Compounded	52 a&b					
table	53					
effective	54					
formula	55 a&b					

Copyright © 1982 by Allyn and Bacon, Inc. Reproduction of this material is restricted to use with *A Guidebook for Teaching Consumer Mathematics*, by Peter A. Pascaris.

MORTGAGE LOANS

I. Describe what is meant by *amortization*. Explain the method of determining interest (formula and when applied) and how the account and number of monthly payments is determined.

II. Determine the amount of interest paid and the total cost of homes that have the terms and monthly mortgage payments indicated.

   A. A $50,000 home, no money down and these monthly payments.

   1) $1118.25 for 5 years.
   2) $724.60 for 10 years.
   3) $559.29 for 20 years.
   4) $523.95 for 30 years.
   5) $514.35 for 40 years.

   B. An $85,000 home, $10,000 down and these monthly payments.

   6) $1706.49 for 5 years.
   7) $1119.84 for 10 years.
   8) $878.69 for 20 years.
   9) $829.65 for 30 years.
   10) $817.14 for 40 years.

III. Use the table in Figure 10-2 and determine a) the monthly payment, b) the amount of interest paid, and c) the total paid for the homes purchased under the following terms.

   1) A $30,000 home, no money down, 12% interest rate, for 30 years.

   2) A $25,000 home, $20,000 down, 12% interest rate, for 10 years.

   3) A $66,000 home, $6000 down, 12% interest rate, for 35 years.

   4) A $100,000 home, $25,000 down, 12% interest rate, for 40 years.

   5) An $85,000 home, $40,000 down, 12% interest rate, for 24 years.

Copyright © 1982 by Allyn and Bacon, Inc. Reproduction of this material is restricted to use with *A Guidebook for Teaching Consumer Mathematics*, by Peter A. Pascaris.

REPRODUCTION PAGE 85

PROPERTY TAXES

PART ONE: Determine the property tax rate (expressed in mills) for the given data.

1) Taxes needed $820,500
   Assessed valuation $24,700,000

2) Taxes needed $1,430,000
   Assessed valuation $28,200,000

3) Taxes needed $3,590,000
   Assessed valuation $70,900,000

4) Taxes needed $653,500
   Assessed valuation $14,200,000

5) Taxes needed $8,250,000
   Assessed valuation $1,904,500,000

6) Taxes needed $375,400
   Assessed valuation $6,275,000

7) Taxes needed $2,850,000
   Assessed valuation $59,700,000

8) Taxes needed $5,168,200
   Assessed valuation $121,300,000

9) Taxes needed $7,391,400
   Assessed valuation $1,329,260,000

10) Taxes needed $253,176
    Assessed valuation $4,138,201

PART TWO: Determine the property taxes for the following homes assessed at the value shown and taxed at the given rate.

11) Find the taxes for these homes taxed at the rate determined in item number 1 above.

   a) $20,000   b) $23,000   c) $31,000   d) $40,000   e) $46,000

12) Find the taxes for these homes taxed at the rate determined in item number 5 above.

   a) $13,000   b) $26,000   c) $33,000   d) $38,000   e) $42,000

13) Find the taxes for these homes taxed at the rate determined in item number 8 above.

   a) $17,000   b) $24,000   c) $32,000   d) $39,000   e) $44,000

14) Find the taxes for these homes taxed at the rate determined in item number 9 above.

   a) $15,000   b) $25,000   c) $34,000   d) $37,000   e) $51,000

15) Find the taxes for these homes taxed at the rate determined in item number 10 above.

   a) $20,000   b) $26,000   c) $32,000   d) $37,000   e) $44,000

Copyright © 1982 by Allyn and Bacon, Inc. Reproduction of this material is restricted to use with *A Guidebook for Teaching Consumer Mathematics*, by Peter A. Pascaris.

APPENDIX D

Feedback Form

Your comments about this book will be very helpful to us in planning other books in the *Guidebook for Teaching* Series and in making revisions in *A Guidebook for Teaching Consumer Mathematics*. Please tear out the form that appears on the following page and use it to let us know your reactions to *A Guidebook for Teaching Consumer Mathematics*. The authors promise a personal reply. Mail the form to:

Mr. Peter A. Pascaris
c/o Longwood Division
Allyn and Bacon, Inc.
470 Atlantic Ave
Boston, Massachusetts 02210

A GUIDEBOOK FOR TEACHING CONSUMER MATHEMATICS

Your school: _____
Address: _____
City and state: _____
Date: _____

Mr. Peter A. Pascaris
c/o Longwood Division
Allyn and Bacon, Inc.
470 Atlantic Avenue
Boston, Massachusetts 02210

Dear Peter:

My name is _____ and I wanted to tell you what I thought of your book *A Guidebook for Teaching Consumer Mathematics*. I liked certain things about the book, including:

I do, however, feel that the book could be improved in the following ways:

There were some other things that I wish the book had included, such as:

Here is something that happened in my class when I used an idea from your book:

                        Sincerely yours,

                        _____